# CONTROL SYSTEMS ENGINEER
# TECHNICAL REFERENCE HANDBOOK

BY CHUCK CORNELL, PE, CAP, PMP

**International Society of Automation**
**P.O. Box 12277**
**Research Triangle Park, NC 27709**

**Notice**

The information presented in this publication is for the general education of the reader. Because neither the author nor the publisher has any control over the use of the information by the reader, both the author and the publisher disclaim any and all liability of any kind arising out of such use. The reader is expected to exercise sound professional judgment in using any of the information presented in a particular application.

Additionally, neither the author nor the publisher have investigated or considered the effect of any patents on the ability of the reader to use any of the information in a particular application. The reader is responsible for reviewing any possible patents that may affect any particular use of the information presented.

Any references to commercial products in the work are cited as examples only. Neither the author nor the publisher endorses any referenced commercial product. Any trademarks or trade names referenced belong to the respective owner of the mark or name. Neither the author nor the publisher make any representation regarding the availability of any referenced commercial product at any time. The manufacturer's instructions on use of any commercial product must be followed at all times, even if in conflict with the information in this publication.

**Library of Congress Cataloging-in-Publication Data**

Cornell, Chuck.
 Control systems engineer technical reference handbook / by Chuck Cornell.
  p. cm.
 Includes bibliographical references.
 ISBN 978-1-937560-47-8 (pbk.)
 1. Systems engineering--Handbooks, manuals, etc. 2. Reliability (Engineering)--Handbooks, manuals,
etc. 3. Machinery--Maintenance and repair--Handbooks, manuals, etc. I. Title.
 TA168.C648 2012
 629.8--dc23
                         2012020506

# DEDICATION

*This book is dedicated to my "big brother" Bob who consistently provides a good example for me to follow and for being someone that I have always aspired to emulate.*

*In addition, this book is dedicated to my late sons Adam Jude Cornell and Charles Thomas Cornell who were the light of my life and still continue to be the source of my inspiration to be the best that I can be.*

# Contents

# PREFACE

The information in this book was prepared so that it can serve a dual purpose. The first purpose is to provide a study aid for the Control Systems Engineering Professional Engineering Exam that is presided over by the NCEES. The second purpose is to provide a technical reference for future use by the instrumentation / automation professional.

Where the author cites any references to commercially available products, it is for reference only and is by no means an endorsement by the author of any commercially available product.

Sample problems presented in this book are not meant to influence the reader on specific problems that may be on the exam, but rather to reinforce the technical material that has been presented to the reader.

# ABOUT THE AUTHOR

Charles Cornell is a Senior Process Control Systems Engineering Manager with more than 30 years of engineering experience in automation, instrumentation, and electrical. He is a licensed professional engineer in the state of North Carolina in both Control Systems and Electrical Power. He is also an ISA Certified Automation Professional (CAP), as well as a PMI certified Project Management Professional (PMP).

Charles is a member of ISA, NSPE, PMI and the CSE PAKS Committee.

# ACKNOWLEDGMENTS

The author wishes to thank the following individuals, whose help was invaluable:

**Robert Cornell,** who provided subject-matter expert input with regard to Review Material.

**Mark Stone,** who provided subject-matter expert input with regard to Types of Programming.

**Robert Harding,** who provided subject-matter expert input with regard to Networks.

**Tony Spivey,** who provided subject-matter expert input with regard to the OSI Model.

**Brian Kovatch,** who provided subject-matter expert input with regard to Batch Control and Advanced Process Control.

**Robin Lowery,** who provided subject-matter expert input with regard to Analytical Measurement.

**Robert Peters,** who provided subject-matter expert input with regard to Motors and VFDs.

**James Potts,** who verified that the sample problems were solved correctly.

**Diana L. Hughes,** who painstakingly proofread the document for grammar and spelling.

# ACRONYMS

**AC:**      Alternating Current

**A/D:**      Analog to Digital

**ANSI:**      American National Standards Institute (www.ansi.org)

**API:**      American Petroleum Institute (www.api.org)

**AS-i:**      Actuator Sensor Interface (www.as-interface.net)

**ASME:**      American Society of Mechanical Engineers (www.asme.org)

**ATEX:**      ATmospheres EXplosibles (French acronym) (www.ce-mark.com/atexdir.html)

**CEMF:**      Counter Electro Motive Force

**CSA:**      Canadian Standards Association (www.csa-international.org)

**D/A:**      Digital to Analog

**DC:**      Direct Current

**D/P:**      Differential Pressure

**ECA:**      Electronic Components Association

**EEMUA:**      Engineering Equipment and Materials User Association (www.eemua.org)

**EIA:**      Electronics Industries Association (now part of ECA)

**EMF:**      Electro Motive Force (as in voltage)

**EMI:**      Electromagnetic Interference

**ERP:**      Enterprise Resource Planning

**GC:**      Gas Chromatograph

**HART:**      Highway Addressable Remote Transducer (www.hartcomm.org)

**HPLC:**      High Pressure Liquid Chromatography

**HTTP:**      Hypertext Transfer Protocol

**I/P:**      Current to Pneumatic

**IEC:**　　International Electrotechnical Commission (www.iec.ch)

**IEEE:**　　Institute of Electrical and Electronic Engineers (www.ieee.org)

**FM:**　　Factory Mutual

**IS:**　　Intrinsic Safety

**ISA:**　　International Society of Automation (www.isa.org)

**ISO:**　　International Organization for Standardization (www.iso.org)

**MES:**　　Manufacturing Execution Systems

**NCEES:**　　National Council of Examiners for Engineering and Surveying (www.ncees.org)

**NEC:**　　National Electrical Code (NFPA 70)

**NEMA:**　　National Electrical Manufacturers Association (www.nema.org)

**NFPA:**　　National Fire Protection Association (www.nfpa.org)

**NIC:**　　Network Interface Controller

**NRC:**　　Nuclear Regulatory Commission (www.nrc.gov)

**OLE:**　　Object Linking and Embedding

**OPC:**　　OLE for Process Control (www.opcfoundation.org)

**OSHA:**　　Occupational Safety & Health Administration (www.osha.gov)

**OSI:**　　Open System Interconnection

**PD:**　　Positive Displacement

**PFD:**　　Probability of Failure on Demand

**PMI:**　　Project Management Institute (www.pmi.org)

**RF:**　　Radio Frequency

**RFI:**　　Radio Frequency Interference

**RG:**　　Radio Guide

**RJ:**　　Registered Jack

**RRF:**　　Risk Reduction Factor

**RTD:** Resistance Temperature Detector

**RTU:** Remote Terminal Unit

**SAMA:** Scientific Apparatus Makers Association[1]

**SFC:** Sequential Function Chart

**SI:** Système international d'unités *(metric system)*

**SIL:** Safety Integrity Level

**SIS:** Safety Instrumented System

**SMTP:** Simple Mail Transfer Protocol

**SNMP:** Simple Network Management Protocol

**STP:** Shielded Twisted Pair

**T/C:** Thermocouple

**TCP/IP:** Transmission Control Protocol / Internet Protocol

**TIA/EIA:** Telecommunications Industry Association / Electronics Industries Association

**USB:** Universal Serial Bus

**UTP:** Unshielded Twisted Pair

**VOM:** Volt – Ohm – Milliamp Meter

**WBF:** World Batch Forum (www.wbf.org)

---

1. SAMA is an obsolete association that is referenced within this document should the reader come across this terminology within legacy documents.

# COMMON ELECTRICAL DEFINITIONS

**Admittance 'Y':** The inverse of the impedance (Z). The SI unit of admittance is the siemens

$$Y = \frac{1}{Z}$$

**Ampere 'A':** SI unit of electric current. It is named after André-Marie Ampère, French mathematician and physicist.

$$A = \frac{C}{s}$$

**Conductivity 'σ':** Measure of a material's ability to conduct an electric current. When an electrical potential difference is placed across a conductor, its movable charges flow, giving rise to an electric current. The conductivity σ is defined as the ratio of the current density J to the electric field strength E

$$J = \sigma E$$

**Conductance 'siemens or mho':** Measures how easily electricity flows along a certain path through an electrical element (reciprocal of resistance).

$$G = \frac{1}{R}$$

**Coulomb 'C':** SI derived unit of electric charge, and is approximately equal to the charge of $6.24151 \times 10^{18}$ protons or $-6.24151 \times 10^{18}$ electrons. It is named after Charles-Augustin de Coulomb.

$$C = A \cdot s = F \cdot V$$

**Coulomb's Law:** Law of physics describing the electrostatic interaction between electrically charged particles. *The magnitude of the electrostatic force between two point electric charges is directly proportional to the product of the magnitudes of each of the charges and inversely proportional to the square of the distance between the two charges.*

$$F = k\frac{Q_1 Q_2}{r^2} \quad and \quad k = \frac{1}{4\pi\varepsilon_0} = \frac{c^2\mu_0}{4\pi} = c^2\left(10^{-7}H \times m^{-1}\right) = 8.987 \times 10^9 N \cdot m^2 \cdot c^{-2}$$

*Where: F = Electrostatic force      Q = Charge*

*r = Distance between the 2 charges      k = Proportional constant*

**Current Density 'J':** Vector quantity whose magnitude is the ratio of the magnitude of current flowing in a conductor to the cross-sectional area perpendicular to the current flow and whose direction points in the direction of the current. Commonly expressed in amperes per square centimeter or amperes per square inch.

$$J = \frac{I}{Area}$$

**Farad 'F':** A farad is the charge in coulombs a capacitor will accept for the potential across it to change 1 volt. The term farad is named after the English physicist Michael Faraday.

$$F = \frac{C}{V} = \frac{A \cdot s}{V} = \frac{C^2}{J} = \frac{J}{V^2} = \frac{W \cdot s}{V^2} = \frac{s}{\Omega}$$

**Flux (magnetic)'Φ':** Measure of the magnetic field strength existing on a two dimensional surface, such as one side of a magnet. The SI unit of magnetic flux is the weber (Wb). One *weber* is equal to $1 \times 10^8$ magnetic field lines.

**Flux Density 'B':** Amount of magnetic flux per unit area of a section, perpendicular to the direction of flux. SI unit for flux density is webers per square meter (Wb/m$^2$). One Wb/m$^2$ equals one tesla.

$$B = \frac{\Phi}{A}$$

**Gauss 'G':** The cgs unit of measurement of a magnetic field B (which is also known as the "magnetic flux density," or the "magnetic induction"), named after the German mathematician and physicist Karl Friedrich Gauss. One gauss is defined as one maxwell per square centimeter; it equals $1 \times 10^{-4}$ tesla (SI unit for B).

**Henry:** a Henry is the SI unit of inductance. It is named after Joseph Henry, the American scientist who discovered electromagnetic induction.

$$H = \frac{Wb}{A} = \frac{V \cdot S}{A} = \frac{J \cdot s^2}{C^2} = \Omega \cdot s$$

**Impedance 'Z':** Describes a measure of opposition to alternating current (AC). Z = R +jX

**Ohm 'Ω':** SI unit of electrical resistance, named after Georg Simon Ohm.

$$\Omega = \frac{E}{I} = \frac{J}{s \cdot A^2} = \frac{J \cdot s}{C^2}$$

**Ohm's Law:** States that the current through a conductor between two points is directly proportional to the potential difference or voltage across the two points, and inversely proportional to the resistance between them provided the temperature remains constant.

$$I = \frac{E}{R} = \frac{J}{s \cdot A^2} = \frac{J \cdot s}{C^2}$$

**Permeability:** The degree of magnetization of a material in response to a magnetic field.

**Reactance 'X':** The opposition of a circuit element to a change of current, caused by the build-up of electric or magnetic fields in the element. Those fields act to produce counter-EMF that is proportional to either the rate of change (time derivative), or accumulation (time integral), of the current. An ideal resistor has zero reactance, while ideal inductors and capacitors consist entirely of reactance with zero resistance.

$$X_L = 2\pi f L \quad X_C = \frac{1}{2\pi f C}$$

**Resistivity '$\Omega \bullet m$':** Measure of how strongly a material opposes the flow of electric current.

$$\rho = \frac{E}{J} = R\frac{A}{\ell}$$

**Volt 'V or E':** SI derived unit of electromotive force, commonly called "voltage". It is named in honor of the Italian physicist Alessandro Volta, who invented the voltaic pile, possibly the first chemical battery.

$$V = \frac{W}{A} = \frac{J}{A \cdot s} = \frac{N \cdot m}{C} = \frac{J}{C}$$

**Watt:** Derived unit of power in the International System of Units (SI), named after the Scottish engineer James Watt. The unit measures the rate of energy conversion. It is defined as one joule per second.

$$W = \frac{J}{s} = \frac{N \cdot m}{s} = \frac{kg \cdot m^2}{s^3} = \frac{V^2}{\Omega} = A^2\Omega$$

# 1. CSE PE EXAM – GENERAL INFORMATION

Reference the NCEES website (www.ncees.org) for detailed information regarding the format and scheduling of the CSE PE Exam. The CSE PE Exam is an 8-hour open book exam that is offered only once a year in October.

For ease of study and reference, the chapters of this book will be aligned to the categories of the CSE PE exam. Some topics will be covered in greater detail than other topics.

**Measurement** comprises approximately 20% of the exam and includes the following subtopics:

- Sensor Technologies and Characteristics

- Material Compatibility

- Calculations (pressure drop, flow element sizing, level including D/P, unit conversions and linearization)

- Installation Details

**Signals, Transmission and Networking** comprises approximately 15% of the exam and includes the following subtopics:

- Signals

    - Pneumatic, electronic, optical, hydraulic, digital, analog and buses
    - Transducers (A/D, D/A, I/P)
    - IS barriers
    - Grounding, shielding, segregation, AC coupling
    - Basic signal circuit design (2-wire, 4-wire, loop powering, buses, etc.)
    - Circuit calculations (voltage, current, impedance)

- Transmission

    - Different communication systems architecture (fiber optic, coaxial cable, wireless, buses, paired conductors) and protocols (TCP/IP, OPC, etc.)
    - Distance considerations versus transmission medium

- Networking (routers, bridges, switches, firewalls, gateways, etc.)

**Final Control Elements** comprises approximately 20% of the exam and includes the following subtopics:

- Valves

    - Types (globe, ball, butterfly, etc.)
    - Characteristics (linear, low noise, equal %, shutoff class)
    - Calculations

- Fluid dynamics (cavitation, flashing, choked flow, etc.)
- Material selection based upon process characteristics (erosion, corrosion, pressure, temperature, etc.)
- Accessories (limit switches, solenoids, positioners, etc.)
- Environmental constraints (fugitive emissions, packing, bellows, etc.)
- Installation practices

- Pressure Relieving Devices

- Motor Controls

- Other Final Control Elements (self-regulating devices, solenoids, etc.)

**Control Systems** comprises approximately 22% of the exam and includes the following subtopics:

- Basic Documentation

- Theory

- Implementation

**Safety Systems** comprises approximately 15% of the exam and includes the following subtopics:

- Basic Documentation

- Theory

- Implementation

**Codes, Standards and Regulations** comprises approximately 8% of the exam and may include questions concerning the following:

- ANSI, API, ASME, IEC, IEEE, ISA, NEC, NEMA, NFPA, OSHA

# 2. MISCELLANEOUS REVIEW MATERIAL

## 2.1 MATHEMATICS

This section only intends to provide a high-level overview of the various math concentrations, not specific in-depth coverage.

### 2.1.1 SI Prefixes

The SI prefixes are derived from a Greek, Latin, Italian and Danish names that precedes an SI unit of measure. This name indicates a decade multiplier or divider.

**Table 2-1. SI Prefixes**

| Prefix | Symbol | Value |
|--------|--------|-------|
| Exa | E | $10^{18}$ |
| Peta | P | $10^{15}$ |
| Tera | T | $10^{12}$ |
| Giga | G | $10^{9}$ |
| Mega | M | $10^{6}$ |
| Kilo | k | $10^{3}$ |
| Hecto | h | $10^{2}$ |
| Deca | da | $10^{1}$ |
| Deci | d | $10^{-1}$ |
| Centi | c | $10^{-2}$ |
| Milli | m | $10^{-3}$ |
| Micro | μ | $10^{-6}$ |
| Nano | n | $10^{-9}$ |
| Pico | p | $10^{-12}$ |
| Femto | f | $10^{-15}$ |
| Atto | a | $10^{-18}$ |

cgs units (centimeter, gram, second)
mks units (meter, kilogram, second)

### 2.1.2 Algebra

The part of mathematics in which letters and other symbols are used to represent numbers and quantities in formulae and equations.

*Quadratic Equations*

$$ax^2 + bx + c = 0$$

$$r_1, r_2 = \frac{-b \pm \sqrt{b^2 - 4ac}}{2a} \qquad a(x - r_1)(x - r_2) = 0$$

The variables $r_1$ and $r_2$ can be real or imaginary depending on the coefficients a, b and c.

- If $b^2-4ac > 0$, then there are two different REAL roots.

  - This is an indication of an over-damped system.

- If $b^2-4ac = 0$, then there are two identical REAL roots.

  - This is an indication of a critically-damped system.

- If $b^2-4ac < 0$, then there are two complex conjugate roots with the following:

  Real part: $\dfrac{-b}{2a}$

  Imaginary part: $\pm\sqrt{\dfrac{c}{a}-\left(\dfrac{b}{2a}\right)^2}$

  - This is an indication of an under-damped system.

## *Exponentiation*

$$x^a x^b = x^{a+b}$$
$$x^{-a} = \frac{1}{x^a}$$
$$x^{\frac{1}{a}} = \sqrt[a]{x}$$
$$\left(x^a\right)^b = x^{ab}$$
$$x^{\frac{b}{a}} = \sqrt[a]{x^b} = \left(\sqrt[a]{x}\right)^b$$

## *Logarithms*

If $b^x = y$, then $x = \log_b y$.

Example:

$$If\ y = 10^x, then\ x = \log_{10} y \quad OR \quad If\ e^x = y, then\ x = \log_e y = \ln y$$
$$\log 100 = \log_{10} 2 \quad \log 0.01 = \log_{10} -2$$

Constants:

$$\log_a 1 = 0$$
$$\log_a a = 1$$

Other Identities:

$$\log_b y^a = (a)\log_b y$$
$$a^x = b^{(x\log b^a)}$$
$$\log_a y = (\log_b y)(\log_a b)$$
$$\log_b xy = \log_b x + \log_b y$$
$$\log(x = jy) = \log(\sqrt{x^2 + y^2}) + j(\log e) \times \tan^{-1}\frac{y}{x}$$

Antilog:

The antilog function is the inverse of the log function:

$$\text{antilog}\, 2 = 10^2 \quad \text{antilog} - 2 = 0.01$$

## *Matrix Mathematics*

A matrix is a rectangular array of numbers.

**Addition and Subtraction:** Matrices MUST be of the same size in order for addition/subtraction to work.

$$\begin{bmatrix} 0 & 1 & 2 \\ 9 & 8 & 7 \end{bmatrix} + \begin{bmatrix} 6 & 5 & 4 \\ 3 & 4 & 5 \end{bmatrix} = \begin{bmatrix} 0+6 & 1+5 & 2+4 \\ 9+3 & 8+4 & 7+5 \end{bmatrix} = \begin{bmatrix} 6 & 6 & 6 \\ 12 & 12 & 12 \end{bmatrix}$$

$$\begin{bmatrix} -1 & 2 & 0 \\ 0 & 3 & 6 \end{bmatrix} - \begin{bmatrix} 0 & -4 & 3 \\ 9 & -4 & -3 \end{bmatrix} = \begin{bmatrix} -1-0 & 2-(-4) & 0-3 \\ 0-9 & 3-(-4) & 6-(-3) \end{bmatrix} = \begin{bmatrix} -1 & 6 & -3 \\ -9 & 7 & 9 \end{bmatrix}$$

$$\begin{bmatrix} -3 & x \\ 2y & 0 \end{bmatrix} + \begin{bmatrix} 4 & 6 \\ -3 & 1 \end{bmatrix} = \begin{bmatrix} 1 & 7 \\ -5 & 1 \end{bmatrix}$$

*First simplify the left side of the equation:*

$$\begin{bmatrix} -3 & x \\ 2y & 0 \end{bmatrix} + \begin{bmatrix} 4 & 6 \\ -3 & 1 \end{bmatrix} = \begin{bmatrix} -3+4 & x+6 \\ 2y-3 & 0+1 \end{bmatrix} = \begin{bmatrix} 1 & 7 \\ -5 & 1 \end{bmatrix}$$

$$\therefore x + 6 = 7 \quad \text{so } x = 1$$
$$\text{and } 2y - 3 = -5 \quad \text{so } y = -1$$

**Multiplication:** Matrix multiplication is not commutative: the order in which matrices are multiplied is important. To multiply matrices, their ranks[2] must be compatible. In the matrix example shown below a (2x3) matrix is multiplied by a (3x2). Their product is a (2x2) matrix. To check for rank compatibility simply write the ranks as (M x N) x (N x Q). The matrices may be multiplied together ONLY if the (N) values are equal. If the (N) values are indeed equal then the resultant matrix will have a rank of (M x Q).

---

2.  Rank of a matrix is defined as, the maximum number of linearly independent column vectors in the matrix, OR the maximum number of linearly independent row vectors in the matrix.

---

Example:

Multiply the ROWS of Matrix A by the COLUMNS of Matrix B.

$$\begin{bmatrix} 1 & 0 & -2 \\ 0 & 3 & -1 \end{bmatrix} \times \begin{bmatrix} 0 & 3 \\ -2 & -1 \\ 0 & 4 \end{bmatrix} = \begin{bmatrix} (1\times0)+(0\times(-2))+((-2)\times0) & (1\times3)+(0\times(-1))+((-2)\times4) \\ (0\times0)+(3\times(-2))+((-1)\times0) & (0\times3)+(3\times(-1))+((-1)\times4) \end{bmatrix} =$$

*Matrix A*    *Matrix B*

$$\begin{bmatrix} (0+0+0) & (3-0-8) \\ (0-6+0) & (0-3-4) \end{bmatrix} = \begin{bmatrix} 0 & -5 \\ -6 & -7 \end{bmatrix}$$

**Division:** There is NO such operation as matrix division. You MUST multiply by a reciprocal. *Not all matrices may be inverted because there is no inverse of zero and you cannot divide by zero.*

*If* $A = \begin{bmatrix} 8 & 3 \\ 5 & 2 \end{bmatrix}$    *Let* $A^{-1} = \begin{bmatrix} w & y \\ x & z \end{bmatrix}$

*Then* $AA^{-1} = \begin{bmatrix} 8w+3x & 8y+3z \\ 5w+2x & 5y+2z \end{bmatrix} = \begin{bmatrix} 1 & 0 \\ 0 & 1 \end{bmatrix}$

*The resulting equations* $\quad \begin{matrix} 8w+3x=1 & 8y+3z=0 \\ 5w+2x=0 & 5y+2z=1 \end{matrix}$

*Have the solution w=2; x=-5; y=-3; z=8*

$$\therefore \quad A^{-1} = \begin{bmatrix} 2 & -3 \\ -5 & 8 \end{bmatrix}$$

## 2.1.3 Trigonometry

The branch of mathematics dealing with the relations of the sides and angles of triangles and with the relevant functions of any angles.

$$\frac{r}{2\pi r}(360°) = \left(\frac{180}{\pi}\right)°$$

$$1 \cdot \text{radian} = \left(\frac{180}{\pi}\right)° = 57.3°$$

$$\sin\theta = \frac{\text{opposite}}{\text{hypotenuse}} = \frac{a}{c}$$

$$\csc\theta = \frac{\text{hypotenuse}}{\text{opposite}} = \frac{c}{a} = \frac{1}{\sin A}$$

$$\cos\theta = \frac{\text{adjacent}}{\text{hypotenuse}} = \frac{b}{c}$$

$$\sec\theta = \frac{\text{hypotenuse}}{\text{adjacent}} = \frac{c}{b} = \frac{1}{\cos A}$$

$$\tan\theta = \frac{\text{opposite}}{\text{adjacent}} = \frac{a}{b} = \frac{\sin A}{\cos A}$$

$$\cot\theta = \frac{\text{adjacent}}{\text{opposite}} = \frac{b}{a} = \frac{\cos A}{\sin A}$$

### *Trigonometric Identities*

$$\sin^2 x + \cos^2 x = 1$$

$$1 + \tan^2 x = \sec^2 x$$

$$1 + \cot^2 x = \csc^2 x$$

$$\sin 2x = 2\sin x \bullet \cos x$$

$$\tan 2x = \frac{2\tan x}{1 - \tan^2 x}$$

$$\cos 2x = 2\cos^2 x - 1 = \cos^2 x - \sin^2 x = 1 - 2\sin^2 x$$

$$\sin(x + y) = \sin x \cos y + \cos x \sin y$$

$$\sin(x - y) = \sin x \cos y - \cos x \sin y$$

$$\cos(x + y) = \cos x \cos y - \sin x \sin y$$

$$\cos(x - y) = \cos x \cos y - \sin x \sin y$$

$$2\sin x \sin y = \cos(x - y) - \cos(x + y)$$

$$2\cos x \cos y = \cos(x - y) + \cos(x + y)$$

$$2\sin x \cos y = \sin(x + y) + \sin(x - y)$$

## 2.1.4 Calculus

A form of mathematics that deals with the finding and properties of derivatives and integrals of functions, by methods originally based on the summation of very small differences. The two main types are differential *calculus* and integral *calculus.*

### *Differential Calculus*

Differentiation is the procedure used to take the derivative of a function. The derivative is a measure of how a function changes as its input changes, as indicated by a tangent line to a curve on the graph (Figure 2-1). For commonly used derivative formulas, reference Table 2-2.

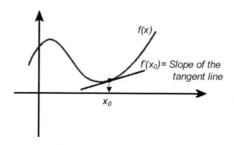

**Figure 2-1. Derivative**

The derivative method is used to compute the rate at which a dependent output y changes with respect to a change in the independent input x. This rate of change is called the **derivative** of y with respect to x (i.e., the dependence of y upon x means that y is a <u>function</u> of x). If x and y are <u>real numbers</u>, and if the <u>graph</u> of y is plotted against x, the derivative measures the <u>slope</u> of this graph at each point. This functional relationship is often denoted y = $f$(x), where $f$ denotes the function.

The simplest case is when y is a <u>linear function</u> of x, meaning that the graph of y against x is a straight line. In this case, y = $f$(x) = m x + b, for real numbers m and c, and the slope m is given by:

$$'m = \frac{change\ in\ y}{change\ in\ x} = \frac{\Delta y}{\Delta x}$$

The idea is to compute the rate of change as the limiting value of the ratio of the differences $\Delta y / \Delta x$ as $\Delta x$ becomes:

$$\frac{dy}{dx}$$

## Differentiation Rules

**Constant Rule:** if $f$(x) is constant, then f′ = 0

**Sum Rule:** for all functions $f$ and g and all real numbers a and b. (af + bg)′ = af′ +bg′

**Product Rule:** for all functions $f$ and g. (fg)′ = f′g + fg′

**Quotient Rule:** for all functions $f$ and g where g ≠ 0.

$$\left(\frac{f}{g}\right)' = \frac{f'g - fg'}{g^2}$$

**Power Rule:** if f′(x)=x″, then f′(x) = nx$^{(n-1)}$

**Chain Rule:** If f(x) = h(g(x)), then F′(x) = h′(g(x)) * g′(x)

Examples:

Find $\frac{dy}{dx}$ if $y = \frac{3x+5}{2x-3}$

$$\frac{dy}{dx} = \frac{3(2x-3) - 2(3x+5)}{(2x-3)^2} =$$

$$\frac{6x - 9 - 6x - 10}{(2x+3)^2} = \frac{-19}{(2x+3)^2}$$

Find f′(3) if $f(x) = x^2 - 8x + 3$
f′(x) = 2x − 8
f′(3) = (2)(3) − 8 = −2

if $y = \frac{4}{x+2}$ find $y'$ at (2,1)

y′ = 0(x + 2) − 1(4)

at (2,1); $y' = \frac{-4}{(2+2)^2} = -\frac{4}{16} = -\frac{1}{4}$

Alternate solution to the above example (right):

$$y = 4(x+2)^{-1} \Rightarrow y' = -4(x+2)^{-1-1} = \frac{-4}{(x+2)^2} \Rightarrow y'(2,1)) = \frac{-4}{(2+2)^2} = -\frac{1}{4}$$

## Table 2-2. Table of Derivatives

Power of x

$$\frac{d}{dx}C = 0 \qquad \frac{d}{dx}x = 1 \qquad \frac{d}{dx}x^n = nx^{(n-1)}$$

Exponential / Logarithmic

$$\frac{d}{dx}e^x = e^x \qquad \frac{d}{dx}b^x = b^x \ln(b) \qquad \frac{d}{dx}\ln(x) = \frac{1}{x}$$

Trigonometric

$$\frac{d}{dx}\sin x = \cos x \qquad \frac{d}{dx}\csc x = -\csc x \cdot \cot x$$

$$\frac{d}{dx}\cos x = -\sin x \qquad \frac{d}{dx}\sec x = \sec x \cdot \tan x$$

$$\frac{d}{dx}\tan x = \sec^2 x \qquad \frac{d}{dx}\cot x = -\csc^2 x$$

Inverse Trigonometric

$$\frac{d}{dx}\sin^{-1} x = \frac{1}{\sqrt{x^2-1}} \qquad \frac{d}{dx}\csc^{-1} x = \frac{-1}{|x|\sqrt{x^2-1}}$$

$$\frac{d}{dx}\cos^{-1} x = \frac{-1}{\sqrt{1-x^2}} \qquad \frac{d}{dx}\sec^{-1} x = \frac{1}{|x|\sqrt{x^2-1}}$$

$$\frac{d}{dx}\tan^{-1} x = \frac{1}{1+x^2} \qquad \frac{d}{dx}\cot^{-1} x = \frac{-1}{1+x^2}$$

Hyperbolic

$$\frac{d}{dx}\sinh x = \cosh x \qquad \frac{d}{dx}\csc hx = -(\coth x \csc hx)$$

$$\frac{d}{dx}\cosh x = \sinh x \qquad \frac{d}{dx}\sec hx = -(\tanh x \sec hx)$$

$$\frac{d}{dx}\tanh x = 1 - \tanh^2 x \qquad \frac{d}{dx}\coth x = 1 - \coth^2 x$$

## Integral Calculus

A common application of integration is to find the *average value* of a function.

For a function u, the average value from x = a to x = b is:

$$\bar{u} = \frac{1}{b-a}\int_a^b u\, dx$$

It is defined informally to be the net signed <u>area</u> of the region in the *xy*-plane bounded by the <u>graph</u> of *f*, the *x*-axis, and the vertical lines $x = a$ and $x = b$ (Figure 2-2). For commonly used integral formulas, reference Table 2-3.

The term *integral* may also refer to the notion of the <u>antiderivative</u>, a function F whose <u>derivative</u> is the given function *f*.

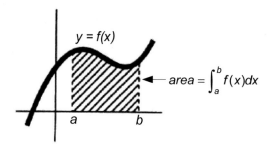

**Figure 2-2. Average Value of a Function**

Product Rule: $d(uv) = udv + vdu$

Quotient Rule: $d\left(\dfrac{u}{v}\right) = \dfrac{vdu - udv}{v^2}$

Chain Rule: $(u \times v)' = (u' \times v)v'$

## Table 2-3. Table of Integrals

$\int f(x)dx = F(x) + C$

$\int kf(x)dx = k \int f(x)dx$  $\qquad \int [f(x) \pm g(x)]dx = \int f(x)dx \pm \int g(x)dx$

$\int kdx = kx + C$  $\qquad \int x^n dx = \dfrac{x^{(n-1)}}{n+1} + C; n \neq 1$

$\int \sin x dx = -\cos x + C$  $\qquad \int \cos x dx = \sin x + C$

$\int \sec^2 x dx = \tan x + C$  $\qquad \int \csc^2 dx = -\cot x + C$

$\int \sec x \tan x dx = \sec x + C$  $\qquad \int \csc x \cot x dx = -\csc x + C$

$\int e^x dx = e^x + C$  $\qquad \int \dfrac{dx}{x} = \ln|x| + C$

$\int \tan x dx = -\ln|\cos x| + C$  $\qquad \int \cot x dx = \ln|\sin x| + C$

$\int \sec x dx = \ln|\sec x + \tan x| + C$  $\qquad \int \csc x dx = -\ln|\sec x + \tan x| + C$

$\int \dfrac{dx}{\sqrt{a^2 - x^2}} = \sin^{-1}\dfrac{x}{a} + C$  $\qquad \int \dfrac{dx}{a^2 + x^2} = \dfrac{1}{a}\tan^{-1}\dfrac{x}{a} + C$

$\int \dfrac{dx}{x\sqrt{x^2 - a^2}} = \dfrac{1}{a}\sec^{-1}\dfrac{x}{a} + C$

Examples:

$\int -8dx = -8x + C$

$\int 3x^2 dx = x^3 + C$

$\int (6x^2 + 5x - 3)dx = \dfrac{6x^3}{3} + \dfrac{5x^2}{2} - 3x + 2 = 2x^3 + 2.5x^2 - 3x + C$

---

## 2.1.6 Differential Equations

A differential equation is a mathematical expression combining a function and one or more of its derivatives.

*First Order (Linear):* A first order differential equation is an equation that contains a first, but no higher, derivative of an unknown function. It can be written as a sum of products of multipliers of the function and its derivatives. If the multipliers are scalar then the differential equation is said to have constant coefficients. If the function or one of its derivatives is raised to some power the equation is said to be non-linear. The form for the first order linear equation is $y' + p(x)y = q(x)$ where p and q are continuous functions of $x$.

*Second Order (Linear):* A second order differential equation is an equation that involves the second derivative of an unknown function, but no derivative of a higher order:

$$y'' + p(x)y' + q(x) = r(x) \text{ where p, q and r are continuous functions of } x.$$

*Homogeneous:* The sum of the derivatives terms is equal to zero.

*Non-Homogeneous:* The sum of the derivative terms is equal to non-zero.

## 2.1.7 Laplace Transforms

The Laplace transform[3] is an important tool for solving systems of linear differential equations with constant coefficients. The strategy is to transform the more difficult differential equations into simple algebra problems where solutions are more easily obtained.

This book will not attempt to teach Laplace transforms, but rather just provide a table of commonly used transforms (Table 2-4).

## 2.1.8 Bode Plot

A Bode plot (Figure 2-3)[4] is a graphical representation of a system, used to evaluate its stability and performance. It is a combination of two graphs (plots) on log (logarithmic) paper and consists of a Bode phase plot and a Bode magnitude (gain) plot. They are both drawn as functions of frequency where each cycle represents a factor of ten in frequency. The Bode magnitude plot is a graph where the frequency is plotted along the x-axis and the resultant gain (represented as decibels – dB) at that frequency is plotted along the y-axis. The Bode phase plot is a graph where the frequency is again plotted along the x-axis and the phase shift of that frequency is plotted along the y-axis.

A "passband" is the range of frequencies or wavelengths that can pass through a circuit without being attenuated.

A "stopband" is a band of frequencies, between specified limits, through which a circuit does not allow signals to pass.

---

3.  Named for French mathematical astronomer Pierre-Simon Laplace.
4.  Image derived from Wikipedia.

---

## Table 2-4. Laplace Transforms

| $f(t)$ | $\mathcal{L}[f(t)]$ |
|---|---|
| $1$ | $\dfrac{1}{s}$ |
| $e^{at}$ | $\dfrac{1}{s-a}$ |
| $\sin at$ | $\dfrac{a}{s^2 + a^2}$ |
| $\cos at$ | $\dfrac{s}{s^2 + a^2}$ |
| $t\sin at$ | $\dfrac{2as}{\left(s^2 + a^2\right)^2}$ |
| $t\cos at$ | $\dfrac{s^2 - a^2}{\left(s^2 + a^2\right)^2}$ |
| $e^{at}\sin bt$ | $\dfrac{b}{\left(s-a\right)^2 + b^2}$ |
| $e^{at}\cos bt$ | $\dfrac{s-a}{\left(s-a\right)^2 + b^2}$ |
| $\dfrac{t^n}{n!}; n \in N$ | $\dfrac{1}{s^{n+1}}$ |
| $e^{at} \times \dfrac{t^n}{n!}; n \in N$ | $\dfrac{1}{\left(s-a\right)^{n+1}}$ |

## 2.1.9 Nyquist Plot

A Nyquist[5] plot (Figure 2-4) exhibits a relationship to the Bode plots of the system. If the Bode phase plot is plotted as the angle θ, and the Bode magnitude plot is plotted as the distance r, then the Nyquist plot of a system is the polar representation of the Bode plot.[6]

*Nyquist Stability Criteria*

This is a test for system stability. The criteria states that the number of unstable closed-loop poles (zeroes) is equal to the number of unstable open-loop poles (zeroes) plus the number of encirclements of the origin of the Nyquist plot of the complex function D(s) (aka the Argument Principle[7]).

---

5.   Named after Harry Theodore Nyquist of Bell Labs.
6.   Image derived from www.math.uic.edu.
7.   Developed by Augustin Louis Cauchy.

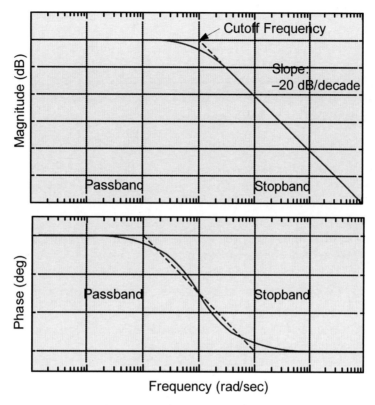

**Figure 2-3. Bode Plot**

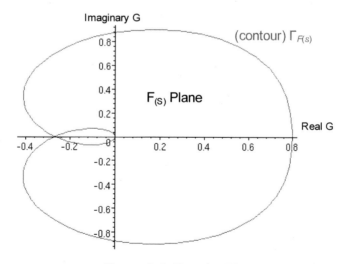

**Figure 2-4. Nyquist Plot**

- A feedback control system is stable if, and only if, the contour $\Gamma_{F(s)}$ in the F(s) plane does not encircle the (-1, 0) point when P (number of poles) is 0 (i.e., the point –(1, 0) is used because that is where a unit circle drawn with its center at the origin (0,0) crosses the real-axis which means that point is -180° from the origin). Therefore, the feedback control system is stable when the unit circle crossing point is at a frequency lower than -180°.

- A feedback control system is unstable when the same unit circle crossing point on the real-axis is at a frequency higher than -180°.

## 2.2 THERMODYNAMICS

Thermodynamics is the study of the effects of work, heat, and energy on a system. There are three laws:

**First Law (Law of Conservation):** Energy can be changed from one form to another, but it cannot be created or destroyed. $\Delta U = Q - W$ (U = internal energy; Q = heat added to system; W = work done by system).

**Second Law (Law of Entropy):** Energy spontaneously disperses from being localized to becoming spread out if it is not hindered from doing so (i.e., heat is transferred from high temperature to low temperature regions).

**Third Law:** This law is an extension of the second law: as temperature approaches absolute zero, the entropy of a system approaches a constant.

### 2.2.1 Terminology

**Exothermic:** A type of chemical reaction that releases energy in the form of heat, light, or sound. This type of reaction may occur spontaneously.

**Endothermic:** A type of chemical reaction that must absorb energy in order to proceed. This type of reaction does not occur spontaneously because work must be done in order to get this reaction to occur.

**Entropy (s):** Entropy is a measure of how much heat must be rejected to a lower temperature receiver at a given pressure and temperature (measured in BTU/lbm–°R). Heat Q released by a system into its surroundings is indicated by a negative quantity ($Q° < °0$); when a system absorbs heat from its surroundings, Q is indicated by a positive value ($Q° > °0$). An example of this heat rejection is to place a glass of hot liquid into a colder environment, this results in a flow of heat from the glass to the environment's surrounding atmosphere until an equilibrium is reached.

$$s = \frac{Q}{T} \quad \text{Q = heat content of the system} \quad \text{T = Temperature of the system (°R)}$$

**Enthalpy (h) (inherent heat):** Enthalpy is measured in British thermal units per pound (mass), or BTU/lbm, and represents the total energy content of the system (the enthalpy SI unit is J/kg). It expresses the internal energy and flow work, or the total potential energy and kinetic energy contained within a substance.

$$h = U + PV \quad \text{(U = internal energy; P = pressure; V = volume)}$$

**Adiabatic process:** A thermodynamic process in which there is no transfer of heat between the process and the surrounding environment. An adiabatic process is generally obtained by surrounding the entire system with a strong insulating material or by carrying out the process so quickly that there is no time for significant heat transfer to take place.

**Isothermal process:** A thermodynamic process in which no temperature change occurs ($\Delta T = 0$). Note that heat transfer can occur without causing a change in temperature of the working fluid.

## 2.2.2 Heat Addition and Temperature

When heat is added to a material, one of two things will occur: the material will change temperature or the material will change state. When a substance is below the temperature at a given pressure required to change state, the addition of *sensible heat* will raise the temperature of the substance. Sensible heat applied to a pot of water will raise its temperature until it boils. Once the substance reaches the necessary temperature at a given pressure to change state, the addition of *latent heat* causes the substance to change state. Adding latent heat to the boiling water does not get the water any hotter, but changes the liquid (water) into a gas (steam).

One can state that a certain amount of heat is required to raise the temperature of a substance one degree. This energy is called the *specific heat capacity ($\Delta Q = mc\Delta T$)*. The specific heat capacity of a substance depends upon the volume and pressure of the material, except for water, the specific heat capacity is 1 BTU/lbm-°F (1kcal/kg-°C in SI units) and remains constant. This means that if we add 1 BTU of heat to 1 lbm of water, the temperature will rise 1°F. The specific heat value and the specific heat capacity value for water are equal.

## 2.2.3 Mollier Steam Diagram

A Mollier diagram (Figure 2-5) can be used to determine enthalpy versus entropy of water and steam with the pressure identified on the y-axis in a log scale, and enthalpy identified on the x-axis. Other properties identified on the Mollier diagram are constant temperature, density and entropy lines. The Mollier diagram[8] is useful when analyzing the performance of adiabatic steady-flow processes.[9]

### *How To Read A Mollier Diagram*

**Example:** Superheated steam at 700psi and 680°F is expanded at constant entropy to 140psi.

1.  Locate point 1 at the intersection of the 700psi and 680°F line – then read h (h = 1333 BTU/lbm).

2.  Follow the entropy line downward vertically to the 140psi line and read h (h = 1178 BTU/lbm)

    $\therefore$ h = 1178 – 1333 = –155 BTU/lbm.

## 2.2.4 Psychrometric Chart

A psychrometric chart (Figure 2-6) is graphical representation of the thermodynamic properties of moist air. The chart is used to determine the state of an air-water vapor mixture when at least two properties are known.

### *How to Use a Psychrometric Chart*

**Example:** Assume dry bulb temperature = 78°F and wet bulb temperature = 65°F.

---

8.  Mollier diagrams are named after Richard Mollier, a professor at Dresden University who pioneered the graphical display of the relationship of temperature, pressure, enthalpy, entropy and volume of steam and moist air.
9.  Reference www.chemicalogic.com for a blank example of this Mollier Diagram.

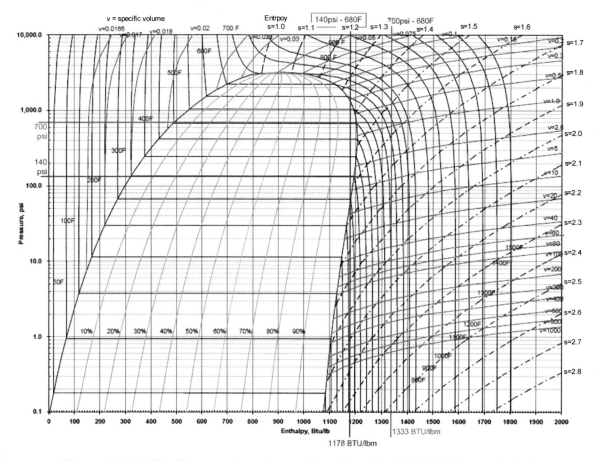

**Figure 2-5. Mollier Diagram (based upon the Scientific IAPWS-95 formulation)**

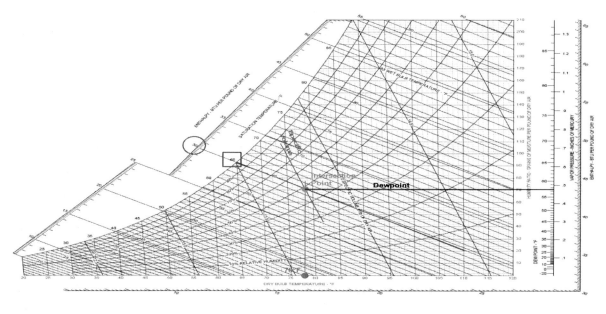

**Figure 2-6. Psychrometric Chart (*Linric Company*)**

- First locate 78°F on the DB Temperature Scale at the bottom of the chart.

- Then locate 65°F WB on the saturation curve scale.

- Extend a vertical line from the 78°F DB point and a diagonal line from the 65°F WB point to intersect the vertical DB line.

- The point of intersection of the two lines indicates the condition of the given air. As a result:

  - Enthalpy of air = 30 BTU/lb
  - Specific volume of air = 13.7 ft$^3$/lb
  - Dewpoint = 57.5°F

## 2.2.5 Properties of Water

### Table 2-5. Properties of Water

| Temperature Of Water °F | Saturation Pressure psia | Specific Volume ft³/lb | Weight Density lb/ft³ | Weight lb/gal |
|---|---|---|---|---|
| 32 | 0.08859 | 0.016022 | 62.414 | 8.3436 |
| 40 | 0.12163 | 0.016019 | 62.426 | 8.3451 |
| 50 | 0.17796 | 0.016023 | 62.410 | 8.3430 |
| 60 | 0.25611 | 0.016033 | 62.371 | 8.3378 |
| 70 | 0.36292 | 0.016050 | 62.305 | 8.3290 |
| 80 | 0.50683 | 0.016072 | 62.220 | 8.3176 |
| 90 | 0.69813 | 0.016092 | 62.116 | 8.3037 |
| 100 | 0.94924 | 0.016130 | 61.996 | 8.2877 |
| 110 | 1.2750 | 0.016165 | 61.862 | 8.2698 |
| 120 | 1.6927 | 0.016204 | 61.7132 | 8.2498 |
| 130 | 2.2230 | 0.016247 | 61.550 | 8.2280 |
| 140 | 2.8892 | 0.016293 | 61.376 | 8.2048 |
| 150 | 3.7184 | 0.016343 | 61.188 | 8.1797 |
| 160 | 4.7414 | 0.016395 | 60.994 | 8.1537 |
| 170 | 5.9926 | 0.016451 | 60.787 | 8.1260 |
| 180 | 7.5110 | 0.016510 | 60.569 | 8.0969 |
| 190 | 9.340 | 0.016572 | 60.343 | 8.0667 |
| 200 | 11.526 | 0.016637 | 60.107 | 8.0351 |
| 210 | 14.123 | 0.016705 | 59.862 | 8.0024 |
| 212 | 14.696 | 0.016719 | 59.812 | 7.9957 |
| 220 | 17.186 | 0.016775 | 59.613 | 7.9690 |
| 240 | 24.968 | 0.016926 | 59.081 | 7.8979 |
| 260 | 35.427 | 0.017098 | 58.517 | 7.8226 |
| 280 | 49.200 | 0.017264 | 57.924 | 7.7433 |
| 300 | 67.005 | 0.01745 | 57.307 | 7.6608 |

| 350 | 134.604 | 0.01799 | 55.586 | 7.4308 |
|-----|---------|---------|--------|--------|
| 400 | 247.259 | 0.01864 | 53.684 | 7.1717 |
| 450 | 422.55 | 0.01943 | 51.467 | 6.8801 |
| 500 | 680.86 | 0.02043 | 48.948 | 6.5433 |
| 550 | 1045.43 | 0.02176 | 45.956 | 6.1434 |
| 600 | 1543.2 | 0.02364 | 42.301 | 5.6548 |
| 650 | 2208.4 | 0.02674 | 37.397 | 4.9993 |
| 700 | 3094.3 | 0.03662 | 27.307 | 3.6505 |

Saturation Pressure: $P = 10^{\left[ A - \left( \frac{B}{(C+T)} \right) \right]}$

*A, B and C are the values of the Antoine constants A, B and C for the temperature T from NIST.*

Density: $\rho = P/RT$ *(P in Pascal; T in Kelvin)*

Specific Volume: $1/\rho$

### *Linear Interpolation in Tables*

Linear interpolation is a convenient way to fill in holes in tabular data. The formula for linear interpolation is:

$$d = d_1 + \left( \frac{g - g_1}{g_2 - g_1} \right) (d_2 - d_1)$$

**Example:** Find the saturation pressure for water at 575°F.

From Table 2-5:
  d1 = 1045.43 psia
  d2 = 1543.2 psia
  g = 575°F
  g1 = 550°F
  f2 = 600°F

$$\therefore \quad d = 1045.3 + \left( \frac{575 - 550}{550 - 600} \right) (1045.43 - 1543.2) =$$
$$1045.3 + (-0.5)(-497.77) = 1045.3 + 248.89 = 1294.19 \text{psia}$$

## 2.2.6 Properties of Saturated Steam

Saturated steam occurs when steam and water are in equilibrium. Once the water's boiling point is reached, the water's temperature ceases to rise and stays the same until all the water is vaporized. As the water goes from a liquid state to a vapor state it receives energy in the form of "latent heat of vaporization." As long as there is some liquid water left, the steam's temperature is the same as the liquid water's temperature. This type of steam is called saturated steam. Industries normally use saturated steam for heating, cooking, drying and other processes.

Table 2-6 lists the properties of saturated steam at different pressures.

## Table 2-6. Properties of Saturated Steam[10]

| Pressure psig | Saturated Temperature ºF | Specific Volume V ft³/lb | | Heat Content BTU/lb | | Latent Heat of Vaporization ($h_{fg}$) BTU/lb |
| --- | --- | --- | --- | --- | --- | --- |
| | | Saturated Liquid ($V_f$) | Saturated Vapor ($V_g$) | Saturated Liquid ($h_f$) | Saturated Vapor ($h_g$) | |
| 0 | 212 | 0.0167 | 26.8 | 180 | 1150 | 970 |
| 1 | 215 | 0.0167 | 24.3 | 183 | 1151 | 967 |
| 2 | 218 | 0.0167 | 23.0 | 186 | 1153 | 965 |
| 3 | 222 | 0.0168 | 21.8 | 190 | 1154 | 963 |
| 4 | 224 | 0.0168 | 20.7 | 193 | 1155 | 961 |
| 5 | 227 | 0.0168 | 19.8 | 195 | 1156 | 959 |
| 6 | 230 | 0.0168 | 18.9 | 198 | 1157 | 958 |
| 7 | 232 | 0.0169 | 18.1 | 200 | 1158 | 956 |
| 8 | 235 | 0.0169 | 17.4 | 203 | 1158 | 955 |
| 9 | 237 | 0.0169 | 16.7 | 205 | 1159 | 953 |
| 10 | 239 | 0.0169 | 16.1 | 208 | 1160 | 952 |
| 11 | 242 | 0.0169 | 15.6 | 210 | 1161 | 950 |
| 12 | 244 | 0.0170 | 15.0 | 212 | 1161 | 949 |
| 13 | 246 | 0.0170 | 14.5 | 214 | 1162 | 947 |
| 14 | 248 | 0.0170 | 14.0 | 216 | 1163 | 946 |
| 15 | 250 | 0.0170 | 13.6 | 218 | 1164 | 945 |
| 16 | 252 | 0.0170 | 13.2 | 220 | 1164 | 943 |
| 17 | 254 | 0.0170 | 12.8 | 222 | 1165 | 942 |
| 18 | 255 | 0.0170 | 12.5 | 224 | 1165 | 941 |
| 19 | 257 | 0.0171 | 12.1 | 226 | 1166 | 940 |
| 20 | 259 | 0.0171 | 11.1 | 227 | 1166 | 939 |
| 25 | 267 | 0.0171 | 10.4 | 236 | 1169 | 933 |
| 30 | 274 | 0.0172 | 9.4 | 243 | 1171 | 926 |
| 35 | 281 | 0.0173 | 8.5 | 250 | 1173 | 923 |
| 40 | 287 | 0.0173 | 7.74 | 256 | 1175 | 919 |
| 45 | 292 | 0.0174 | 7.14 | 262 | 1177 | 914 |
| 50 | 298 | 0.0174 | 6.62 | 267 | 1178 | 911 |
| 55 | 302 | 0.0175 | 6.17 | 272 | 1179 | 907 |
| 60 | 307 | 0.0175 | 5.79 | 277 | 1181 | 903 |
| 65 | 312 | 0.0176 | 5.45 | 282 | 1182 | 900 |
| 70 | 316 | 0.0176 | 5.14 | 286 | 1183 | 897 |
| 75 | 320 | 0.0176 | 4.87 | 290 | 1184 | 893 |
| 80 | 324 | 0.0177 | 4.64 | 294 | 1185 | 890 |
| 85 | 327 | 0.0177 | 4.42 | 298 | 1186 | 888 |
| 90 | 331 | 0.0178 | 4.24 | 301 | 1189 | 887 |
| 95 | 334 | 0.0178 | 4.03 | 305 | 1189 | 884 |
| 100 | 338 | 0.0178 | 3.88 | 308 | 1190 | 882 |
| 105 | 341 | 0.0179 | 3.72 | 312 | 1191 | 877 |
| 110 | 343 | 0.0179 | 3.62 | 314 | 1191 | 877 |
| 115 | 347 | 0.0180 | 3.44 | 318 | 1192 | 872 |
| 120 | 350 | 0.0180 | 3.34 | 321 | 1193 | 872 |

| | | | | | | |
|---|---|---|---|---|---|---|
| 125 | 353 | 0.0180 | 3.21 | 324 | 1193 | 867 |
| 130 | 355 | 0.0180 | 3.12 | 327 | 1194 | 867 |
| 135 | 358 | 0.0181 | 3.02 | 329 | 1194 | 864 |
| 140 | 361 | 0.0181 | 2.92 | 332 | 1195 | 862 |
| 145 | 363 | 0.0181 | 2.84 | 335 | 1196 | 860 |
| 150 | 366 | 0.0182 | 2.75 | 337 | 1196 | 858 |
| 155 | 368 | 0.0182 | 2.67 | 340 | 1196 | 854 |
| 160 | 370 | 0.0182 | 2.60 | 342 | 1196 | 854 |
| 165 | 373 | 0.0183 | 2.53 | 345 | 1197 | 852 |
| 170 | 375 | 0.0183 | 2.47 | 347 | 1197 | 850 |
| 175 | 378 | 0.0183 | 2.40 | 350 | 1198 | 848 |
| 180 | 380 | 0.0184 | 2.34 | 352 | 1198 | 846 |
| 185 | 382 | 0.0184 | 2.29 | 355 | 1199 | 844 |
| 190 | 384 | 0.0184 | 2.23 | 357 | 1199 | 842 |
| 200 | 388 | 0.0185 | 2.14 | 361 | 1199 | 838 |
| 210 | 392 | 0.0185 | 2.05 | 365 | 1200 | 835 |
| 220 | 396 | 0.0186 | 1.96 | 369 | 1200 | 831 |
| 230 | 399 | 0.0186 | 1.88 | 373 | 1201 | 828 |
| 240 | 403 | 0.0187 | 1.81 | 377 | 1201 | 824 |
| 250 | 406 | 0.0187 | 1.75 | 380 | 1201 | 821 |
| 260 | 410 | 0.0188 | 1.68 | 384 | 1201 | 817 |
| 270 | 413 | 0.0188 | 1.63 | 387 | 1202 | 814 |
| 280 | 416 | 0.0189 | 1.57 | 391 | 1202 | 811 |
| 290 | 419 | 0.0190 | 1.52 | 394 | 1202 | 807 |
| 300 | 421 | 0.0190 | 1.47 | 397 | 1202 | 805 |

## 2.2.7 Properties of Superheated Steam

As opposed to saturated steam, when all the water is vaporized any subsequent addition of heat will raise the steam's temperature. Steam heated beyond the saturated steam level is called superheated steam. Superheated steam is used almost exclusively for turbines. Turbines have a number of stages. The exhaust steam from the first stage is directed to a second stage on the same shaft, and so on. This means that saturated steam would get wetter and wetter as it went through the successive stages. This is due to the fact that saturated steam has a greater volume of water in it as the pressure gets lower. Not only would this situation promote water hammer[11], but the water particles would cause severe erosion within the turbine.

There is a good reason why superheated steam is not as suitable for process heating as is saturated steam. Superheated steam has to cool to saturation temperature before it can condense to release its enthalpy of evaporation. The amount of heat given up by superheated steam as it cools to saturation temperature is relatively small in comparison to its enthalpy of evaporation. Thus, if the steam has a large degree of superheat, it may take a relatively long time to cool, during which time the steam will release very little energy.

---

10. There are many free online tools available to calculate these values based upon the pressure in psig, www.spiraxsarco.com is just one example of a website that makes one of these calculation tools available free of charge.

11. Water hammer (aka fluid hammer) is a pressure surge that occurs when the fluid in motion is suddenly stopped or forced to change directions abruptly. This pressure wave then propagates throughout the equipment causing noise and vibration.

Table 2-7 lists the properties of superheated steam at various pressures.

### Table 2-7. Properties of Superheated Steam[12]

| Pressure | | Sat. Temp. | | Total Temperature ºF | | | | | | | | | | |
|---|---|---|---|---|---|---|---|---|---|---|---|---|---|---|
| psia | psig | | | 350 | 400 | 500 | 600 | 700 | 800 | 900 | 1000 | 1100 | 1300 | 1500 |
| 15 | 0.3 | 213.03 | V | 31.939 | 33.963 | 37.985 | 41.986 | 45.978 | 49.964 | 53.946 | 57.926 | 61.905 | 69.858 | 77.807 |
| | | | $h_g$ | 1216.2 | 1239.9 | 1287.3 | 1335.2 | 1383.8 | 1433.2 | 1483.4 | 1534.5 | 1586.5 | 1693.2 | 1803.4 |
| 20 | 5.3 | 227.96 | V | 23.900 | 25.428 | 28.457 | 31.466 | 34.465 | 37.458 | 40.447 | 43.435 | 46.420 | 52.388 | 58.352 |
| | | | $h_g$ | 1215.4 | 1239.2 | 1286.9 | 1334.9 | 1383.5 | 1432.9 | 1483.2 | 1534.3 | 1586.3 | 1693.1 | 1803.3 |
| 30 | 15.3 | 250.34 | V | 15.859 | 16.892 | 18.929 | 20.945 | 22.951 | 24.952 | 26.949 | 28.943 | 30.936 | 34.918 | 38.896 |
| | | | $h_g$ | 1213.6 | 1237.8 | 1286.0 | 1334.2 | 1383.0 | 1432.5 | 1482.8 | 1534.0 | 1586.1 | 1692.9 | 1803.2 |
| 40 | 25.3 | 267.25 | V | 11.838 | 12.624 | 14.165 | 15.685 | 17.195 | 18.699 | 20.199 | 21.697 | 23.194 | 26.183 | 29.168 |
| | | | $h_g$ | 1211.7 | 1236.4 | 1285.0 | 1333.6 | 1382.5 | 1432.1 | 1482.5 | 1533.7 | 1585.8 | 1692.7 | 1803.0 |
| 50 | 35.3 | 281.02 | V | 9.424 | 10.062 | 11.306 | 12.529 | 13.741 | 14.947 | 16.150 | 17.350 | 18.549 | 20.942 | 23.332 |
| | | | $h_g$ | 1209.9 | 1234.9 | 1284.1 | 1332.9 | 1382.0 | 1431.7 | 1482.2 | 1533.4 | 1585.6 | 1692.5 | 1802.9 |
| 60 | 45.3 | 292.71 | V | 7.815 | 8.354 | 9.400 | 10.425 | 11.438 | 12.446 | 13.450 | 14.452 | 15.452 | 17.448 | 19.441 |
| | | | $h_g$ | 1208.0 | 1233.5 | 1283.2 | 1332.3 | 1381.5 | 1431.3 | 1481.8 | 1533.2 | 1585.3 | 1692.4 | 1802.8 |
| 70 | 55.3 | 302.93 | V | 6.664 | 7.133 | 8.039 | 8.922 | 9.793 | 10.659 | 11.522 | 12.382 | 13.240 | 14.952 | 16.661 |
| | | | $h_g$ | 1206.0 | 1232.0 | 1282.2 | 1331.6 | 1381.0 | 1430.9 | 1481.5 | 1532.9 | 1585.1 | 1692.2 | 1802.6 |
| 80 | 65.3 | 312.04 | V | 5.801 | 6.218 | 7.018 | 7.794 | 8.560 | 9.319 | 10.075 | 10.829 | 11.581 | 13.081 | 14.577 |
| | | | $h_g$ | 1204.0 | 1230.5 | 1281.3 | 1330.9 | 1380.5 | 1430.5 | 1481.1 | 1532.6 | 1584.9 | 1692.0 | 1802.5 |
| 90 | 75.3 | 320.28 | V | 5.128 | 5.505 | 6.223 | 6.917 | 7.600 | 8.277 | 8.950 | 9.621 | 10.290 | 11.625 | 12.956 |
| | | | $h_g$ | 1202.0 | 1228.9 | 1280.3 | 1330.2 | 1380.0 | 1430.1 | 1480.8 | 1532.3 | 1584.6 | 1691.8 | 1802.4 |
| 100 | 85.3 | 327.82 | V | 4.590 | 4.935 | 5.588 | 6.216 | 6.833 | 7.443 | 8.050 | 8.655 | 9.258 | 10.460 | 11.659 |
| | | | $h_g$ | 1199.9 | 1227.4 | 1279.3 | 1329.6 | 1379.5 | 1429.7 | 1480.4 | 1532.0 | 1584.4 | 1691.6 | 1802.2 |
| 120 | 105.3 | 341.27 | V | 3.7815 | 4.0786 | 4.6341 | 5.1637 | 5.6813 | 6.1928 | 6.7006 | 7.2060 | 7.7096 | 8.7130 | 9.7130 |
| | | | $h_g$ | 1195.6 | 1224.1 | 1277.4 | 1328.2 | 1378.4 | 1428.8 | 1479.8 | 1531.4 | 1583.9 | 1691.3 | 1802.0 |
| 140 | 125.3 | 353.04 | V | - | 3.4661 | 3.9526 | 4.4119 | 4.8588 | 5.2995 | 5.7364 | 6.1709 | 6.6036 | 7.4652 | 8.3233 |
| | | | $h_g$ | - | 1220.8 | 1275.3 | 1326.8 | 1377.4 | 1428.0 | 1479.1 | 1530.8 | 1583.4 | 1690.9 | 1801.7 |
| 160 | 145.3 | 363.55 | V | - | 3.0060 | 3.4413 | 3.8480 | 4.2420 | 4.6295 | 5.0132 | 5.3945 | 5.7741 | 6.5293 | 7.2811 |
| | | | $h_g$ | - | 1217.4 | 1273.3 | 1325.4 | 1376.4 | 1427.2 | 1478.4 | 1530.3 | 1582.9 | 1690.5 | 1801.4 |
| 180 | 165.3 | 373.08 | V | - | 2.6474 | 3.0433 | 3.4093 | 3.7621 | 4.1084 | 4.4508 | 4.7907 | 5.1289 | 5.8014 | 6.4704 |
| | | | $h_g$ | - | 1213.8 | 1271.2 | 1324.0 | 1375.3 | 1426.3 | 1477.7 | 1529.7 | 1582.4 | 1690.2 | 1801.2 |
| 200 | 185.3 | 381.80 | V | - | 2.3598 | 2.7247 | 3.0583 | 3.3783 | 3.6915 | 4.0008 | 4.3077 | 4.6128 | 5.2191 | 5.8219 |
| | | | $h_g$ | - | 1210.1 | 1269.0 | 1322.6 | 1374.3 | 1425.5 | 1477.0 | 1529.1 | 1581.9 | 1689.8 | 1800.9 |
| 220 | 205.3 | 389.88 | V | - | 2.1240 | 2.4638 | 2.7710 | 3.0642 | 3.3504 | 3.6327 | 3.9125 | 4.1905 | 4.7426 | 5.2913 |
| | | | $h_g$ | - | 1206.3 | 1266.9 | 1321.2 | 1373.2 | 1424.7 | 1476.3 | 1528.5 | 1581.4 | 1689.4 | 1800.6 |
| 240 | 225.3 | 397.39 | V | - | 1.9268 | 2.2462 | 2.5316 | 2.8024 | 3.0661 | 3.3259 | 3.5831 | 3.8385 | 4.3456 | 4.8492 |
| | | | $h_g$ | - | 1202.4 | 1264.6 | 1319.7 | 1372.1 | 1423.8 | 1475.6 | 1527.9 | 1580.9 | 1689.1 | 1800.4 |
| 260 | 245.3 | 404.44 | V | - | - | 2.0619 | 2.3289 | 2.5808 | 2.8256 | 3.0663 | 3.3044 | 3.5408 | 4.0097 | 4.4750 |
| | | | $h_g$ | - | - | 1262.4 | 1318.2 | 1371.1 | 1423.0 | 1474.9 | 1527.3 | 1580.4 | 1688.7 | 1800.1 |
| 280 | 265.3 | 411.07 | V | - | - | 1.9037 | 2.1551 | 2.3909 | 2.6194 | 2.8437 | 3.0655 | 3.2855 | 3.7217 | 4.1543 |
| | | | $h_g$ | - | - | 1260.0 | 1316.8 | 1370.0 | 1422.1 | 1474.2 | 1526.8 | 1579.9 | 1688.4 | 1799.8 |
| 300 | 285.3 | 417.35 | V | - | - | 1.7665 | 2.0044 | 2.2263 | 2.4407 | 2.6509 | 2.8585 | 3.0643 | 3.4721 | 3.8764 |
| | | | $h_g$ | - | - | 1257.7 | 1315.2 | 1368.9 | 1421.3 | 1473.6 | 1526.2 | 1579.4 | 1688.0 | 1799.6 |
| 320 | 305.3 | 423.31 | V | - | - | 1.6462 | 1.8725 | 2.0823 | 2.2843 | 2.4821 | 2.6774 | 2.8708 | 3.2538 | 3.6332 |
| | | | $h_g$ | - | - | 1255.2 | 1313.7 | 1367.8 | 1420.5 | 1472.9 | 1525.6 | 1578.9 | 1687.6 | 1799.3 |

| | | | | | | | | | | | | | | |
|---|---|---|---|---|---|---|---|---|---|---|---|---|---|---|
| 340 | 325.3 | 428.99 | V | - | - | 1.5399 | 1.7561 | 1.9552 | 2.1463 | 2.3333 | 2.5175 | 2.7000 | 3.0611 | 3.4186 |
| | | | $h_g$ | - | - | 1252.8 | 1312.2 | 1366.7 | 1419.6 | 1472.2 | 1525.0 | 1578.4 | 1687.3 | 1799.3 |
| 360 | 345.3 | 434.41 | V | - | - | 1.4454 | 1.6525 | 1.8421 | 2.0237 | 2.2009 | 2.3755 | 2.5482 | 2.8898 | 3.2279 |
| | | | $h_g$ | - | - | 1250.3 | 1310.6 | 1365.6 | 1418.7 | 1471.5 | 1542.4 | 1577.9 | 1686.9 | 1798.8 |
| 380 | 365.3 | 439.61 | V | - | - | 1.3606 | 1.5598 | 1.7410 | 1.9139 | 2.0825 | 2.2484 | 2.4124 | 2.7366 | 3.0572 |
| | | | $h_g$ | - | - | 1247.7 | 1309.0 | 1364.5 | 1417.9 | 1470.8 | 1523.8 | 1577.4 | 1686.5 | 1798.5 |
| 400 | 385.3 | 444.60 | V | - | - | 1.2841 | 1.4763 | 1.6499 | 1.8151 | 1.9759 | 2.1339 | 2.2901 | 2.5987 | 2.9037 |
| | | | $h_g$ | - | - | 1245.1 | 1307.4 | 1363.4 | 1417.0 | 1470.1 | 1523.3 | 1576.9 | 1686.2 | 1798.2 |

V = specific volume, ft³/lb

$h_g$ = total heat of steam, BTU/lb

## 2.3 STATISTICS

*Degrees of Freedom (df)*

df is the number of values that are free to vary in the final calculation of a statistic:

df = n – 1 where n = number of samples.

**Standard Deviation ($\sigma$)** is a measure of the spread of the data about the mean value. Reference Figure 2-7 for the normal distribution curve between standard deviations.

$$\sigma = \sqrt{\frac{\Sigma(x-\bar{x})^2}{n-1}}$$   $x$ = sample value; $\bar{x}$ = mean value; $n$ = # of samples; $n-1 = df$

**Example:**
Consider a population consisting of the following values:

2, 4, 4, 4, 5, 5, 7, 9

There are eight data points in total, with a <u>mean</u> (or average) value of 5:

$$\frac{2+4+4+4+5+5+5+7+9}{8} = \frac{40}{8} = 5$$

To calculate the population standard deviation, we compute the difference of each data point from the mean, and square the result:

$$(2-5)^2 = (-3)^2 = 9$$

$$(4-5)^2 = (-1)^2 = 1$$

$$(4-5)^2 = (-1)^2 = 1$$

---

12. There are many free online tools available to calculate these values based upon the pressure and super-heat temperature, www.spiraxsarco.com is just one example of a website that makes one of these calculation tools available free of charge.

$$(4-5)^2 = (-1)^2 = 1$$

$$(5-5)^2 = (-0)^2 = 0$$

$$(5-5)^2 = (-0)^2 = 0$$

$$(7-5)^2 = (-2)^2 = 4$$

$$(9-5)^2 = (-4)^2 = 16$$

Next we average these values and take the square root, which gives the standard deviation:

$$\sqrt{\frac{9+1+1+1+0+0+4+16}{8-1}} = \sqrt{\frac{32}{7}} = 2.138$$

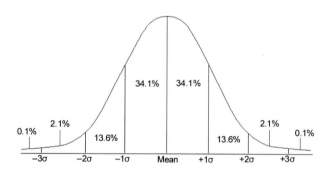

**Figure 2-7. Percentages in Normal Distribution between Standard Deviations**

### Six Sigma

Six Sigma comes from the notion that if one has six standard deviations between the process mean and the nearest specification limit, there will be practically no items that fail to meet specifications. To achieve Six Sigma, a process must not produce more than 3.4 defects per one million opportunities.

### 1.5 Sigma Shift

It has been shown that in the long term, processes usually do not perform as well as they do in the short term. As a result of this performance, the number of sigmas that will fit between the process mean and the nearest specification limit is likely to drop over time, as compared to an initial short-term study. To account for this real-life increase in process variation over time, an empirically-based 1.5 sigma shift is introduced into the calculation.

According to this concept, a process that fits Six Sigmas between the process mean and the nearest specification limit in a short-term study will, in the long term, only fit 4.5 sigmas. Either the process mean will move over time, or the long-term standard deviation of the process will be greater than that observed in the short term, or possibly both. Therefore, the widely accepted definition of a Six Sigma process is one that produces 3.4 Defective Parts per Million Opportunities (DPMO).

This definition is based on the fact that a process that is normally distributed will have 3.4 defective parts per million beyond a point that is 4.5 standard deviations above or below the mean. So the 3.4 DPMO of a "Six Sigma" process corresponds in fact to 4.5 sigmas, that is, 6 sigmas minus the 1.5 sigma shift introduced to account for long-term variation. This is done to prevent underestimation of the defect levels likely to be encountered in real-life operation.

Table 2-8 gives long-term DPMO values corresponding to various short-term sigma levels.

### Table 2-8. Long-Term DPMO Values

| ΣLevel | DPMO | % Defective | % Yield |
|---|---|---|---|
| 1 | 691,462 | 69% | 31% |
| 2 | 308,538 | 31% | 69% |
| 3 | 66,807 | 6.7% | 93.3% |
| 4 | 6,210 | 0.62% | 99.38% |
| 5 | 233 | 0.023% | 99.977% |
| 6 | 3.4 | 0.00034% | 99.99966% |

## 2.4 BOOLEAN LOGIC OPERATIONS[13]

**AND** Gate: All inputs must be true for output to be true

| In1 | In2 | Out |
|---|---|---|
| 0 | 0 | 0 |
| 0 | 1 | 0 |
| 1 | 0 | 0 |
| 1 | 1 | 1 |

**NAND** Gate: All inputs must be false for output to be true

| In1 | In2 | Out |
|---|---|---|
| 0 | 0 | 1 |
| 0 | 1 | 1 |
| 1 | 0 | 1 |
| 1 | 1 | 0 |

---

13. Developed by George Boole (1815-1864).

**OR** Gate: Any input can be true for output to be true

| In1 | In2 | Out |
|-----|-----|-----|
| 0 | 0 | 0 |
| 0 | 1 | 1 |
| 1 | 0 | 1 |
| 1 | 1 | 1 |

**NOR** Gate: If any input is true the output will be false

| In1 | In2 | Out |
|-----|-----|-----|
| 0 | 0 | 1 |
| 0 | 1 | 0 |
| 1 | 0 | 0 |
| 1 | 1 | 0 |

**XOR** Gate: All the inputs must be different for the output to be true

| In1 | In2 | Out |
|-----|-----|-----|
| 0 | 0 | 0 |
| 0 | 1 | 1 |
| 1 | 0 | 1 |
| 1 | 1 | 0 |

**S-R Flip-Flop:** Latch Circuit

| S | R | Q | $\overline{Q}$ |
|---|---|---|---|
| 0 | 0 | Keep output state | Keep output state |
| 0 | 1 | 0 | 1 |
| 1 | 0 | 1 | 0 |
| 1 | 1 | Unstable condition | Unstable condition |

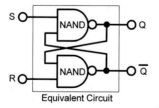
Equivalent Circuit

## 2.5 CONVERSION FACTORS

### Table 2-9. Common Conversion Factors

| Unit | = | Unit |
|---|---|---|
| Gallon | 8.34 | Lbs Water @ 60°F |
| Density of Water | 62.4 Lbs/Ft$^3$ | |
| Density of Air | 0.07649 Lbs/Ft$^3$ | |
| SG Water @ 60°F | 1 | |
| MW of Air | 29 | |
| SG of Liquid | MW of Liquid ÷ 18.02 | |
| SG of Gas | MW of Gas ÷ 29 | |

### Table 2-10. Distance Factors

| Multiply | By | To Obtain |
|---|---|---|
| Inch | 2.54 | Centimeter |
| Centimeter | 0.3937 | Inch |
| Foot | 0.3048 | Meter |
| Meter | 3.28083 | Foot |

### Table 2-11. Volume Factors

| Multiply | By | To Obtain |
|---|---|---|
| Gallon | 0.13368 | Ft$^3$ |
| Gallon | 0.003754 | M$^3$ |
| Gallon | 3.7853 | Liter |
| Liter | 0.2642 | Gallon |
| Liter | 0.03531 | Ft$^3$ |
| Liter | 0.001 | M$^3$ |
| Ft$^3$ | 7.481 | Gallon |
| Ff$^3$ | 28.3205 | Liter |
| Ft$^3$ | 0.028317 | M$^3$ |
| M$^3$ | 35.3147 | Ft$^3$ |
| M$^3$ | 3.28083 | Gallon |
| M$^3$ | 1000 | Liter |

### Table 2-12. Mass Factors

| Multiply | By | To Obtain |
|---|---|---|
| Pound | 0.4536 | Kilogram |
| Kilogram | 2.2046 | Pound |

## Table 2-13. Force Factors

| Multiply | By | To Obtain |
|---|---|---|
| Newton | 0.22481 | Pound-Force |
| Pound-Force | 4.4482 | Newton |

## Table 2-14. Energy Factors

| Multiply | By | To Obtain |
|---|---|---|
| BTU | 778.17 | Ft-Lbf |
| BTU | 1.055 | KJoules |
| BTU/Hr | 0.293 | Watt |
| HP | 0.7457 | Kilowatt |
| HP | 2545 | BTU/Hr |

## Table 2-15. Temperature Factors

| Unit | Use Equation | To Obtain |
|---|---|---|
| °F | $(°F - 32)*1.8$ | °C |
| °F | $(°F + 459.67)/1.8$ | °K |
| °F | $(°F + 459.67)$ | °R |
| °C | $(°C \times 1.8) + 32$ | °F |
| °C | $°C + 273.15$ | °K |
| °C | $(°C \times 1.8) + 32 + 459.67$ | °R |
| °K | $(°K \times 1.8) - 459.67$ | °F |
| °K | $°K - 273.15$ | °C |
| °K | $°K \times 1.8$ | °R |
| °R | $°R - 459.67$ | °F |
| °R | $(°R - 32 - 459.67)/1.8$ | °C |
| °R | $°R/1.8$ | °K |

## Table 2-16. Pressure Factors

| Multiply | By | To Obtain |
|---|---|---|
| Atmosphere | 1.01295 | Bar |
| Atmosphere | 29.9213 | Inches Hg |
| Atmosphere | 760 | mm Hg |
| Atmosphere | 406.86 | Inches WC * |
| Atmosphere | 14.696 | PSI |
| Atmosphere | $1.01295 \times 10^5$ | $N/M^2$ or Pa |
| Bar | 0.9872 | Atm |
| Bar | 29.54 | Inches Hg |
| Bar | 750.2838 | mm Hg |
| Bar | 401.65 | Inches WC |
| Inches WC | 0.03612 | PSI |
| Inches WC | 0.07354 | Inches Hg |
| Inches WC | 1868.1 | mm Hg |
| Inches WC | 248.9 | $N/M^2$ or Pa |
| Inches WC | 0.001868 | Micron or mtorr |
| PSI | 27.68 | Inches WC |
| PSI | 2.036 | Inches Hg |
| PSI | 51.71 | mm Hg |
| PSI | 0.068046 | Atm |
| PSI | 0.068948 | Bar |
| PSI | 6892.7 | $N/M^2$ or Pa |
| Micron or mtorr | 0.0005353 | Inches WC |
| $N/M^2$ or Pa | 0.004018 | Inches WC |
| $N/M^2$ or Pa | 0.00014508 | PSI |

\* WC indicates water column

## Table 2-17. Viscosity

| Multiply | By | To Obtain |
|---|---|---|
| cSt | $0.999 g/cm^3$ | cP |
| cP | $1/0.999 g/cm^3$ | cSt |

Kinematic viscosity (stoke) = Absolute viscosity (poise)/S.G.
Dynamic viscosity (cP) = 0.001 Pa-s

## 2.6 EQUATIONS/LAWS/FORMULAS

*Pressure*

$$P = \frac{F}{A}$$

F = Force applied
A = Area

*Boyle's Law*

$$P_1 V_1 = P_2 V_2$$

Boyle's law states that *at constant temperature,* the <u>absolute pressure</u> and the volume of a gas are inversely proportional. The law can also be stated in a slightly different manner: that the product of absolute pressure and volume is always constant.

P = Pressure in PSIA
V = Volume in feet$^3$

*Charles's Law*

$$\frac{V_1}{T_1} = \frac{V_2}{T_2} \quad \text{OR} \quad V_1 T_2 = V_2 T_1$$

These expressions may be combined into the form of PV/T = constant for a fixed mass of gas.

Charles's law states that *at constant pressure,* the volume of a given mass of an ideal gas increases or decreases by the same factor as its temperature on the absolute temperature scale (i.e., the gas expands as the temperature increases, the temperature is the average of molecular motion, therefore, the molecular motion will increase with a corresponding temperature increase, thus causing the gas to expand.).

T = Temperature in °R *(Note the absolute temperature scale)*
V = Volume in feet$^3$

*Gay-Lussac's Law*

$$\frac{P_1}{T_1} = \frac{P_2}{T_2} \quad \text{OR} \quad P_1 T_2 = P_2 T_1$$

The pressure of a fixed mass and fixed volume of a gas is directly proportional to the gas's temperature.

T = Temperature in °R
P = Pressure in PSIA

## Ideal Gas Law (for compressible fluids)

$$PV = RT$$

R = Gas Constant (Value = 1544/MW)
P = Pressure in PSIA
V = Volume in feet$^3$
T = Temperature in °R
MW = Molecular Weight

## Pascal's Law

Pascal's Law states that a change in the pressure of an enclosed incompressible fluid is conveyed undiminished to every part of the fluid and to the surfaces of its container. *Note: this is the principle used in the pressure factor table (Table 2-16) to convert between pressure and inches-WC, mm-Hg, etc.*

$$\Delta P = \rho g(\Delta h)$$

$\Delta P$ = Hydrostatic pressure
$\rho$ = Mass density
g= Gravitation constant
$\Delta h$ = Difference in elevation between the two points within the fluid column

## Bernoulli's Equation

The Bernoulli equation states that as the speed of a moving fluid increases, the pressure within the fluid decreases:

$$\frac{P_1 V_1}{T_1} = \frac{P_2 V_2}{T_2}$$

P + 1/2 $\rho v^2$ + $\rho gh$ = Constant
P = Pressure in PSIA
$\rho$ = Mass Density
g = Gravitation constant
h = Height above reference level
v = Velocity

This form of the Bernoulli equation ignores viscous effects. If the flow rate is high, or the flowing material has a very low viscosity, Bernoulli's equation should not be used. For example, liquid flowing in a pipe may be more accurately described with Poiseuille's equation[14] to account for flow rate, viscosity and pipe diameter. Poiseuille's equation states:

---

14. Also known as the Hagen-Poiseuille Law that states that the volume flow of an incompressible fluid through a circular tube is equal to $\pi/8$ times the pressure differences between the ends of the tube, times the fourth power of the tube's radius divided by the product of the tube's length and the dynamic viscosity of the fluid.

$$Q = \frac{\Delta P \pi r^4}{8 \mu \ell}$$

Q = Volumetric flow rate, in$^3$/sec
$\ell$ = Tube length, inches
r = Tube radius, inches
$\mu$ = Viscosity, lb•sec/in$^2$

## Volumetric Flow Rate

$$Q = AV \quad Q_{(gpm)} = 3.12 \, A_{(sq\ in)} \times V_{(ft/sec)}$$

Q = Volumetric Flow Rate
A = Cross Sectional Area of the Pipe
V = Velocity of the Fluid

## Darcy's Formula (general formula for pressure drop)

$$h = \frac{fLV^2}{2Dg}$$

h = Pressure drop in feet of fluid
L = Length of pipe (feet)
V = Velocity of the fluid (ft/sec)
g = Acceleration of gravity (32.2 ft/sec$^2$)
D = Pipe ID (feet)
f = The Darcy-Weisbach friction factor f = 16 ÷ Re (Re = Reynolds number)

## Velocity of Exiting Fluid

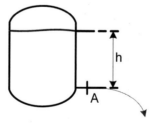

$$V = \sqrt{2gh} \quad Q = A\sqrt{2gh}$$

V = Velocity of the fluid (ft/sec)
g = Gravitation constant (32.2 ft/sec$^2$)
h = Height above reference level (in feet)
A = Area of opening (in sq ft) (the smaller the area, the greater the fluid velocity)
Q = Volumetric flow rate (ft$^3$/sec)

## Convert Actual Cubic Feet per Minute (ACFM) to Standard Cubic Feet per Minute (SCFM)

$$ACFM = SCFM \left[ \frac{14.7}{P_a} \times \frac{T_a}{519.67} \right] \quad \text{equivalent to} \quad \frac{P_1 V_1}{T_1} = \frac{P_2 V_2}{T_2}$$

$P_a$ = Actual pressure (PSIA)

$T_s$ = Standard temperature (519.67°R) NOTE: °R =60°F+459.67 (convert from °F to °R)

$T_a$ = Actual temperature (°R)

## Joule-Thomson (Kelvin) Effect/Coefficient

When the pressure of a non-ideal (real) gas changes from high to low (such as through a valve), a change of temperature occurs, proportional to the pressure difference across the restriction. The Joule-Thomson coefficient ($\mu_{JT}$) is the change of temperature per unit change of pressure.

The rate of change of temperature T with respect to pressure P in a Joule-Thomson process (that is, at constant enthalpy H) is the *Joule-Thomson (Kelvin) coefficient*. This coefficient can be expressed in terms of the gas's volume V, its heat capacity at constant pressure $C_p$, and its coefficient of thermal expansion $\alpha$ as:

$$\mu_{JT} \equiv \left(\frac{\partial T}{\partial P}\right)_H = \frac{V}{C_p}(\alpha T - 1)$$

Where:

V   = Volume of gas

$C_p$ = The gas' heat capacity at constant pressure

$\alpha$   = The gas' coefficient of thermal expansion

H   = Enthalpy constant

$\partial T$ = Rate of change of temperature

$\partial P$ = Rate of change of pressure

The value of $\mu_{JT}$ is typically expressed in °C/bar (SI units: °K/Pa)

Table 2-18 defines when the Joule-Thomson effect cools or warms a real gas:

### Table 2-18. Joule-Thomson Effect

| Gas Temperature | $\mu_{JT}$ is: | sign of $\partial P$ | sign of $\partial T$ | ∴ The gas |
|---|---|---|---|---|
| < Inversion Temperature (1) | Positive | Negative (2) | Negative | COOLS |
| > Inversion Temperature (1) | Negative | Negative (2) | Positive | WARMS |

(1) Inversion Temperature: The critical temperature below which a non-ideal (real) gas that is expanded (with a constant enthalpy) will experience a temperature decrease.

(2) When a gas expands the pressure is always lower; therefore $\partial P$ is always negative.

## Mass Flow – Gas Equations

Substitute Q for V/t:     Substitute for Q:

$$w = \frac{m}{t} = \frac{M}{10^3 R}\left(\frac{V}{t}\right)\left(\frac{p}{T}\right) \qquad w = \frac{MQ}{10^3 R}\left(\frac{p}{T}\right) \qquad Q = k\sqrt{D}; k = \frac{Mk_f}{10^3 R}$$

Simplified:  $w = k\sqrt{D\left(\dfrac{p}{T}\right)}$

w = Mass flow rate (kg/sec)
Q = Volume flow rate (m³/sec)
p = Absolute pressure (pascal)
T = Absolute temperature (Kelvin)
M = MW (g/mol)
R = Universal gas constant = 8.314 J ÷ (°K x mol)
D = Flowmeter D/P (pascal)
k = Mass flow proportionality constant
$k_f$ = Flowmeter proportionality constant
$M = \rho AV$
M = Mass flow rate (lbs/sec)
A = Cross sectional area (ft²)
ρ = Fluid density (lbs/ft³)
V = Velocity (ft/sec)

*Density will vary in inverse proportion to temperature, and in direct proportion to pressure.*

## Surface Area Formulas

- Sphere: $4\pi r^2$

- Right Circular Cone: $\pi r^2 + \pi rs$

- Right Circular Cylinder: $2\pi rh + 2\pi r^2$

- Pyramid: Area of Base + Area of the (4) Triangular Sides

## Volume Formulas

- Sphere:

  $\dfrac{4}{3}\pi r^3$

- Right Circular Cone:

  $\dfrac{1}{3}\pi r^2 h$

- Right Circular Cylinder:

  $\pi r^2 h$

- Pyramid:

  $\dfrac{1}{3}A \bullet h$  (A = Area of base)

---

# 3. MEASUREMENT

## TOPIC HIGHLIGHTS

- Measurement sensor technologies: temperature, flow, level, vibration, pH, etc.
- Sensor characteristics
- Calculations
- Sample temperature problems

## 3.1 TEMPERATURE MEASUREMENT SENSORS

Temperature may be measured by a variety of techniques. All of the techniques infer temperature by measuring some change in a physical characteristic.

Table 3-1 compares the three primary types of contact temperature measurement sensors: thermocouples, resistance temperature detectors (RTDs) and thermistors.

### Table 3-1. Comparison of Contact Temperature Sensors

| Relative Advantages of Contact Temperature Sensors | | | |
|---|---|---|---|
| **Quality** | **T/Cs** | **RTDs** | **Thermistors** |
| Temp Range | -400 to 4200°F | -200 to 1475°F | -176 to 392°F |
| Accuracy | < RTD | > T/C | > T/C & RTD |
| Ruggedness | Highly Rugged | Sensitive to Shock | NOT Rugged |
| Linearity | Highly NON-Linear | Somewhat NON-Linear | Highly NON-Linear |
| Drift | Subject to Drift | < T/C | < T/C |
| Cold Junction | Required | None | None |
| Compensation Response | Fast | Relatively Slow | Faster than RTD |
| Cost | Low, except for noble metals | > T/C | Low |

### 3.1.1 Thermocouple (T/C)

A thermocouple is a junction between two dissimilar metals that produces a voltage related to a temperature difference. These devices are based upon the Seebeck effect[15]. Changes in the temperature at the T/C junction induce a change in voltage between the other ends of the T/C wire. As the temperature rises at the junction of the T/C, the voltage output of the T/C also rises. This voltage rise, however, is not linear with the temperature change. T/C voltage tables are available from many of the T/C vendors.

The following criteria are generally used in selecting a thermocouple:

---

15. Seebeck effect: named for scientist Thomas Johann Seebeck (1770-1831). A circuit made of two dissimilar metals produces electricity if the two places where the metals connect are held at different temperatures.

- Temperature range required (range requirements have an effect on the combination of metals to be used for its construction).

- Response time required (this is a time constant, which is the time at which the thermocouple's voltage output reaches 63.2% of its final value in response to a step change). The main factors that affect thermocouple response time are:

  - Thermocouple bead size (typically two times the diameter of the wire that makes up the thermocouple). Note, the bead is where the two dissimilar metals are joined together.
  - Conducting medium (air, water, etc.)

- Environmental conditions (corrosive, presence of electrical noise, etc.)

- Durability

- Grounding (whether or not a grounded junction is required)

- Installation requirements (insertion length, connection type, etc.)

## *Thermocouple Junction Options*

Sheathed (enclosed) thermocouple probes are typically constructed in three available junction types (Figure 3-1):

- **Grounded:** The tip of a grounded junction thermocouple has the thermocouple wires physically attached (welded) to the inside of the sheath wall. This type of construction results in good heat transfer from the fluid to be measured through the sheath wall to the thermocouple junction (which provides faster response to temperature changes). This type of junction is recommended for measuring static or flowing corrosive gases or liquids and for high pressure applications.

- **Ungrounded:** The thermocouple junction is physically detached from the sheath. The response time for an ungrounded junction is slower than for the grounded junction type, however, the thermocouple junction is electrically isolated from the sheath. The sheath for an ungrounded junction thermocouple is filled with an insulating material, typically magnesium oxide (MgO).

- **Exposed:** The tip of the thermocouple extends beyond the sheath, which allows for faster and more accurate results but makes the junction vulnerable to damage. The sheath insulation is sealed where the junction exits the sheath in order to prevent liquid or gas penetration into the sheath.

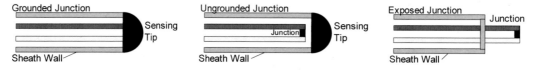

**Figure 3-1. Thermocouple Junction Types**

**Table 3-2. Thermocouple Types (Reference Figure 3.2 for common curves)**

| T/C Type | Names of Materials | Insulation Colors | Useful Range |
|---|---|---|---|
| B | Platinum 30% Rhodium (+)<br>Platinum 6% Rhodium (-) | + Grey<br>- Red<br>Overall: Grey | 2500–3100°F<br>(1370–1700°C) |
| C | W5Re Tungsten 5% Rhenium (+)<br>W26Re Tungsten 26% Rhenium (-) | + White/Red (Glass Braid)<br>- Red (Glass Braid)<br>Overall: White/Red (Glass Braid) | 3000–4200°F<br>(1650–2315°C) |
| E* | Chromel (+)<br>Constantan (-) | + Purple<br>- Red<br>Overall: Purple | 200–1650°F<br>(95–900°C) |
| J* | Iron (+)<br>Constantan (-) | + White<br>- Red<br>Overall: Black | 200–1400°F<br>(95–760°C) |
| K* | Chromel (+)<br>Alumel (-) | + Yellow<br>- Red<br>Overall: Yellow | 200–2300°F<br>(95–1260°C) |
| N | Nicrosil (+)<br>Nisil (-) | + Orange<br>- Red<br>Overall: Orange | 1200–2300°F<br>(650–1260°C) |
| R | Platinum 13% Rhodium (+)<br>Platinum (-) | + Black<br>- Red<br>Overall: Green | 1600–2640°F<br>(870–1450°C) |
| S | Platinum 10% Rhodium (+)<br>Platinum (-) | + Black<br>- Red<br>Overall: Green | 1800–2640°F<br>(980–1450°C) |
| T* | Copper (+)<br>Constantan (-) | + Blue<br>- Red<br>Overall: Red | -330–660°F<br>(-200–350°C) |

\* Indicates most commonly used types of thermocouples.

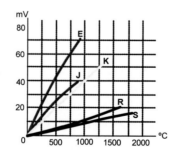

**Figure 3-2. Thermocouple Output vs. Temperature**

**Thermopile:** A device that converts heat energy into electrical energy. It is made up of thermocouple junction pairs connected electrically in series.

*Thermocouple R-A-S-S Rule (Cold Junctions)*

For every thermocouple there is a "cold junction" effect when the leads are connected to a measurement device or a simulated voltage input device. Thermocouple measurement devices will typically automatically add or subtract the cold junction effect with internal circuitry. However, common VOMs and voltage sources may not have the built-in circuitry, therefore compensation needs to be made (reference Figure 3-3).

*Note: The "cold junction" nomenclature originated from the use of ice as a reference point.*

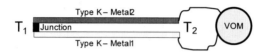

**Figure 3-3. Measuring the Thermocouple Voltage**

- **Receive** – **Add** the cold junction effect (i.e., the temperature at the sensing junction end is higher than the temperature at the cold junction end)

$$T_1 = \left( \frac{\mu V}{\text{Seebeck Coefficient}} \right) + T_2 = \left( \frac{\mu V}{xx \mu V / \,^\circ C} \right) + T_2$$

- **Send** – **Subtract** the cold junction effect (i.e., the temperature at the sensing junction end is lower than the temperature at the cold junction end)

$$T_1 = \left( \frac{\mu V}{\text{Seebeck Coefficient}} \right) - T_2 = \left( \frac{\mu V}{xx \mu V / \,^\circ C} \right) - T_2$$

## 3.1.2 Resistance Temperature Detector (RTD)

An RTD is a type of temperature sensor that contains a resistor that changes its resistance value as its temperature changes. The resistor is typically constructed of a pure metal, such as platinum, copper or nickel, or an alloy of these. These elements were chosen because of their linear resistance values and their ability to be drawn into a fine wire.

As the temperature rises at the sensing point, the resistance value also increases. The resistance rise versus the temperature rise in an RTD is more linear than that of the voltage output of a thermocouple. As you can see from the curves in Figure 3-4, the response of the platinum RTD is the most linear, making it the most popular choice of RTD material. However, you must consider the temperature limitations of all the materials involved, where they are applied, and the temperatures to which each will be exposed.

*RTD Construction*

**Wire Wound:** A typical RTD consists of a fine platinum wire wrapped around some form of non-conducting core such as ceramic or glass. Since the RTD wire is usually quite fragile it is covered with a protective coating (sheath).

---

**TEMPERATURE MEASUREMENT**

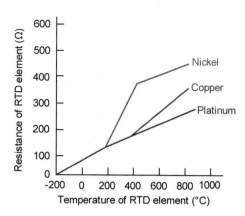

**Figure 3-4. Temperature Response of RTD Materials**

**Thin Film:** This type consists of a ceramic substrate that has a thin film of platinum deposited on it in a resistance pattern. The thin film is then heat bonded to the substrate. The resistance is adjusted and finalized by the vendor by the use of laser trimming of the substrate. The final assembly is then coated with glass reinforcement.

**Coiled:** The goal of the coil constructed RTD is to provide strain relief to the thin RTD wire. The elements are constructed with a helical coil and placed within an insulated core. This design allows for the sensing wire to expand and contract.

### RTD Temperature Coefficient of Resistance (α)

This value symbolizes the resistance change factor per degree of temperature change (i.e., it expresses the sensitivity to temperature changes of the wire used in the RTD element). The formula for α is as follows:

$$\frac{\Delta R}{R_0} = \alpha(\Delta T) \quad \therefore \alpha = \frac{\Delta R}{(R_0)(\Delta T)} \quad \text{engineering units are: } \Omega/\Omega/°C$$

Some common α values for RTD elements over the range of 0 to 100°C are:

$$\alpha = \frac{(R_{100°C} - R_{0°C})}{(R_{0°C})(100°C - 0°C)}$$

Platinum :
$$\alpha = \frac{(138.51\Omega - 100\Omega)}{(100\Omega)(100°C - 0°C)} = 0.00385\Omega/\Omega/°C \quad \text{IEC Standard (IEC751)}$$

Copper :
$$\alpha = \frac{(12.897\Omega - 9.035\Omega)}{(9.035\Omega)(100°C - 0°C)} = 0.00427\Omega/\Omega/°C$$

Nickel :
$$\alpha = \frac{(200.64\Omega - 120\Omega)}{(120\Omega)(100°C - 0°C)} = 0.00672\Omega/\Omega/°C$$

*RTD Lead Wire Configuration*

Since an RTD is a resistance type sensor, any resistance in the extension wires between the RTD and the instrument will induce an error in the reading. If the extension wires are maintained at a constant temperature, then this error may be compensated for in the calibration of the instrument. However, since it is usually not the case that the extension wires are maintained at a constant temperature, some other form of compensation must be used to reduce the potential for introducing error into the reading due to changing extension wire resistance. This other form of compensation is typically the use of compensation conductors within the extension lead wires (e.g., 3-wire and 4-wire RTDs) (Figure 3.5).

In **2-Wire** loop construction, the sensor resistor measurement will include the lead wire resistance. Therefore, the 2-wire construction is typically used only with high resistance sensors, and when the extension wire length will be very short.

**3-Wire** loop construction is the most common design. The lead wire resistance is factored out as long as all of the lead wires have the same resistance. The 3-wire construction reduces errors to a negligible level in most applications.

**4-Wire** loop construction is typically found in laboratories and other applications where very precise measurement is required. The fourth wire allows the measurement equipment to factor out all of the lead wire resistance and other unwanted resistance from the measurement circuit.

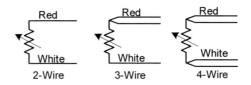

**Figure 3-5. RTD Lead Wire Configurations**

*RTD Accuracy (Table 3.3)*

- **Class A RTD:** Highest RTD element tolerance and accuracy, Class A (IEC-751), α = 0.00385 (accuracy of ±0.15°C at 0°C). This is only used when the Class B RTD accuracy limits are not sufficient.

- **Class B RTD:** Most common RTD element tolerance and accuracy, Class B (IEC-751), α = 0.00385 (accuracy of ±0.3°C at 0°C).

## Table 3-3. RTD Accuracy

| Permissible Deviations from Basic Values | | | | | |
|---|---|---|---|---|---|
| **Class A** | | | **Class B** | | |
| | **Deviation** | | | **Deviation** | |
| Temp. (°C) | ±Ω | ±°C | Temp. (°C) | ±Ω | ±°C |
| -200 | 0.24 | 0.55 | -200 | 0.56 | 1.3 |
| -100 | 0.14 | 0.35 | -100 | 0.32 | 0.8 |
| 0 | 0.06 | 0.15 | 0 | 0.12 | 0.3 |
| 100 | 0.13 | 0.35 | 100 | 0.30 | 0.8 |
| 200 | 0.20 | 0.55 | 200 | 0.48 | 1.3 |
| 300 | 0.27 | 0.75 | 300 | 0.64 | 1.8 |
| 400 | 0.33 | 0.95 | 400 | 0.79 | 2.3 |
| 500 | 0.38 | 1.15 | 500 | 0.93 | 2.8 |
| 600 | 0.43 | 1.35 | 600 | 1.06 | 3.3 |
| 650 | 0.46 | 1.45 | 650 | 1.13 | 3.6 |

### 3.1.3 Thermistor

With a name derived from the combination of "thermal" and "resistor," this is a temperature sensing element composed of semiconductor material that exhibits a large change in resistance in response to a small change in temperature (Figure 3-6). The base semiconductor material is a mixture of metal oxides pressed into a shape (bead, disk, wafer, etc.) with lead wires and then sintered at a high temperature. The thermistor bead is then typically coated with epoxy or glass for further protection.

Unlike thermocouples and RTDs, a thermistor has a negative temperature coefficient. This means that the resistance of the thermistor decreases as the temperature rises.

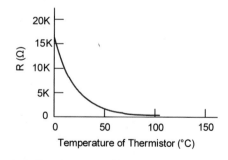

**Figure 3-6. Resistance Change with Temperature**

*Uses of Thermistors*

- Applications that require high resolution over a narrow temperature span

- Where low cost is an absolute must

- Where point sensing or miniaturization is required

### 3.1.4 Temperature Switch

A temperature switch is an On-Off device that is responsive to temperature changes. Types of switch construction include:

- **Filled thermal system:** As the temperature changes, the volume of the fill fluid changes as well (increasing with temperature increase and decreasing with temperature decrease). The change in fluid volume creates higher or lower pressure against some form of piston assembly which, in turn, actuates a microswitch. The fill system can either be direct or remote. Direct systems are composed of a rigid stem filled with fluid that is directly connected to the switch assembly. Remote systems are composed of a fluid filled bulb, typically 3 inches in length, with various lengths of filled armored capillary between the bulb and the switch assembly. Care should be taken when selecting the proper fill fluid for the thermal system. This will ensure that there is no boiling/vaporization or freezing of the fill fluid.

- **Bimetallic:** A bimetallic strip is used to convert temperature change into mechanical movement. The bimetallic strip consists of two dissimilar metals with different thermal expansion coefficients. The strips are joined together throughout their length so that the difference in the thermal expansion capabilities causes the flat strip to bend in one direction or the other (i.e., it bend one way when heated, and the other way when cooled). This mechanical action then causes some form of snap-action or mercury switch to actuate.

### 3.1.5 Temperature Indicator (Thermometer)

A thermometer is very similar in operation to a temperature switch except that instead of actuating a switch, a pointer on a dial is rotated to indicate the temperature being measured. For a bimetallic thermometer, the bimetallic element is formed into a coil to provide for a rotary motion to a central shaft. The "liquid in glass" thermometer utilizes a thermally sensitive liquid (e.g., mercury, ammonia) that is contained in a graduated glass tube and expands or contracts as the temperature increases or decreases.

### 3.1.6 Thermowell

Thermowells are typically used in temperature measurement applications where isolation between a temperature sensor and the environment whose temperature is to be measured must be provided. Thermowells are inserted into the process and, as such, are subject to static and dynamic fluid forces. These dynamic forces govern the thermowell design. Vortex shedding is the major concern for a thermowell, as it is capable of forcing the thermowell into a flow-induced resonance with consequent metal fatigue failure. The latter is particularly significant at high fluid velocities.

The ASME Performance Test Code (PTC 19.3 – Temperature Measurement) is the most widely used basis for thermowell design.

Thermowell construction material is typically chosen based upon the temperature of the process fluid into which the thermowell is being immersed, the corrosion characteristics of the process, and the possibility of erosive conditions.

*Thermowell Shank Types*[16]

- **Step Shank** (has an outer diameter of 1/2 inch at the end of the thermowell immersion length to provide a quicker response time)

- **Straight Shank** (same size all along the immersion length)

- **Tapered Shank** (the outside diameter of the thermowell decreases gradually along the immersion length)

- **Built-up Step Shank** (typically used for high velocity application or long insertion lengths)

**Lagging:** The lagging extension of a thermowell is often referred to as the thermowell's T length (Figure 3-7). The lagging extension is located on the cold side of the process connection and is usually an extension of the hex length of the thermowell. Typically, the T length enables the probe and thermowell to extend through insulation or walls.

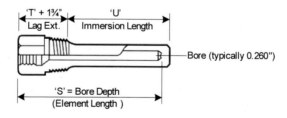

**Figure 3-7. Thermowell T Length**

*Temperature Sensor/Thermowell Installation*

- *Pipes larger than 3 inches in diameter:* Thermowells are usually best sized to extend to the center of the pipe (Figure 3-8).

---

16. Images obtained from www.ashcroft.com.

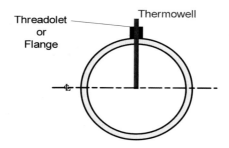

**Figure 3-8. Thermowell Installation (1)**

- *Pipes smaller than 3 inches in diameter:* There are two generally accepted methods for thermowell installation in pipes less than 3 inches in diameter.

    - Using a piping tee fitting to allow the tip of the thermowell to extend to the center of the main run pipe (Figure 3-9).

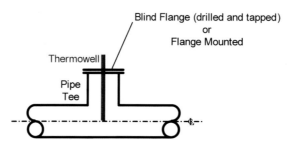

**Figure 3-9. Thermowell Installation (2)**

    - Using a piping elbow (Figure 3-10). When a piping elbow is used, the tip of the temperature sensor/thermowell must be facing opposite the flow direction of the fluid to be measured.

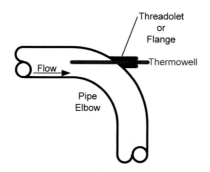

**Figure 3-10. Thermowell Installation (3)**

## 3.2 PRESSURE MEASUREMENT SENSORS

*Terminology (see Figure 3-11)*

**Absolute Pressure (psia):** Pressure that is measured against a full vacuum

**Gauge Pressure (psig):** Pressure that is measured relative to the surrounding atmosphere

**Vacuum:** Pressure that is lower than atmospheric (may be partial or full)

**Differential Pressure (D/P or ΔP):** The difference between two pressure readings

**Compound Pressure:** Combined gauge pressure indication with vacuum indication

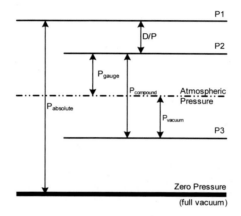

**Figure 3-11. Relationship of Pressure Terms**

## 3.2.1 Manometer

A basic manometer consists of a reservoir filled with liquid of known density and a vertical graduated tube. The difference in the two column heights indicates the static pressure (the sum of the readings above and below zero on the graduated scale). Most common types of manometers are U-tube/slack tube (Figure 3-12, U-tube) and inclined. Manometer design may or may not include a fluid reservoir depending on the manufacturer.

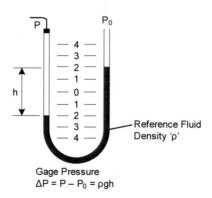

Gage Pressure
$$\Delta P = P - P_0 = \rho g h$$

**Figure 3-12. U-tube Manometer**

### 3.2.2 Bourdon Tube

With a simple bourdon tube, one end of the tube is sealed, and the other end is connected to the source of pressure that is to be measured. The end that pressure is applied to is held stationary. When pressure is applied to the inside of the tube, the tube begins to straighten out, which causes a small amount of movement at the sealed end of the tube. The bourdon tube has three major designs (the size, shape, and material of the bourdon tube depend on the pressure range):

*C-style Bourdon Tube* (Figure 3-13): Most commonly used in direct reading gauges. (Should not be over-pressurized because this may permanently stretch the bourdon tube and render the gauge inaccurate.)

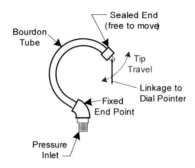

**Figure 3-13. C-style Bourdon Tube**

*Spiral/Helical Bourdon Tube* (Figure 3-14): Very similar in operation to that of the C-style bourdon tube. The spiral/helical design bourdon tube designs allow for more travel of the tube tip. These are typically used in higher pressure applications ≥ 1000 psig.[17]

**Figure 3-14. Spiral (left) and Helical (right) Bourdon Tubes**

### 3.2.3 Pressure Diaphragm

The pressure diaphragm consists of a flexible membrane that separates regions of different pressures. The flexible membrane's degree of deformation is dependent upon the difference in pressure on the opposite sides of the membrane. This deformation is then measured using a variety of technologies such as capacitive, strain gauge, piezoresistive and LVDT.

---

17. Images obtained from www.instrumenttoolbox.com.

*Capacitive:* The movement of the diaphragm between two capacitor plates changes the capacitance, which is detected as an indication of the change in the process pressure.

*Strain Gauge:* A strain gauge is affixed to a diaphragm; when the diaphragm deforms due to a change in pressure, there is a corresponding change in resistance in the strain gauge.

*Piezoresistive:* This type is similar in construction to a strain gauge diaphragm, except that the piezoresistive pressure component is affixed to a diaphragm that is commonly made of silicon. When the diaphragm deforms due to a change in pressure, there is a corresponding change in resistance from the piezoresistive component.

*LVDT (Linear Variable Differential Transformer):* The movement of the diaphragm causes the movement of a core within magnetically coupled coils. There is a corresponding electrical output change as a result of a change in pressure.

### 3.2.4 Pressure Transducer/Transmitter

The pressure transducer/transmitter (Figure 3-15 shows one type) receives the low voltage output from one of the pressure diaphragm technologies described above (and there are others, including mechanical and optical) and converts that low voltage signal into an output that can be interpreted by a control system (4–20 mA, 1-5 VDC, Fieldbus, etc.).[18]

**Figure 3-15. One Type of Pressure Transducer/Transmitter**

### 3.2.5 Diaphragm Seal

Unlike a pressure diaphragm, a diaphragm seal (labeled "metal diaphragm" in Figure 3-16) is a flexible membrane whose purpose is to isolate a pressure sensor from the process fluid.[19]

### 3.2.6 Pressure Sensor Installation Details

*Steam/Liquid Service (the device must be located below process tap location; Figure 3-17)*

*Gas Service (when the instrument is located above the process tap location – preferred; Figure 3-18):*

---

18.  Rosemount 3501N transmitter.
19.  Image obtained from www.machinedesign.com.

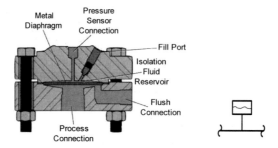

**Figure 3-16. Diaphragm Seal**

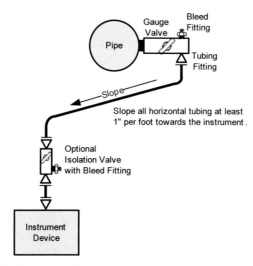

**Figure 3-17. Pressure Sensor Installation for Steam/Liquid Service**

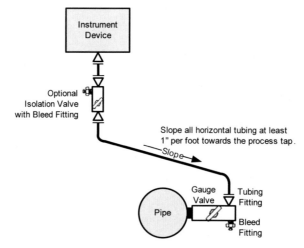

**Figure 3-18. Pressure Sensor Installation for Gas (1)**

*Gas Service (when the instrument is located below the process tap location; Figure 3-19):*

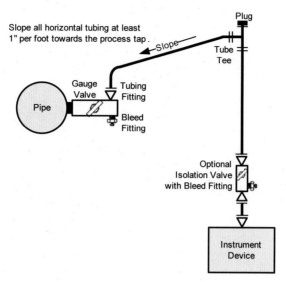

**Figure 3-19. Pressure Sensor Installation for Gas (2)**

## 3.3 VOLUMETRIC FLOW MEASUREMENT SENSORS

*Reynolds Number*

The Reynolds number (Re) is a dimensionless number that is the ratio of inertial forces (Vq) to viscous forces ($\mu/L$) and consequently, it quantifies the relative importance of these two types of forces for given flow conditions.

The Reynolds number is found from the equation:

$$Re = \frac{3160 * Q * G}{D * \rho} \text{ For liquids} \qquad Re = \frac{6316 * Q}{D * \rho} \text{ For gases \& steam}$$

Q: Flow in GPM                   Q: Flow in lb/hr

G: Specific Gravity

D: Pipe ID

$\rho$: Viscosity (in cp)

Flow type you can expect (laminar, transitional, or turbulent) based on the Reynolds number equation result:

Laminar: Re < 2300

Transitional: 2300 < Re < 4000

Turbulent: Re > 4000

## 3.3.1 Sensors Based on Differential Pressure (D/P Producers)

*Orifice Plate:* ⊣‖⊢

### Introduction

**Bernoulli's principle** states that there is a relationship between the pressure of a fluid and the velocity of the fluid: when the velocity increases, the pressure decreases and vice versa.

**Beta Ratio:** The ratio between the orifice bore diameter and the pipe inside diameter (*Beta Ratio (β): d/D*). The beta ratio is stable between 0.20 and 0.70; outside of this range an uncertainty in the flow measurement signal will appear.

**Orifice Plate** is a thin metal plate with a precise hole machined in it.

**Common D/P Ranges** for full scale flow: 50 inches WC or 100 inches WC.

**Vena Contracta:** The point in a flow stream where the diameter of the flow stream is the smallest, typically 0.35 to 0.85 pipe diameters downstream of the orifice plate.

**Vents/Drains:** When vents or drains are installed on an orifice plate, the flow through either the vent or drain should be < 1% of total flow.

- *Vent:* Hole located at the top of the orifice plate to allow entrained gases in a liquid flow to vent past the orifice plate.

- *Drain:* Hole located at the bottom of the orifice plate to prevent the build-up of condensate in a steam flow behind the orifice plate.

The orifice plate is commonly used to measure flow in clean liquid, gas and steam service. The plate may be constructed of any material; however, stainless steel is the most common material. The thickness of the orifice plate is a function of the following characteristics: line size, process pressure, temperature, and the D/P across the plate. The most common orifice plate thickness is 1/8 inch.

As the process fluid passes through the orifice, the fluid converges and the velocity of the fluid increases to a maximum value. Then, after the orifice plate, as the process fluid diverges to fill the entire pipe, the velocity decreases back to its original value. The pressure loss across an orifice plate is permanent, therefore the outlet pressure will always be less than the inlet pressure. The pressure on both sides of the orifice is measured (thru taps), resulting in a D/P that is proportional to the flow rate.

There are two distinct disadvantages of orifices:

- High permanent pressure loss

- Erosion of the machined orifice bore, which will cause progressive inaccuracies in the measured D/P.

## *Orifice Plate Types*

### Concentric Square-Edged

The concentric square-edged orifice plate (Figure 3-20) is the most common of the orifice plate types. The orifice bore is equidistant from (concentric to) the inside diameter of the pipe. The front edge of the bore has a sharp square edge, which provides for negligible friction drag. The process flow through a sharp-edged orifice plate is characterized by a rapid increase in fluid velocity. Concentric square-edged orifices are not recommended for use on multi-phase (i.e., combinations of solids, liquids, or gases) process fluid in a <u>horizontal</u> run of pipe; however, this type is still preferred in a <u>vertical</u> run of pipe.

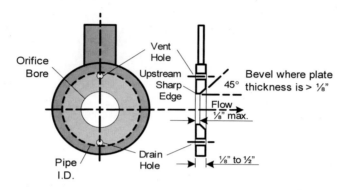

**Figure 3-20. Concentric Square-Edged Orifice Plate**

### Segmental and Eccentric

Segmental and eccentric orifice plates (Figure 3-21) are functionally the same as the concentric orifice plate. With an eccentric orifice plate the circular opening is offset from center. With the segmental orifice plate, the circular section of the segment is concentric with the pipe. The segmental portion of the orifice eliminates the potential for the damming of solid materials on the upstream side of the orifice when the orifice is mounted in a horizontal pipe. If the secondary phase of the process fluid is a liquid, then the opening should be at the bottom of the pipe. If the secondary phase of the process fluid is a gas, then the opening should be at the top of the pipe. These types of orifice plates are typically used with sediment-laden liquids or slurries.

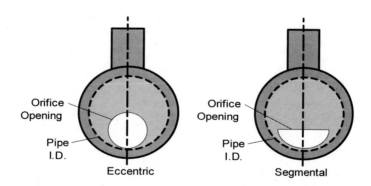

**Figure 3-21. Eccentric and Segmental Orifice Plates**

## Integral Orifice

Similar to an orifice plate, except that it is typically used to measure very small flow rates. Upstream and downstream piping requirements are built into the meter body (Figure 3-22).[20]

**Figure 3-22. Integral Orifice D/P Pressure Flowmeter**

### Orifice Pressure Tap Types/Locations

## Flange Taps

These taps are located 1 inch (centerline) from the upstream face of the orifice plate and 1 inch (centerline) from the downstream face of the orifice plate (Figure 3-23). The standard tap size is 1/2 inch NPT. This is the most common type of pressure tap used in pipe sizes ≥ 2 inches ≤ 6 inches.[21]

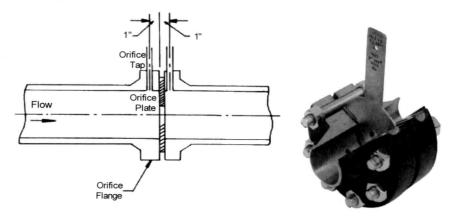

**Figure 3-23. Flange Tap Drawing (left); Flange Tap Cutaway (right)**

## Vena Contracta Taps

These taps are located one pipe diameter upstream and at the minimum pressure point downstream (at the vena contracta) (Figure 3-24). The minimum pressure point will vary with

---

20. Rosemount ProPlate Flowmeter assembly.
21. Image obtained from www.avcovalve.com.

the Beta ratio. This type is only used where process flow rates are very constant and plates are not changed. Pipe sizes > 6-inch lines.

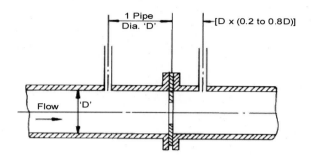

**Figure 3-24. Vena Contracta Tap Location**

### Radius Taps
These taps are very similar to vena contracta pressure taps, except that the lower pressure tap is located 1/2 pipe diameter downstream from the orifice plate (Figure 3-25).

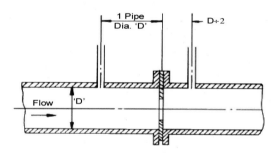

**Figure 3-25. Radius Tap Location**

### Corner Taps
These taps are very similar to flange taps except that the taps are located immediately adjacent to the plate faces, upstream and downstream (Figure 3-26). Pipe size is typically < 2 inches in diameter.[22]

*Orifice Plate Sizing Calculations*

**Useful Orifice Ratio Equations:**

$$\frac{F_2}{F_1} = \sqrt{\frac{\Delta P_2}{\Delta P_1}} \qquad \frac{P_2}{P_1} = \left(\frac{F_1}{F_2}\right)^2 \qquad A_1 V_1 = A_2 V_2$$

**Liquid Orifice:**

---

22. Image obtained from www.qps-orifice.com.

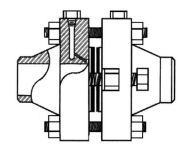

**Figure 3-26. Corner Tap Location**

$$S = \frac{Q_M * G_b}{ND^2 \sqrt{G_F \sqrt{h_M}}}$$

Basic Equation: $Q_M = 5.667 SD^2 \sqrt{\dfrac{h_M}{G_F}}$

$Q_M$ = Maximum flow in GPM

$G_b$ = Base S.G. [(S.G. of liquid @ 60°F (Water @ 60°F = 1)]

N = 5.667 for GPM

D = Pipe ID in inches

$G_F$ = Flowing SG of liquid @ flowing temperature

$h_M$ = Meter differential in inches WC

S = Orifice ratio (reference Spink book for corresponding β). S = 0.598 β$^2$ + 0.01 β$^3$+ [(0.00001947 β$^2$)(10 β$^{4.425}$)] for flange, vena-contracta, radius or corner taps.

**Steam or Gas Orifice**

*Basic Equation Steam*

$$W_{lbs/hr} = 359 SD^2 \sqrt{h_m S_W}$$

*Basic Equation Gas*

$$Q_{scfh} = 218.4 SD^2 \frac{T_{abs}}{P_{abs}} \sqrt{\frac{h_m P_f}{T_f G_f}} \quad T_f = T_{abs} \text{ in } °R$$

$P_{abs}$ = 14.7

$SG_{gas}$ = MW ÷ 29

W = Flow in lbs/hr

SW = Specific Weight of vapor in lbs/ft$^3$ = 1 ÷ Specific Volume

For Steam, reference Table 2.6 (use 1/specific volume)

For Gas, reference table below for some common gas weight densities ($S_W = \rho g$):

$h_M$ = Meter differential in inches WC

D = Pipe ID in inches

S = Orifice ratio (reference Spink[23] book for corresponding β)

*A rule of thumb to use in <u>gas flow</u> is that critical flow[24] is reached when the downstream pipe tap registers an absolute pressure of approximately 50% or less than the upstream pipe tap.*

## Common Gas Weight Densities:

| Gas | Weight Density 'ρ' |
|---|---|
| Air | 0.752 |
| Ammonia | 0.0448 |
| Carbon Monoxide | 0.0727 |
| Carbon Dioxide | 0.1150 |
| Chlorine | 0.1869 |
| HCl | 0.0954 |
| Helium | 0.0104 |
| Hydrogen | 0.00523 |
| Methane | 0.0417 |
| Natural Gas | 0.0502 |
| Nitrogen | 0.0727 |
| Oxygen | 0.0831 |
| Propane | 0.1175 |

## Orifice Plate Installation Details

*Liquid or Steam Service (Horizontal Pipe)*

*Liquid or Steam Service (Vertical Pipe)*

*Gas Service (Horizontal Pipe)*

*Gas Service (Vertical Pipe)*

---

23. Principles and Practice of Flow Meter Engineering by L.K. Spink printed by the Foxboro Company (1978).
24. Critical Flow is also referred to as choked flow which is a condition where the flow will no longer increase even if the downstream pressure decreases because the gas velocity is equal to sonic velocity.

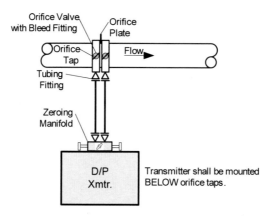

**Figure 3-27. Orifice Plate Installation (1)**

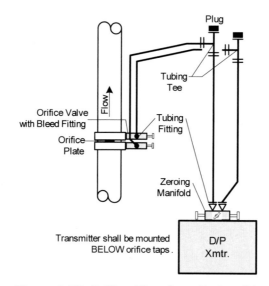

**Figure 3-28. Orifice Plate Installation (2)**

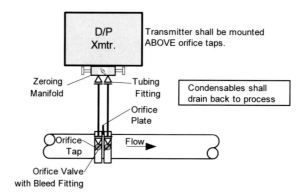

**Figure 3-29. Orifice Plate Installation (3)**

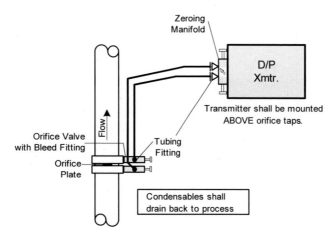

**Figure 3-30. Orifice Plate Installation (4)**

*Venturi Tube*

In this type, a D/P region is created by reducing the cross-sectional process flow area (venturi effect). The venturi tube (Figure 3-31) operates very similar to an orifice plate; however, these meters create less of a permanent pressure loss than that of orifice plates. Venturi tubes are primarily used on larger flows where the use of an orifice plate is impractical.

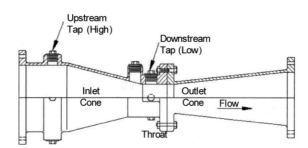

**Figure 3-31. Venturi Tube (Cutaway Drawing and Typical Unit)**

**Venturi Tube Sizing Calculation (Liquid)**

$$Q_m = \frac{CA_{throat}\sqrt{2\rho\Delta P}}{\sqrt{1-B^4}} \qquad Q_v = \frac{Q_m}{\rho}$$

A = Area of Throat

C = Coefficient of Discharge (friction coefficient) this coefficient is developed experimentally by the flow element manufacturer.

ΔP = Differential Pressure

Qm = Mass Flow Rate

Qv = Volumetric Flow Rate

ρ = Density

## Venturi Tube Installation Details

The location of the D/P transmitter for the venturi tube should follow the same rules as for orifice plate installation.

### Cone Flowmeter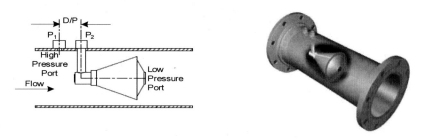

The cone (Figure 3-32) is a D/P type flowmeter. The D/P is created by a cone placed in the center of the pipe. The cone is shaped so that it flattens the process fluid velocity profile within the pipe. The D/P is measured by sensing the difference in pressure upstream of the cone element at the meter wall and the pressure downstream of the cone through the center of the cone element.[25]

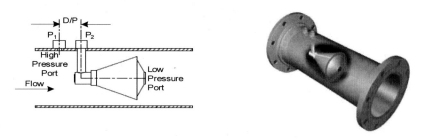

**Figure 3-32. Cone Flowmeter (Drawing and Cutaway View)**

### Cone Sizing Calculation

$$B = \frac{\sqrt{D^2 - d^2}}{D} \qquad k_1 = \frac{\pi}{576}\sqrt{2G_C}\,\frac{D^2 B^2}{\sqrt{1 - B^4}}\,C_F \qquad ACFS = k_1\sqrt{\frac{5.197\,\Delta P}{\rho}}$$

B = Cone Beta Ratio

$K_1$ = Flow Constant

$C_G$ = Gravitational Constant

D = Pipe ID

d = Cone Diameter

$C_F$ = Flowmeter Coefficient (use 1 if unknown)

### Pitot Tube/Annubar

**Pitot Tube:** Consists of a tube with a short 90° bend. The pitot tube (Figure 3-33) is inserted vertically into the process fluid with the short 90° bend directed upstream.

Multi-port Averaging Pitot tube (Annubar): The major difference between an annubar (Figure 3-34) and a pitot tube is that an annubar takes multiple samples across a section of a pipe or duct, versus just one for a basic pitot tube. This allows the annubar to average the differential pressures encountered across the flow velocity profile.

---

25.  McCrometer cone element.

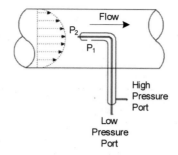

**Figure 3-33. Pitot Tube Installation**

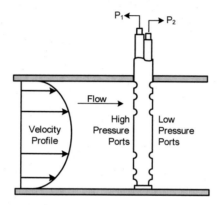

**Figure 3-34. Annubar**

**Pitot/Annubar Sizing Calculation**

*Liquid:*

$$\Delta P = \frac{Q^2 S_f}{K^2 D^4 32.14}$$

$\Delta P$ = D/P in inches WC

Q = Flowrate in GPM

$S_f$ = S.G. at flowing conditions

K = Flow Coefficient (use 1 if unknown)

D = Pipe ID

*Steam or Gas:*

$$\Delta P = \frac{Q^2 (lb/hr)}{K^2 D^4 \rho 128900} \quad \text{or} \quad \Delta P = \frac{Q^2 (scfm) S_s T_R}{K^2 D^4 P 16590}$$

$\Delta P$ = D/P in inches WC

$S_s$ = S.G. at 60°F

K = Flow Coefficient (use 1 if unknown)

D = Pipe ID

$\rho$ = Density (in lb/ft$^3$)

P = Static Line Pressure (in PSIA)

$T_R$ = Temperature in °R

### Elbow Flowmeter

This type of flowmeter (Figure 3-35) is useful when an "order of magnitude" flow measurement is needed. An existing elbow in the piping system may be utilized without the need to install a new flowmeter. When a liquid flows through an elbow, the centrifugal forces cause a pressure difference between the outer and inner sides of the elbow. This is because the process fluid on the outer radius of the elbow must travel at a faster velocity than the process fluid on the inner radius of the elbow. The D/P produced by the elbow meter is very small and thus a highly sensitive D/P sensor is required to measure this difference in pressure.

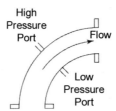

**Figure 3-35. Elbow Flowmeter**

### Variable Area/Rotameter

The rotameter is a special form of D/P producer, wherein the area of the flow restriction is varied to maintain a constant D/P. The rotameter's operation (Figure 3-36) is based on the variable area principle, which states that process fluid flow raises a float in a tapered tube, increasing the area needed for passage of that process fluid. The greater the flow, the higher the float is raised. The volumetric flow rate is directly proportional to the displacement of the float. The float will stabilize within the tube when the upward force exerted by the flowing process fluid equals the downward gravitational force exerted by the weight of the float. The rotameter MUST be oriented vertically and used on clean liquids and gases only, and should not be used in highly viscous applications.

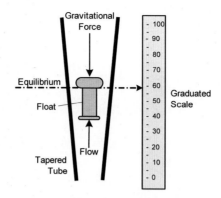

**Figure 3-36. Rotameter**

## Rotameter Sizing Calculation

*Liquid Volumetric Rate:*

$$Q = \frac{(GPM)(\rho)(2.65)}{\sqrt{\rho(\rho_f - \rho)}}$$

$\rho$ = Density of flowing process fluid

$\rho_f$ = Density of float

2.65 = Unit conversion factor

*Gas or Vapor Volumetric Rate:*

$$Q = \frac{(SCFM)(\rho_g std)(10.34)}{\sqrt{\rho_f(\rho_g act)}}$$

$\rho_g std$ = Density of gas at 14.7 psia and 70°F

$\rho_g act$ = Density of flowing process gas or vapor

$\rho_f$ = Density of float

10.34 = Unit conversion factor

### *Target Meter*

With the target meter, a physical target is attached to an extension rod and located directly in the process fluid flow. The force that is exerted upon the target by the flowing fluid is measured via force or strain sensors attached the extension rod (typically a cantilevered arm). This force (drag) measurement is proportional to the process flow rate squared.

## 3.3.2 Electronic Volumetric Flowmeters

*Vortex Shedder*

Vortex shedder flowmeters consist of a bluff body (Figure 3-37) placed in the process fluid flow path. When the flowing process fluid strikes the bluff body, regular alternating vortices (von Karman effect[26]) are formed downstream of the bluff body (similar to wind striking a flag pole which causes the flag to wave in alternating directions). A sensor in turn detects these alternating vortices and generates a flow signal that is based upon the Strouhal number[27]. The higher the frequency of the alternating vortices, the higher the flow rate.

Vortex shedder flowmeters should not be used in flow applications with low Reynolds numbers. In addition, vortex flowmeters should be rigidly supported so as not to allow any pipe vibration to interfere with the vortex sensing device.

**Strouhal Number:**

$$S = \frac{f_{Hz}(h_f)}{V_{avg}}$$

Where:
| | | |
|---|---|---|
| $S$ | = | Strouhal number |
| $f_{Hz}$ | = | Vortex shedding frequency |
| $h_f$ | = | Bluff body width (in feet) |
| $V_{avg}$ | = | Average fluid velocity |

The volumetric flowrate Q of the vortex flowmeter is the product of the average fluid velocity and the cross-sectional area A available for flow:

$$Q = AV_{avg} = \frac{[(A)(f_{Hz})(d)(B)]}{S}$$

Where:
| | | |
|---|---|---|
| $d$ | = | Meter internal diameter |
| $B$ | = | $d - h_f$ |

Since $[(A)(f_{Hz})(d)(B)] = K$ factor for the flow meter, $S = \dfrac{f_{Hz}(h_f)}{V_{avg}}$ the equation reduces to $Q = f_{Hz}K$

Figure 3-37 is a cross-sectional view of different styles of bluff bodies.

**Figure 3-37. Bluff Bodies**

---

26. The von Kármán effect refers to the tendency of any fluid as it flows past an object and reaches the other side to curl or produce eddies. The eddies set up an oscillation that may be reinforced by the natural frequency of the structure.
27. A dimensionless number (Vincenc Strouhal) describing oscillating flow mechanisms.

**Vortex Shedder Installation:** The preferred installation is in a vertical pipe run with the flow direction up. A vortex flowmeter may be installed within a horizontal pipe run as well; however, the minimum straight run of pipe must be observed for proper operation of the flowmeter (15 pipe diameters upstream and 5 pipe diameters downstream).

*Magnetic Flowmeter (Magmeter)*   ⊣M⊢

The magmeter operates on the principle of Faraday's Law[28]. A magmeter is constructed of a piece of pipe with a non-magnetic, insulating liner and a pair of magnetic coils that are placed on opposite sides of the pipe. Electrodes penetrate through the pipe and its lining into the flowing process fluid. The magnitude of the voltage detected by these electrodes is directly proportional to the velocity of the conductor, which in the case of a magmeter is the flowing conductive process fluid, and the distance between the electrodes and the strength of the magnetic field. Within the magmeter, the space between the electrodes is fixed, as is the strength of the magnetic field, leaving only the flowing conductive fluid as the variable component.

The use of magmeters is advantageous when the process fluid contains a high percentage of solids and/or is highly corrosive. Permanent pressure loss is minimal with a magmeter due to the fact that there is no obstruction with the flow path. However, magmeters may not be used on non-conductive fluids (typically $\leq 50\ \mu S/cm$).

The voltage signal generated between the electrodes can be calculated from (see Figure 3-38).

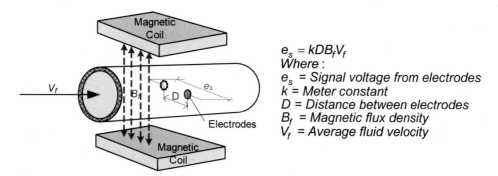

$e_s = kDB_fV_f$
Where:
$e_s$ = Signal voltage from electrodes
$k$ = Meter constant
$D$ = Distance between electrodes
$B_f$ = Magnetic flux density
$V_f$ = Average fluid velocity

**Figure 3-38. Magmeter Schematic and Voltage Output Calculation**

For a circular pipe the volumetric flow rate is as follows:

$$Q = \frac{\pi}{4}KD\frac{e_s}{B_f}$$

**Magmeter Installation:** The preferred installation is in a vertical pipe run with the flow direction up. A magmeter may be installed within a horizontal pipe run as well; however, the minimum straight run of pipe must be observed for proper operation of the flowmeter (5 pipe diameters upstream and 2 pipe diameters downstream).

---

28. Faraday's Law (law of electromagnetic induction): The value of the induced voltage is proportional to the rate of change of magnetic flux.

## Ultrasonic Flowmeter

The ultrasonic flowmeter operates on the principle that the speed at which sound propagates through a process fluid is dependent upon the fluid's density. If the density is maintained constant, then the flowmeter can measure the time of ultrasonic passage (reflection) to determine the velocity of a flowing fluid. There are different types of ultrasonic flowmeters, including Doppler and time-of-flight.

### Doppler

A shift in frequency is the basis upon which all Doppler ultrasonic flowmeters work. The transducer sends an ultrasonic wave into the flowing process fluid. The ultrasonic waves are then reflected back by acoustical discontinuities such as particles or entrained gas. The resulting reflected waves will have a frequency change that varies by the velocity of the moving discontinuities in the flowing process fluid. The meter detects this frequency change (Figure 3-39) and calculates a flow velocity. The flow velocity is directly proportional to the change in frequency.[29]

$$Q_{ft^3/sec} = \left[ \frac{\pi}{4} D^2 \frac{V_{SO}}{\sin\theta} \right] \frac{\Delta f_{Hz}}{\left( f_{Hz} \right)_{cw}}$$

*Where:*
$\Delta f_{Hz}$ = *Doppler frequency shift*
$\left( f_{Hz} \right)_{cw}$ = *Transmitter's fixed frequency*
$\theta$ = *Beam angle*
$D$ = *Pipe internal diameter*
$V_{SO}$ = *Velocity of sound in the liquid*

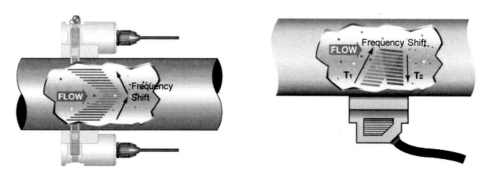

**Figure 3-39. Doppler Ultrasonic Flowmeter, Showing Frequency Shift**

### Time of Flight (Transit Time)

In this design, the time of flight of the ultrasonic wave is measured between two separate transducers, one upstream and one downstream (Figure 3-40). The difference in the elapsed time going with or against the flow determines the fluid velocity. When there is flow, the effect is to increase the frequency of the ultrasonic wave. Like the Doppler effect meter, the time of flight meter detects this frequency change and calculates a flow velocity. The flow velocity is directly proportional to the change in frequency.[30]

---

29. Single and Dual transducer design from Dynasonic.
30. Image obtained from www.dynasonics.com.

---

$$Q_{ft^3/sec} = K \left[ \frac{\pi D^3}{4 \sin 2\theta} \right] \left[ (f_{Hz})_{DN} - (f_{Hz})_{UP} \right]$$

*Where :*

$\left[ (f_{Hz})_{DN} - (f_{Hz})_{UP} \right]$ = *Frequency difference between downstream and upstream*
$\theta$ = *Beam angle*
$D$ = *Pipe internal diameter*
$K$ = *Ultrasonic meter coefficient*

**Figure 3-40. Time of Flight Ultrasonic Flowmeter**

### 3.3.3 Mass Flowmeters

*Coriolis*

A mass flowmeter is differentiated from a volumetric flowmeter in that the output of the meter is proportional with the mass of a process media moving through meter with respect to time as opposed to a volume of process media flowing through a meter with respect to time. This mass measurement is possible because the mass flowmeter also detects the flowing process media's density. The density is measured by the meter using the following principle, when the internal meter flow tube(s) are caused to vibrate at their natural frequency, the mass of the fluid contained in the tube can be deduced. Dividing this mass on the known volume contained within the tube(s) of the tube provides the **density** of the fluid.

This type of flowmeter utilizes the Coriolis effect[31] in its design. When the process fluid is flowing through the tubes (Figure 3-41), the process fluid accelerates on the inlet side of the tube and decelerates on the outlet side of the tube. As a result, the tube is slightly twisted by these opposing forces. The amount of tube twist is proportional to the mass flow rate.

To measure this twist, a magnet is attached to one of the tubes and a magnetic pickup coil is attached to the other tube (at both inlet and outlet locations). The relationship of the sine wave outputs represents the flow rate and the relative position of the tubes. When there is no flow through the tubes, the sine wave outputs from both of the pickup coils are in phase with each other (i.e., there is no twist).

There are two basic tube configurations for the Coriolis flowmeter: straight or curved. In both designs, process fluid is flowing through both tubes.

---

31. Coriolis effect: described by French engineer Gustave-Gaspard Coriolis, the effect of the Coriolis force is an apparent deflection of the path of an object that moves within a rotating coordinate system. The object does not actually deviate from its path, but it appears to do so because of the motion of the coordinate system.

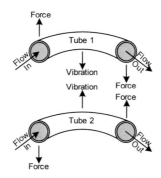

**Figure 3-41. Principle of Coriolis Flowmeter Operation**

$$Q_{mass} = \frac{F}{2\ell\Omega}$$

*Where* :
*F = Force (as in F = ma)*
*ℓ = Fluid element length*
*Ω = Angular frequency*

**Coriolis Flowmeter Installation:** The preferred installation is in a vertical pipe run with the flow direction up. The Coriolis flowmeter may be installed within a horizontal pipe run as well; however, the tube side must never be facing up. When mounted in a horizontal pipe run, the Coriolis flowmeter must be installed with the tubes facing either sideways or down. Note: mounting the tubes facing down in a horizontal pipe line creates a drainage problem, because a pocket of liquid will pool at the U tube area (straight through coriolis meters do not have this problem). To eliminate this pocket of fluid, it may be necessary to blow the line clear.

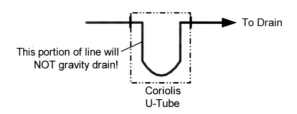

*Thermal Mass*

This type is based on an operational principle that states that the rate of heat absorption by a process flow stream is directly proportional to its mass flow rate. The amount of heat absorbed is measured with two integral temperature sensors within a probe. One of the sensors is heated so that it is maintained at a constant temperature. The other sensor measures the process fluid temperature. As the molecules of the flowing process fluid comes in contact with the heated probe, they will absorb heat. This absorbed heat will then have a tendency to cool the heated probe. The higher the flow rate of the flowing process fluid, the more molecules that will contact the heated probe; thus absorbing more heat. The amount of heat taken away from the heated probe is proportional to the number of molecules of the flowing process fluid (its mass). The molecules (mass) can be measured by the amount of energy that the needs to supply back to the heated probe so that a constant differential temperature above the reference probe. Therefore, the amount of heat energy required to maintain this constant temperature differential is proportional to the process fluid mass flow rate.

### Hot-Wire Anemometer

This type, used for gases, consists of an electrically heated fine-wire element. Tungsten is traditionally used as the wire material because of its strength and its high temperature coefficient of resistance. When it is placed in a moving stream of gas, the wire cools, which causes its electrical resistance to change. The rate of cooling corresponds to the mass flow rate. As the fluid velocity increases, the rate of heat transfer from the heated wire to the flow stream increases, and as it decreases, the heat transfer rate decreases. The most common type of hot-wire anemometer is the constant-temperature type, which is powered by an adjustable current to maintain a constant temperature. The fluid velocity is a function of input current and flow temperature.

## 3.3.4 Mechanical Volumetric Flowmeters

### Turbine Meter

The turbine flowmeter consists of a multi-bladed rotor, mounted at right angles to the process flow stream, that transforms energy from the flow stream into rotational energy. As each turbine blade rotates past the magnetic pickup coil, the magnetic field is broken and a small voltage pulse is generated. The frequency of the pulses is proportional to the velocity of the flow.

**Turbine Meter Installation:** Install in a horizontal pipe run only. The minimum straight run of pipe must be observed for proper operation of the flowmeter (10 pipe diameters upstream and 5 pipe diameters downstream).

**Sizing:**

$$Q = \frac{60f}{K}$$

f = Pulses per second (the 60 factor is used to convert the pulses per second to pulses per minute.

K = Pulses per gallon (this is determined experimentally by the meter manufacturer).

Q = Flow in GPM

### Positive-Displacement (PD) Meter

The basic operating principle of all positive displacement flowmeter designs is to fill a chamber of known volume with fluid and then count the number of repeated fillings and discharges of the volume chambers over a known time interval. Though accuracy is high, the permanent pressure loss associated with positive displacement meters is higher than that of other types of flowmeters. This type is commonly used in custody transfer applications. Coriolis mass flowmeters have made inroads into areas formerly dominated by positive displacement meters due to their accuracy.

Some common types of positive displacement meters are nutating disc ("wobble plate"), oscillating piston, rotary vane, oval gear, and rotating lobe.

Positive displacement meters are NOT to be used with dirty or low viscosity fluids.

**Positive Displacement Meter Installation:** Install in a horizontal pipe run. Due to the weight of the majority of positive displacement meter designs, the meter should be supported independent of the piping system (i.e., do not use the piping system as the only support for the meter).

### 3.3.5 Open Channel Volumetric Flow Measurement

A common method of measuring liquid flow through an open channel is to measure the height of the liquid as it passes over an obstruction such as a flume or weir in the channel. For any open channel that is free flowing, there is a specific relationship between the height of liquid and the flow rate. Whenever a given height occurs, there will always be the same liquid flow rate. Therefore, if the flow rate is known for each height, a height-to-flow rate relationship can be constructed. Ultrasonic detectors are frequently used to detect the change in height of the liquid.

*Weir*

As described above, a weir provides an obstruction in the flow path within an open channel. The liquid surface level downstream of the weir needs to be at least 2 1/2 inches below the bottom of the weir opening.

There are two basic weir designs:

**Rectangular Weir:** The rectangular weir (Figure 3-42) is the most commonly used design. The surface height over the rectangular weir is measured and correlated with the fluid flow rate through the opening in the weir. The fluid flow rate is a function of height over the rectangular weir (measured approximately 4H upstream of the weir).

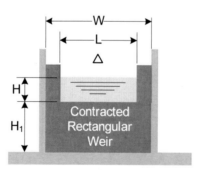

**Figure 3-42. Rectangular Weir**

$$Q = 3.33 \left[ L - (0.2H) \right] H^{3/2}$$

For H < 1.25 ft; H ÷ $H_1$ < 2.4 and W > 3 ft.

Q = Fluid flow rate (ft$^3$/sec)
H = Height of the fluid over the weir (feet)

**V-Notch:** This type (Figure 3-43) operates similarly to a rectangular weir. However, due to the shape of the V-notch design, a small change in flow rate results in a larger change in depth of the fluid over the weir as compared to the rectangular type. This type of design is more accurate in measurement than a rectangular design, but should only be used to measure low fluid flow rates.

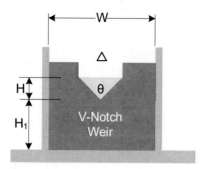

**Figure 3-43. V-Notch Weir**

$$Q = 4.28(C)\left[Tan\left(\frac{\theta}{2}\right)\right](H+k)^{5/2}$$

For H < 1.25 ft; H ÷ $H_1$ < 2.4 and W > 3 ft.

Q = Fluid flow rate ($ft^3$/sec)
H = Height of the fluid over the weir (feet)
θ = Notch angle (degrees) should be between 20 and 100°
C = Effective discharge coefficient
C = 0.6072 − 0.0008745(tanθ) + 6.1039x$10^{-6}$(tanθ)$^2$
K = Head correction factor (feet) Reference Figure 3-44 below for typical curves.
k = 0.01449 − 0.0003396 θ + 3.2982x$10^{-6}$(tanθ)$^2$ − 1.06215x$10^{-8}$(tanθ)$^3$

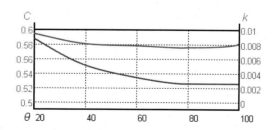

**Figure 3-44. Typical Head Correction Factor Curves**

*Flume*

A flume (Figure 3-45; Table 3-4) operates on the same basic principle as the weir except that the obstruction (referred to as a constriction) is in the horizontal plane rather than the vertical plane. This constriction causes the fluid level to change, and that fluid level in the constriction can be correlated with flow rate. Flumes have the advantage of less pressure loss (head) than that of a weir, but have a more complicated construction. This type is primarily used where flow may contain suspended solids, such as in waste water treatment. The most popular flume design is the Parshall[32] flume.

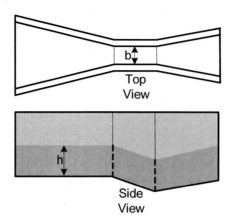

**Figure 3-45. Flume**

Free Flow Equation: $Q = Ch^n$

**Table 3-4. Coefficient Values for Parshall Flume**

| Throat Width b | Coefficient C | Exponent n |
|----------------|---------------|------------|
| 1" | 0.338 | 1.55 |
| 6" | 2.08 | 1.58 |
| 2'-0" | 8.00 | 1.55 |
| 6'-0" | 24.00 | 1.59 |
| 10'-0" | 39.38 | 1.60 |
| 20'-0" | 76.25 | 1.60 |
| 30'-0" | 113.13 | 1.60 |
| 40'-0" | 150.00 | 1.60 |
| 50'-0" | 186.88 | 1.60 |

---

32.  Named after Ralph Parshall, who invented the flume that now carries his name.

---

## 3.3.6 Flowmeter Selection Guide

### Table 3-5. Flowmeter Selection Guide

| Flowmeter | Pipe Size, in. (mm) | Steam | Gases (Vapors) Clean | Gases (Vapors) Dirty | Pressure High | Pressure Low | Liquids Clean | Viscous High | Viscous Low | Dirty | Corrosive | Very Corr. | Slurries Fibrous | Slurries Abrasive | Open Channel | Rangeability (Average) | Minimum Reynolds # | Viscosity Effect | Maximum Temperature °F (°C) | Maximum Pressure PSIG | Permanent Pressure Loss |
|---|---|---|---|---|---|---|---|---|---|---|---|---|---|---|---|---|---|---|---|---|---|
| **Square Root Scale: Maximum Single Range 4:1 (Typical)** | | | | | | | | | | | | | | | | | | | | | |
| **Orifice** | | | | | | | | | | | | | | | | | | | | | |
| Sq-Edged | > 1.5" (40mm) | ✓ | ✓ | X | ✓ | ✓ | ✓ | X | ? | X | ? | X | X | X | X | 4:1 | > 10,000 | High | 1000F (540C) | 4000 | Medium |
| Integral | <= 0.5" (12mm) | ? | ✓ | X | ✓ | ✓ | ✓ | X | ? | X | ? | X | X | X | X | 4:1 | > 10,000 | High | 1000F (540C) | 4000 | Medium |
| Wedge | < 12" (300mm) | ✓ | ✓ | ✓ | ✓ | ✓ | ✓ | ? | ✓ | ✓ | ? | X | ? | ? | X | 4:1 | > 10,000 | High | 1000F (540C) | 4000 | Medium |
| Eccentric | >= 2" (50mm) | ? | ? | ✓ | ✓ | ✓ | ? | X | ? | ? | ? | X | ? | ? | X | 4:1 | > 10,000 | High | 1000F (540C) | 4000 | Medium |
| Segmental | >= 4" (100mm) | ? | ? | ✓ | ✓ | ✓ | ? | X | ? | ? | ? | X | ? | X | X | 4:1 | > 10,000 | High | 1000F (540C) | 4000 | Medium |
| V-Cone | 0.5 - 72" | ✓ | ✓ | ? | ✓ | ✓ | ✓ | ? | ✓ | ✓ | ? | X | X | ? | X | 10:1 | > 8000 | Medium | 1000F (540C) | 4000 | Medium |
| Target | >= 0.5" (12mm) | ? | ✓ | ? | ✓ | ✓ | ✓ | ? | ✓ | ✓ | ? | X | X | X | X | 10:1 | > 2000 | High | 500F (260C) | 15000 | Medium |
| Venturi | >= 2" (50mm) | ✓ | ✓ | ? | ✓ | ✓ | ✓ | ? | ? | ✓ | ? | X | ✓ | ? | X | 4:1 | > 75,000 | High | 1000F (540C) | 4000 | Low |
| Pitot | >= 3" (75mm) | X | ✓ | X | ✓ | ✓ | ✓ | X | ? | X | ? | X | X | X | X | 3:1 | > 100,000 | Low | 1000F (540C) | 4000 | Low |
| Avg Pitot | >= 1" (25mm) | X | ✓ | SD | ✓ | ✓ | ✓ | X | ? | SD | ? | X | X | X | X | 3:1 | > 15,000 | Low | 300F (149C) | 275 | Low |
| Elbow | >= 2" (50mm) | X | ✓ | ? | ✓ | ✓ | ✓ | X | ? | ✓ | ? | X | X | X | X | 3:1 | > 10,000 | Low | 1000F (540C) | 4000 | Very Low |
| **Linear Scale Range 10:1 or Better** | | | | | | | | | | | | | | | | | | | | | |
| Magmeter | 0.1 - 72" | X | X | X | X | X | ✓ | ? | ✓ | ✓ | ✓ | ✓ | ✓ | ✓ | ? | 40:1 | > 4,500 | None | 360F (180C) | 1500 | None |
| **PD Meter** | | | | | | | | | | | | | | | | | | | | | |
| Gas | < 12" (300mm) | X | ✓ | X | ? | ✓ | X | ✓ | X | X | X | X | X | X | X | 10:1 | NA | NA | 250F (120C) | 1400 | High |
| Liquid | < 12" (300mm) | X | X | X | X | X | ✓ | ✓ | ? | X | ? | X | X | X | X | 10:1 | NA | High | 600F (315C) | 1400 | High |
| **Turbine** | | | | | | | | | | | | | | | | | | | | | |
| Gas | 0.25 - 24" | SD | ✓ | SD | ✓ | ✓ | X | X | ? | X | X | X | X | X | ? | 10:1 | NA | NA | 500F (260C) | 3000 | High |
| Liquid | 0.25 - 24" | X | X | X | X | X | ✓ | X | ? | X | ? | X | X | SD | ? | 10:1 | > 5,000 | High | 500F (260C) | 3000 | High |
| **Ultrasonic** | | | | | | | | | | | | | | | | | | | | | |
| Time of Flight | >= 0.5" (12mm) | X | SD | SD | SD | SD | ✓ | ? | ? | ✓ | ✓ | ✓ | ? | ? | ? | 20:1 | > 10,000 | None | 500F (260C) | Pipe rating | None |
| Doppler | >= 0.5" (12mm) | X | X | X | X | X | X | ? | ? | ✓ | ✓ | ✓ | ✓ | ✓ | X | 10:1 | > 4,000 | None | 500F (260C) | Pipe rating | None |
| Rotameter | <= 3" (75mm) | ? | ✓ | ? | ✓ | ✓ | ✓ | ✓ | ✓ | ✓ | ? | ? | X | X | X | 10:1 | NA | Medium | 400F-Glass 1000F-Metal | 350 - Glass 720 - Metal | Medium |
| Vortex | 1.5 - 16" | ✓ | ✓ | ? | ✓ | ✓ | ✓ | ? | ? | ✓ | ? | ? | ? | X | X | 10:1 | > 10,000 | Medium | 400F (200C) | 1500 | Medium |
| **Mass** | | | | | | | | | | | | | | | | | | | | | |
| Coriolis | 0.25 - 6" | ? | X | ? | ✓ | ✓ | ✓ | ✓ | ✓ | ✓ | ? | ? | ? | ? | X | 10:1 | NA | None | 800F (427C) | 5700 | Low |
| Thermal | >= 0.5" (12mm) | X | ✓ | ? | ✓ | ✓ | ✓ | ? | ? | ? | ? | ? | ? | ? | X | 10:1 | NA | None | 1500F (816C) | Pipe rating | Low |
| **Open Channel** | | | | | | | | | | | | | | | | | | | | | |
| Weir | | X | X | X | X | X | ✓ | X | ✓ | ✓ | ✓ | ✓ | ✓ | ✓ | ✓ | 100:1 | NA | Very Low | | Atmospheric | Very Low |
| Flume | | X | X | X | X | X | ✓ | X | ✓ | ✓ | ? | ? | ? | ? | ✓ | 50:1 | NA | Very Low | | Atmospheric | Very Low |

✓ Designed for this application (generally suitable)
? May be applicable (worth consideration)
X Not Applicable
SD Some designs may be applicable.

## 3.4 LEVEL MEASUREMENT SENSORS

### 3.4.1 Inferential Level Measurement Techniques

*Differential Pressure (D/P) Level Measurement*

Liquid level can be measured (inferred) by measuring a D/P caused by the weight (pressure) of the fluid column in a vessel compared to a known reference pressure measurement. Since the weight of the fluid column is a function of the fluid's specific gravity, the fluid being measured should have stable specific gravity properties in order to accurately measure level using D/P level measurement. The D/P level measurement technique will work in atmospheric, pressurized and vacuum applications. D/P level measurement should not be used on vessels/ containers that are not uniform is shape, for example, conical bottom.

- For atmospheric vessels, the high pressure side is connected to the bottom of the vessel, and the low pressure side (reference) is vented to atmosphere. Reference Figure 3-50 for installation detail.

- For pressurized/vacuum vessels, the high pressure side is connected to the bottom of the vessel, and the low pressure side is connected to the vapor space section (top) of the vessel. Reference Figure 3-51 for installation detail.

The D/P calibration range for a transmitter = Maximum height of liquid column x S.G. of the liquid. Whenever the transmitter is located at a different plane than the lower tap, zero adjustments must be made in the calibration (reference zero elevation/suppression).

- Extended diaphragms (Figure 3-46) may be used on viscous, slurry, or other plugging type applications. The extended diaphragm allows the sensing diaphragm to be mounted flush with the inside of the vessel wall, thereby avoiding the usual pocketing problems encountered by conventional flange mounted transmitters.[33]

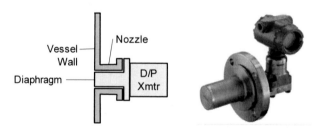

**Figure 3-46. Extended Diaphragm D/P Level (Schematic and Typical Unit)**

- Chemical seals (diaphragm seals) (Figure 3-47) may be used with D/P level devices where the process is corrosive or toxic. Whenever the capillaries for the low and high side are of different lengths, this may introduce an error into the reading for small D/P readings because the thermal expansion will be different for both legs. In addition, whenever chemical seals with capillaries are used in association with the D/P

---

33. Rosemount extended diaphragm assembly.

transmitter, zero adjustments must be made to account for any head associated with the capillaries.[34]

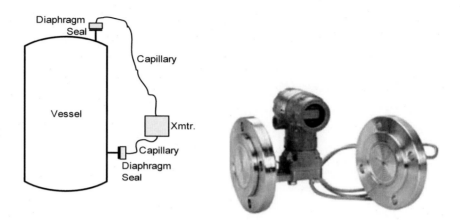

**Figure 3-47. Chemical Seals (Installation Schematic and Typical Unit)**

### Bubbler
This form of D/P level measurement uses a dip tube and a flow of gas, typically nitrogen or instrument air. The amount of gas pressure required to force the gas through the open bottom end of the dip tube, which is at or near the level of the bottom of the vessel, is equal to the hydraulic head of the fluid in the vessel. A change in gas pressure is therefore proportional to a change in fluid level (hydraulic head). Bubblers should only be used in atmospheric vented vessels. Bubblers consume a certain volume of pressurized gas or air during operation, therefore the cost of the compressed gas or air should be calculated into the lifecycle cost of a bubbler level measurement installation.

$$H = L_{DT} \times SG_{fluid}$$

$H$ = Head pressure
$L_{DT}$ = Length of dip tube submerged in process fluid
$SG_{fluid}$ = Specific gravity of the process fluid

### D/P Transmitter Zero Elevation/Suppression
This correction is applicable whenever the D/P sensing cell is at an elevation other than the connecting nozzle (Figure 3-48 a & b). The zero of the D/P cell needs to be elevated or suppressed. Note that two separate zero reference points exist. One zero reference point is the level of the tank that is considered to be zero (lower range value). The other zero reference point is the point at which the D/P cell experiences a zero differential pressure.

*Zero Suppression:* Zero suppression is required if the D/P transmitter is mounted below the position of the capillary/seal high pressure side reference tap. *(Note: If the instrument zero value is a positive value, then this is an indication of zero suppression).*

---

34. Rosemount 1199 diaphragm.

---

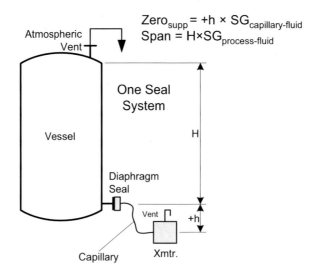

**Figure 3-48. Zero Suppression**

*Zero Elevation:* Zero elevation is required if the D/P transmitter is mounted above the position of the capillary/seal high pressure side reference tap (Figure 3-49). *(Note: If the instrument zero value is a negative value, then this is an indication of zero elevation).*

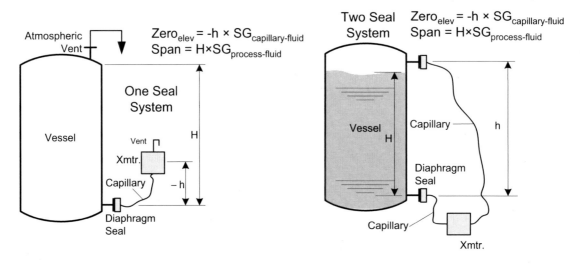

**Figure 3-49. Zero Elevation (a: one seal system; b: two seal system)**

## D/P Level Installation Details

*Atmospheric Vessel:*

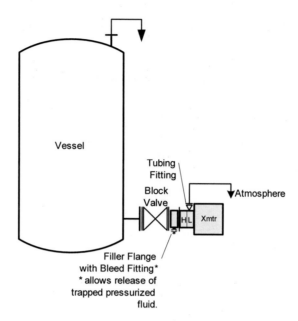

**Figure 3-50. D/P Level Installation (1)**

*Pressurized Vessel:*

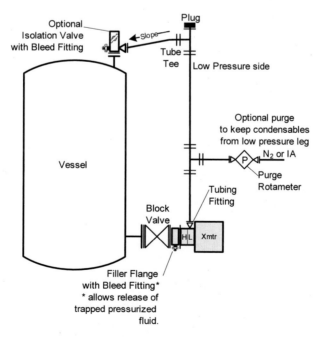

**Figure 3-51. D/P Level Installation (2)**

## Weight Level Measurement

The level of material in a vessel (whether liquid or solid) may also be inferred through the weight of the vessel. To infer level by weight, load cells are typically used. A load cell is a device that converts a weight-force into an electrical signal, usually via strain gauges[35]. Four strain gauges are usually used per load cell (Figure 3-52). Two of the strain gauges are in tension, while the other two are in compression.

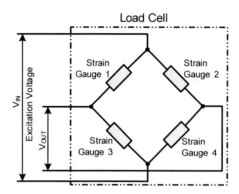

**Figure 3-52. Load Cell**

The electrical output (on the order of a few millivolts) from the load cells is sent to a summing box, which adds all of the outputs together. It then amplifies the resulting output so that the signal may be used by a control system.

### Types of Strain Gauge Load Cells

- *Bending Beam (Cantilever):* This type uses strain gauges that are bonded to the upper and lower parts of the load cell beam to detect bending stress (which results in beam deflection) when the load cell is subjected to force. They are used for lower-end weight capacity applications.

- *Shear Beam (Single/Double) (Figure 3-53):* This type uses strain gauges that are bonded to a reduced cross-sectional area of the load cell beam at an orientation of 45° from each other in order to detect shear stress. They are used for medium to large weight capacity applications.

- *Canister (Tension/Compression) (Figure 3-54):* Not as popular as the beam type load cells in compression applications due to the installation requirement of exact placement (check rods). They may be used as a solution in tension applications.

### Installation

A vessel will typically use three or four load cells, depending on how many support legs there are for the vessel. Care should be taken so as not to introduce any tension or compression into the installation from external connections (e.g., piping, conduit, etc.).

---

35. Strain Gauge: A metallic string will differ in its resistance depending on whether it is stretched or allowed to contract. The longer the string is stretched the higher the resistance. The strain gauge utilizes this principle to detect strain by changes in resistance. Gage factor $Fg = (\Delta R/R) \div (\Delta L/L)$.

**Figure 3-53. Beam Type (Sensortronics)**

**Figure 3-54. Canister Type (Sensortronics)**

### Nuclear Level Measurement

Nuclear level measurement devices, which may also be used for either solids or liquids, consist of two components mounted on opposite sides of a vessel: a gamma radiation source and a detector for the gamma radiation. Reference Figure 3-55 for a basic nuclear level measurement installation.

The process fluid in the vessel will absorb gamma radiation in proportion to its mass. Therefore, when the vessel is full, maximum gamma ray absorption will occur. When the vessel is empty, minimum gamma ray absorption will occur.

Gamma radiation sources (Cesium 137 and Cobalt 60) are used because they have sufficient penetrating power to pass through metal vessel walls.

Cesium 137 is the most common of the sources used. It has a half-life of 30 years. Due to its half life, it will most likely not require replacement of the Cesium 137 source within level transmitter's lifetime. Cesium 137 decays by emission of beta particles and gamma rays to Barium 137m.

Cobalt 60 has a higher penetrating capacity, however, its half-life is only 5.27 years. It ultimately decays to non-radioactive nickel.

The gamma ray detector may be either the scintillation counter type or the ion chamber type.

*Scintillation counter (aka photon detector):* When the gamma rays strike the material contained within the scintillation counter, energy is transferred to the bound electrons, causing them to leave the valence band. This, in turn, causes light particles (photons) to be produced. These photons are then converted into an electrical current which is proportional to the amount of gamma rays striking the scintillation counter.

*Ion chamber:* An ion chamber is a tube filled with an inert gas that also contains an electrode in the center of the tube. Voltage is applied to this tube, then when the gamma rays strike the tube, they ionize the inert gas and produce a small electrical current proportional to the amount of gamma rays striking the tube.

Some gamma ray detectors may be susceptible to x-rays and cause false low level readings (i.e., if the detector is located near an area where pipe x-ray activities are taking place, it may be necessary to shield the detector from the x-ray signals).

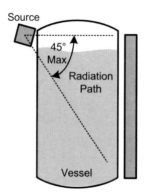

**Figure 3-55. Basic Nuclear Level Installation.**

Nuclear level measurement may be used when tank penetrations are not permitted (e.g., they present risk to human life).

The nuclear sources are regulated by the NRC (Nuclear Regulatory Commission). Any site that uses nuclear level measurement sources must have a named Radiation Safety Officer on their site who is the point of contact with the NRC. In addition, the nuclear sources require an NRC license. Paperwork is required to track the sources from receipt to disposal.

### 3.4.2 Visual Level Measurement Techniques

*Level Gauge*

A level gauge is a device that allows the liquid level within a vessel to be visually inspected. The level gauge is mounted parallel to the side of the vessel. As the level of the liquid within the vessel fluctuates, the level of liquid within the level gauge fluctuates accordingly.

**Tubular Glass Level Gauge:** This type is not recommended for use in industrial areas due to glass breakage potential, unless it is of armored design.

**Flat Glass Level Gauge:** Flat glass gauges come in two designs: transparent and reflex.

*Transparent (Figure 3-56):* This type consists of two pieces of flat, polished glass assembled on opposite sides of a chamber, allowing the liquid level to be viewed through the gauge. The use of a backlight (illuminator) is usually recommended with a transparent glass level gauge. Flat glass gauges are frequently used in water/steam level applications (e.g., steam drum).[36]

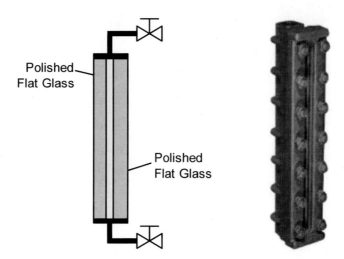

**Figure 3-56. Transparent Flat Glass Level Gauge**

*Reflex (Figure 3-57):* Reflex glass level gauges work on the principles of light refraction and reflection. They consist of prisms that are molded with grooves at 90° and polished in order to provide a black/silver bicolor indication of the fluid level. Light striking the glass in the vapor phase is reflected back, thus appearing a shiny silvery color. Light striking the glass that is covered with liquid is refracted and not reflecting back, the glass therefore appears black. It does not need an external illumination source except at nighttime.[37]

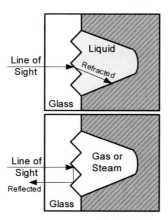

**Figure 3-57. Reflex Glass Level Gauge**

---

36. Image obtained from tycoflowcontrol.com.
37. Cesare Bonetti reflex gauge glass.

## Magnetic Flag Indicator

This type is used in liquid level measurement applications where level gauge glass may present an unsafe environmental condition if it is broken. Magnetic flag indicators (Figure 3-58) consist of a non-magnetic metal cage that contains an internal float that holds one or more magnets and rides on the liquid level. The magnets within the float are magnetically coupled to an external indicator. The external indicator consists of a series of wafers (flags) that contain a permanent magnet and have one color on the front and a contrasting color on the back. As the magnet in the float passes by, the wafers (flags) are flipped over, indicating the level. All of the wafers (flags) below the float must be of one color and all of the wafers (flags) above the float must be of the contrasting color. The magnetic level gauge may also be fitted with point level switches or continuous level transmitters.

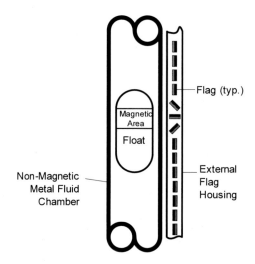

**Figure 3-58. Magnetic Flag Level Indicator**

## Tape Float (Tank Gauging)

The tape float design (Figure 3-59) consists of a tape connected to a float at one end and some form of counterweight at the other end. The purpose of the counterweight is to keep the tape in tension. The float rides on the surface of the process fluid within the vessel. Changes in the fluid level result in the up and down movement of the counterweight. Marking a graduated scale on the outside of the vessel next to the counterweight enables direct viewing of the level within the vessel based upon the position of the counterweight. This is a high maintenance, low accuracy device. When used, it is typically on a very large atmospheric tank.

## 3.4.3 Electrical Properties Level Measurement Techniques

### Capacitance

With capacitance level measurement, the capacitive electrode, process fluid, and vessel wall form an electrical capacitor. As the fluid level in the vessel changes, the dielectric value of the material between the capacitive electrode and vessel wall (air or process fluid) also changes. This change in dielectric value causes a change in the capacitance value that is proportional to the level of the fluid in the vessel. Capacitance level should not be used in applications where

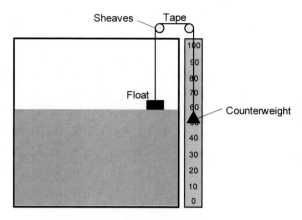

**Figure 3-59. Tape Float Level Gauge**

coating or buildup may occur on the electrode. Traditionally, it is used more in point level measurement applications than in continuous level measurement applications.

### Conductivity

The process fluid to be measured must have a minimum conductivity for this technology to function properly. (Figure 3-60) The minimum conductivity level is typically around 10 µS/cm. When a conductive process fluid contacts the electrode, which is immersed in the fluid, an electrical circuit is established. A low current flows to ground via the process fluid to the electrode. This low current is detected by the instrument and converted into a form that can be used by a control system. When the electrode is installed in conductive vessels, the vessel wall forms the ground; in nonconductive vessels, an additional grounded electrode is required. The process fluid being measured MUST be consistently conductive to the same degree for this technology to work. Traditionally, conductivity is used more in point level applications than in continuous level applications.[38]

**Figure 3-60. Conductivity Type Level Sensor**

### Magnetostrictive

This type is similar in operation to a float level device. It consists of a magnetostrictive[39] wire (Figure 3-61) within a protective sleeve, and a float mechanism that has a permanent magnet inside. The float follows the liquid level and travels vertically up and down on the protective sleeve called a wave guide, The point at where the magnet is on the waveguide produces a twist on the waveline (reference Figure 3-61). The magnetostrictive wire is pulsed with a low voltage

---

38. Image obtained from www.roxspur.com.
39. Magnetostriction is a property of ferromagnetic materials that causes them to change their shape or dimensions during the process of magnetization.

signal several times a second. This pulse results in a magnetic field that moves down the wire. When this magnetic field pulse reaches the float containing the permanent magnet, the magnetic fields react with each other. This magnetic field interaction causes a torsion ultrasonic wave that is then reflected back to the source. The level device tracks the time it takes for this torsion wave to travel back up the wire, and then calculates the position of the float which, in turn, reflects the process fluid level in the vessel.[40]

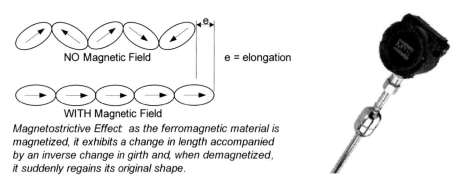

NO Magnetic Field            e = elongation

WITH Magnetic Field

*Magnetostrictive Effect: as the ferromagnetic material is magnetized, it exhibits a change in length accompanied by an inverse change in girth and, when demagnetized, it suddenly regains its original shape.*

**Figure 3-61. Magnetostrictive Effect**

### 3.4.4 Float/Buoyancy Level Measurement Techniques

*Displacer*

A displacer level works on Archimedes' Principle[41]. Within the displacer chamber (cage) (Figure 3-62), a weighted float displaces its own weight of the process fluid in which it floats. This weighted float is connected to an operating mechanism (torque tube and rod) which, when sensing changes in the buoyant force of the fluid, converts this force into a measurement of level. When the liquid level is below the displacer, the full weight of the displacer is measured by the transducer. As the liquid level rises, the apparent weight of the displacer decreases. This difference in tension on the torque tube is proportional to the level. Displacers should only be used for low viscosity, clean fluids.[42]

*Ball Float*

Ball float devices (Figure 3-63) operate on a similar principle as the displacer level device except that they operate based upon the float principle rather than Archimedes' Principle. This is due to the fact that the ball float is hollow and floats on top of the fluid service, whereas, Archimedes' Principle is about a solid float displacing its own weight in the fluid. This type of level device is used almost exclusively for point level applications, and not for continuous level applications. May be horizontal side mounted or vertical top mounted. The vertical top mounted design may be single-point or custom multi-point (multi-point design may be applicable to sump pump level control applications).[43]

---

40.  MTS Sensor magnetostrictive probe.
41.  Archimedes' Principle states that a body wholly or partially immersed in a fluid is buoyed up by a force equal to the weight of the fluid displaced.
42.  Fisher (Emerson) displacer level instrument.
43.  Mobrey float switch (Transcat).

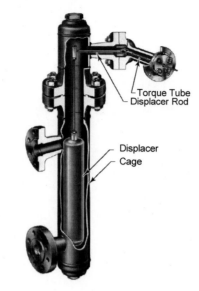

**Figure 3-62. Displacer Level Sensor**

**Figure 3-63. Ball Float Level Sensor**

## 3.4.5 Time of Flight Level Measurement Techniques

*Ultrasonic*

An ultrasonic level transmitter operates by generating an ultrasonic pulse (acoustic wave) and measuring the time it takes for an echo to return. The echo signal received back is greatly affected by the angle between the sensor and the material reflecting the acoustic wave (Figure 3-64). Therefore, it is necessary that the surface of a process fluid and that of the sensor be kept parallel when installing the sensor.

The ultrasonic pulse needs a medium to transmit the pulse; therefore, this technology will <u>not</u> work in vacuum service. Ultrasonic technology may be used for both point and continuous level measurement.

Ultrasonic level measurement is also used in conjunction with open channel flow measurement.

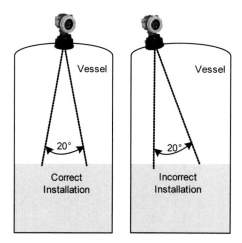

**Figure 3-64. Ultrasonic Level Transmitter Installation**

*Radar (Microwave)*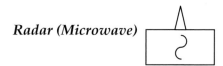

Radar level measurement works on the principle of measuring the time it takes a microwave pulse to travel to the surface to be measured and the resulting echo to return. Unlike ultrasonic units, radar units may be used in vacuum applications and the pulses travel at the speed of light vs the speed of sound for ultrasonic. They are available in two design types: antenna and guided wave. The antenna type design concentrates the microwave pulse into a narrow beam. The guided wave type design uses a rod or cable to guide the microwave pulses into a very concentrated area.

**Note:** When guided wave radar level measurement is used in the food, beverage, and pharma/biotech industries, you must ensure that the device is capable of being set up to tune out false high level signals when the vessel is being filled via the spray ball assembly.

*Laser*

Laser devices (Figure 3-65) measure level by detecting how long it takes for an infrared light pulse to travel to the process surface and back again (Figure 3-66) (similar to ultrasonic and radar). Most laser level devices also contain a visible laser beam, in addition to the infrared beam, which is utilized for sighting purposes. Laser measurement is used almost exclusively with solids level measurement (e.g., silos). Silos are often in excess of 60 feet high, and relatively narrow (10–15 feet), and may exhibit a steep angle of repose (reference Figure 3-66). Consequently, ultrasonic devices may struggle with false echoes, and the low dielectric value of many solids often causes a poor reflective surface for radar.[44]

---

44. K-Tek Corp. laser level transmitter.

---

**Figure 3-65. Laser Level Sensor**

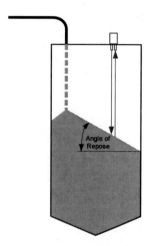

**Figure 3-66. Laser Level Measurement**

## 3.4.6 Miscellaneous Level Measurement (Switch) Techniques

### *Rotating Paddle*

In this technique, a small motor keeps a paddle wheel in motion at very slow speed (Figure 3-67). When the level rises to the paddle wheel, the motor stalls and causes a switch to activate. Typically, power to the motor is also shut off during the stalled condition to prevent the flow of continuous locked rotor current. The rotating paddle is used exclusively for point level applications, not continuous level applications. This type of level switch is ideal for use on solids.[45]

**Figure 3-67. Rotating Paddle Level Sensor**

---

45.   K-Tek rotating paddle switch.

### Radio Frequency

This device employs a radio frequency (RF) balanced impedance bridge circuit to detect if the probe is in contact with the material that is to be sensed. When material comes in contact with the probe, the bridge becomes unbalanced. A radio frequency signal is emitted by the electronics in the head of the level switch. The probe serves as an antenna to detect the frequency. This type is used almost exclusively for point level applications.

### Thermal

The operation of the thermal level switch is similar to that of a thermal flow switch. Thermal level sensors (Figure 3-68) detect the difference in thermal conductivity between the process fluid and air (or another gas or vapor). When the probe is covered by process fluid, there is a different heat transfer rate than when the probe is uncovered, which is interpreted by the control system as point level.[46]

**Figure 3-68. Thermal Level Sensor**

### Vibratory

This device (Figure 3-69) operates similar to a tuning fork where the principle of either resonance frequency or oscillation and is available in either a rod or fork type design. Either design uses the damping of vibration when the vibrating rod or fork is covered with process fluid. These are typically used for point level applications, and not for continuous level applications. This type of level sensor is ideal for use on solids.[47]

**Figure 3-69. Vibratory Point-Level Sensor**

## 3.5 ANALYTICAL MEASUREMENT SENSORS

This section will only briefly touch on different forms of analytical measurement, as this area is too broad and would require a book of its own to describe in detail.[48]

---

46. K-Tek thermal level switch.
47. Endress & Hauser vibratory level sensor.

### 3.5.1 Combustible Gas Analyzers

*Catalytic Bead (Electrocatalytic) Type*

This design typically involves the use of two beads (Figure 3-70) contained within a sensing head. One bead is an "active" bead, meaning that it is coated with a catalyst and oxidizes any combustible gas present in the atmosphere. The other bead is considered a reference bead, which is used to provide a baseline signal. This bead is coated with an inert material such as glass. As the gas oxidizes on the active bead, the bead temperature increases. This increase in temperature is proportional to the concentration of gas in the atmosphere and causes the resistance of the active bead to rise. When the active bead resistance is compared to the reference bead resistance (via a bridge circuit), a small voltage differential is obtained. This small voltage differential signal is then amplified for use with a control system. This type of sensor should not be mounted in a location where it is subject to mechanical shock or vibration as this could cause damage to the fine platinum wire contained within the active bead.

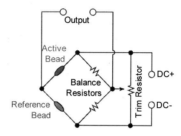

**Figure 3-70. Catalytic Bead CG Sensor**

**Contamination and Poisoning:** Contamination involves having the beads coated by some foreign material (oil, grease, paint, etc.). Poisoning involves the catalyst having a strong reaction which results in heavy absorption of the material into the catalyst. Unfortunately the only way to detect contamination or poisoning is to check the probe with calibration gases.

*Infrared Type*

Many gases absorb infrared (IR) radiation at certain wavelengths as it passes through the volume of gas. This design consists of a light source and a light detector that measures the light intensity at two specific wavelengths. One wavelength is the absorption (active) wavelength, while the other wavelength (reference wavelength) is outside of the absorption wavelength. When a volume of gas is present between the light source and detector, if the amount of light at the active wavelength is reduced while the amount of light at the reference wavelength stays the same, the gas concentration can be determined from the differences in the amount of light transmitted at those wavelengths. IR detectors may be either single-source (Figure 3-71) or open-path (Figure 3-72). Unlike catalytic bead technology, they are not subject to poisoning. Poisoning is where the sensor loses its sensitivity, and may become totally unresponsive because it has come into contact with chemicals such as lubricants, chemicals that contain silicon, chlorine and heavy metals.[49]

---

48. Two good sources for analytical measurement technical date are: *Analytical Instrumentation*, R.E. Sherman, Editor, ISBN 978-1-55617-581-7, ISA and *Instrument Engineers' Handbook, 4th Edition, Vol. 1: Process Measurement and Analysis*, Béla Lipták, Editor, ISBN 0-8493-1083-0, ISA/CRC Press.

**Figure 3-71. IR Detector (Single Source)**

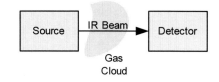

**Figure 3-72. IR Detector (Open Path)**

### 3.5.2 Dew Point

*Dew Point Temperature*

The dew point is the temperature at which water vapor starts to condense out of the air. It is the temperature at which air becomes completely saturated. Above this temperature the moisture will stay in the air.

If the dew-point temperature is close to the air temperature, the relative humidity is high. If the dew point is well below the air temperature, the relative humidity is low.

If moisture condenses on a cold bottle from the refrigerator, the dew-point temperature of the air is above the temperature in the refrigerator.

The dew point temperature can be measured by filling a metal can with water (at room temperature) and ice cubes. Immediately stir the water-ice solution with a thermometer and watch the outside of the can. When the water vapor in the air starts to condense on the outside of the can, the temperature on the thermometer is close to the dew point of the air. Do not allow the water and ice to reach an equilibrium temperature close to 32°F before stirring.

The dew point is given by the saturation line in the psychrometric chart.

*Chilled Mirror Dewpoint Sensor*

By chilling a mirrored surface to a point at which the vapor pressure of the water on the surface is at equilibrium with the water vapor pressure in the gas sample above the mirror and then measuring the temperature of this surface, the dew point can be calculated. The mirror must be constructed of a material with good thermal conductivity and plated with an inert metal to

---

49. MSA Ultima Gas Monitor.

prevent tarnishing and oxidation. A beam of light (LED) is aimed at the mirror surface and a photodetector then monitors the reflected light (Figure 3-72).

When the gas sample flows over the chilled mirror, dew droplets will form on the surface of the mirror. These dew droplets will cause the reflected light to be scattered. As the amount of reflected light to the photodetector decreases, the output of the detector will decrease proportionally. By accurately measuring the surface temperature of the mirror at this point the dew point temperature may be determined.

The chilled mirror sensor may also be used with gas samples. The principle is the same. However, it should not be used in applications where the gas may react with water (e.g. chlorine, ammonia) or where the gas vapor dew point is higher than the water dew point. The gas vapor may contain water, but the gas vapor's dewpoint may not be higher than water because the water vapor would condense for the gas vapor.

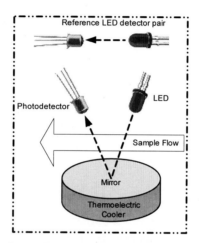

**Figure 3-72. Chilled Mirror Dewpoint Sensor**

## 3.5.3 Humidity Sensors

*Terminology*

**Relative Humidity:** Relative humidity is the amount of moisture in the gas compared to what the gas can "hold" at that temperature.

$$RH \% = \frac{\text{actual vapor density}}{\text{saturation vapor density}} \times 100\%$$

**Dry Bulb Temperature:** The dry bulb temperature, usually referred to as gas temperature, is the gas temperature that is most commonly used. When people refer to the temperature of the gas, they are normally referring to its dry bulb temperature. It is called "dry bulb" because the gas temperature is indicated by a thermometer that is not affected by the presence of moisture in the gas.

Dry bulb temperature can be measured using a normal thermometer freely exposed to the gas, but shielded from radiant heat and liquid water. The temperature is usually given in °C or °F. The SI unit is °K, where 0°K is equal to -273°C.

The dry bulb temperature is an indicator of heat content and is shown along the bottom axis of a psychrometric chart. Constant dry bulb temperatures will appear as vertical lines in the psychrometric chart.

**Wet Bulb Temperature:** The wet bulb temperature is the temperature of adiabatic saturation. This is the temperature indicated by a moistened thermometer bulb that is exposed to gas flow.

Wet bulb temperature can be measured by using a thermometer with the bulb wrapped in wet fabric, such as muslin. The adiabatic evaporation of water from the fabric and its cooling effect is indicated by a "wet bulb temperature" lower than the "dry bulb temperature" of the gas.

The rate of evaporation from the wet bandage on the bulb, and the difference between the dry bulb temperature and wet bulb temperature, depends on the humidity of the gas. The evaporation rate is reduced when the gas contains more water vapor.

The wet bulb temperature is always lower than the dry bulb temperature except at 100% relative humidity (the gas is at the saturation line), where they are identical.

Combining the dry bulb and wet bulb temperature in a psychrometric diagram or Mollier chart gives the humidity level of the gas. Lines of constant wet bulb temperatures run diagonally from the upper left to the lower right in the psychrometric diagram.

*Capacitive Type Humidity Sensor*

This type of sensor (Figure 3-73) consists of a substrate (e.g., silicon) on which a thin film of polymer or metal oxide is deposited. The substrate is placed between two conductive electrodes. The incremental change in the dielectric constant of a capacitive humidity sensor is close to being directly proportional to the relative humidity of the surrounding environment. The is because the dielectric material absorbs the water vapor from the surrounding environment until it reaches equilibration with the surrounding environment. The dry dielectric coefficient of the substrate is much lower than that of water, as such the capacitance increases as it absorbs water. The typical uncertainty is ±2% within the RH range of 5%–95%.

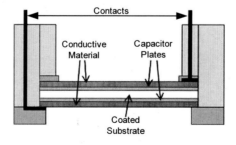

**Figure 3-73. Capacitive Type Humidity Sensor**

## Hygroscopic Type (Resistance) Humidity Sensor

This type of sensor uses materials whose electrical resistance changes in proportion (exponentially inverse) to the amount of moisture within the surrounding atmosphere (Figure 3-74).

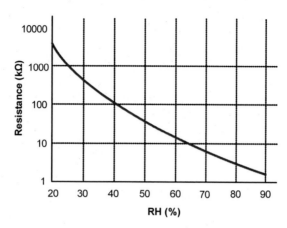

**Figure 3-74. Hygroscopic Type (Resistance) Humidity Sensor**

## 3.5.4 Electrical Conductivity Analyzers

### Conductive Method

The conductivity (also called conductance) of a liquid may be measured by placing two or more electrodes, with a known distance between them, into the liquid and then applying a DC voltage (Figure 3-75). The reciprocal of the resistance between the electrodes as a result of the liquid is the conductance of the liquid.[50]

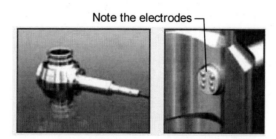

**Figure 3-75. Electrical Conductivity Sensor**

### Inductive Method

This technique consists of two coils potted in a synthetic material, a current generator, and a current receiver. When the coils (Figure 3-76) are placed into contact with a liquid and an AC current is generated within the primary coil, an induced current is created within the secondary coil and sensed by the current receiver. The strength of the induced current in the secondary coil is dependent upon the ion concentration of the liquid and thus its conductivity.[51]

---

50.  Optek Conductivity Probe.

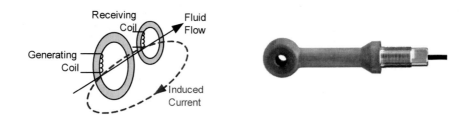

**Figure 3-76. Inductive Conductivity Sensor**

## 3.5.5 pH/ORP Analyzers

pH is an abbreviation for the "**p**otential of **H**ydrogen." In the process environment, pH is usually an important parameter to be measured and controlled. The pH of a solution is an indication of how acidic or basic the solution is. The pH scale is from 0–14, with 0–6 being an acid and 8– 14 being a base, and 7 being neutral (e.g., pure water). The pH measurement is a measurement of the hydrogen ion concentration of the solution. A low pH values equates to a very large concentration of hydrogen ions, while a high pH value equates to a small number of hydrogen ions.

- A very acidic water solution would contain a hydrogen ion concentration of approximately $1 \times 10^0$ moles. (Hence the pH value of 0)

- A very basic water solution would contain a hydrogen ion concentration of approximately $1 \times 10^{-14}$ moles. (Hence the pH value of 14, negative logarithm)

- Pure neutral water solution would contain a hydrogen ion concentration of approximately $1 \times 10^{-7}$ moles. (Hence the pH value of 7)

*pH Method*

A pH sensor (Figure 3-77) consists of three components: a measuring electrode, a reference electrode and a temperature sensor. The pH loop is, essentially, a battery setup with the measuring electrode being the positive terminal and the reference electrode being the negative terminal. When the probe is immersed in a solution, the measuring electrode potential changes according to the concentration of hydrogen ions (approximately 59.16 mV per pH unit), while the reference electrode potential does not change at all. The difference in the potential of the two electrodes is in relation to the pH of the process fluid. The output of the measuring electrode will change with temperature, so a temperature element is necessary to compensate the signal for process temperature. *Note: A glass type pH probe tip should never be permitted to be out of solution for an extended period of time. The tip must be protected with a solution prior to being put into service.*

Another type of pH sensor is semi-conductor based and utilizes an ISFET (Ion Selective Field Effect Transistor). As the ion concentration within the fluid to be measured contacts the ISFET, the current through the transistor will change accordingly with the ion concentration. The advantage of this type of probe over the glass probe is that it is not easily broken.[52]

---

51. Endress+Hauser Conductivity Probe.
52. Endress & Hauser Non-Glass type probe.

---

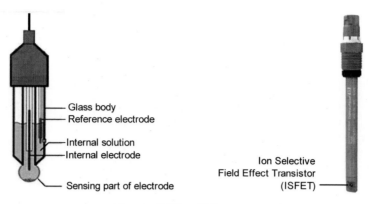

Figure 3-77. pH Sensor

*ORP Method*

**ORP** is an acronym for "**O**xidation-**R**eduction **P**otential." ORP is the tendency of a solution to acquire electrons (negative voltage) and thereby have its electrical potential reduced. Oxygen is the most common oxidizer. Rust forms by iron donating an electron to oxygen. The oxygen gains electrons and is thereby reduced. This electrochemical process is evident in the form of rust (metal cations).

An ORP sensor consists of two components: a measuring electrode and a reference electrode. The measuring electrode is constructed so that it has a positive electrical characteristic, while the reference electrode is constructed so that has a negative electrical characteristic. The reference electrode is surrounded by an electrolytic solution so that it produces a very small constant voltage. The voltage of the measuring electrode will vary in proportion to the reduction potential of the process fluid. The difference in the potential of the two electrodes is in relation to the ORP of the process fluid.

## 3.5.6 Dissolved Oxygen Analyzers

The amount of oxygen that a given volume of water can hold is a function of the atmospheric pressure at the water-air interface; the temperature of the water; and the amount of other dissolved substances. Dissolved oxygen (DO) is defined as the physical distribution of oxygen in water. A DO analyzer will determine the amount of oxygen (by volume) present in a unit volume of water. This is an important measurement in waste water treatment facilities, because the DO content is an indication of the water's ability to support aquatic life. Low DO content also promotes growth of harmful bacteria.

There are two DO measurement techniques: polarographic (aka Clark cell) and luminescent. The Clark cell (Figure 3-78) consists of an anode, a cathode, an electrolyte solution (typically KCl), and a gas-permeable membrane. The membrane permits oxygen to pass through. The oxygen is then reduced by the cathode, which results in a current flow between the anode and cathode. This current flow is proportional to the oxygen content of the electrolyte solution. This technique is affected by temperature, therefore temperature compensation is required to attain high accuracy.

The second measurement technique is *luminescence*. This device consists of a sensor coated with a luminescent material. Light from a blue LED is transmitted through the water to the sensor

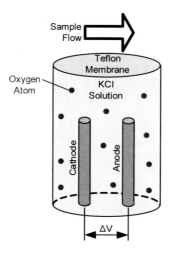

**Figure 3-78. Clark Cell**

surface (Figure 3-79). The light excites the luminescent material, causing it to emit red light. The time it takes the transmitted blue to be sent and the red light to be received back is correlated to the oxygen content of the water. Note: In the presence of oxygen, the luminescence material lifetime is reduced. Check with the manufacturer for recommended replacement intervals.

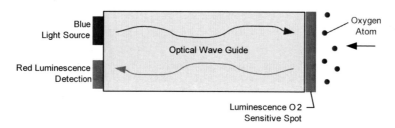

**Figure 3-79. Luminescence DO Sensor**

### 3.5.7 Oxygen Content (in Gas)

*Paramagnetic Oxygen Analyzer*

The paramagnetic[53] oxygen analyzer is based upon the principle that oxygen is a paramagnetic material. This design is known as the "dumbbell" design. This is due to the double sphere design (Figure 3-80). Two spheres filled with nitrogen are suspended on a taut piece of wire in a non-uniform magnetic field so that the assembly resembles a dumbbell. The dumbbell is allowed to move freely while it is suspended by the wire. There are two opposing coils on each side of the spheres to create magnetic fields. When the sample containing oxygen is passed through the sensor, the oxygen molecules are attracted to the stronger of the two magnetic fields. This, in turn, causes the dumbbell assembly to rotate. The resulting deviation is detected by a light source, reflecting mirror, and light-receiving element. The output of the light receiving element is then amplified. for use in a transmitter that is calibrated to read out oxygen content.

---

53. A paramagnetic material is a material in which an induced magnetic field is parallel and proportional to the intensity of the magnetizing field but is much weaker than in ferromagnetic material.

---

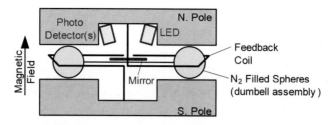

**Figure 3-80. Paramagnetic Oxygen Analyzer**

*Zirconium Oxide Oxygen Analyzer*

This design is based upon the Nernst principle.[54]

This oxygen measurement device utilizes the fact that zirconium oxide conducts oxygen ions when heated above 1112°F (600°C). The zirconium oxide tube (Figure 3-81) is fitted with platinum electrodes on the interior and exterior to provide for a catalytic surface for the exchange of oxygen molecules and ions. As the oxygen molecules come into contact with the platinum electrodes, they become ionized and are transported across the zirconium tube. This results in a charge being developed and sets up a potential difference between the electrodes that is proportional to the log of the ratio of oxygen concentrations on each side of the zirconium oxide.

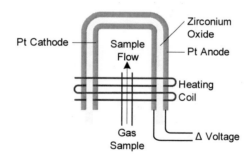

**Figure 3-81. Zirconium Oxide Oxygen Analyzer**

Because of the high temperatures associated with this type of device, care should be taken to not locate this device within a hazardous classified area, unless it is specifically rated for the application.

Other $O_2$ content analyzers include the ambient temperature electrochemical oxygen sensor and the polarographic oxygen sensor.

### 3.5.8 Turbidity Analyzers

There are two techniques used to measure turbidity (Figure 3-82) within a process fluid: light absorption and light scattering.

---

54. Nernst Principle: The Nernst voltage generated by a $ZrO_2$ electrochemical cell (Nernst cell) is determined by the relation of the partial pressures of oxygen at the platinum electrodes on both sides of the cell.

## Units of Measure for Turbidity

- JTU: Jackson Turbidity Unit (now considered obsolete). The lower the value, the clearer the fluid. Water containing 100ppm of silica (diatomaceous earth) has a turbidity of 21.5JTU.

- NTU: Nephelometric Turbidity Unit. Water containing 1mg of finely divided silica per liter has a turbidity of 1NTU at a measurement angle of 90°.

- FTU: Formazin Turbidity Unit (ISO unit), same as NTU except that the measurement angle may be any angle.[55]

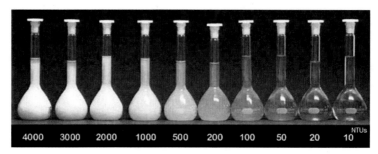

**Figure 3-82. Turbidity**

### Light Absorption

Light absorption-based photometers[56] typically incorporate a focused beam of light and project this beam through the process fluid (Figure 3-83). They work much the same way as a laboratory spectrophotometer. Process-scale absorption based photometers are set at a fixed wavelength or bandwidth, rather than having the ability to operate at selected wavelengths. NOTE: wavelength is a very specific frequency while bandwidth defines a range of frequencies.

Three-dimensional solids will exhibit absorbance at virtually all frequencies of light; however, with short wavelength infrared (Near Infrared, NIR), solids contributing to turbidity can be isolated from visible color as well as from other dissolved constituents. This bandwidth is just beyond visible red light, in the 700–1100 nanometer range. The attenuation of the NIR light transmittance is used as an indicator of the concentration of solids. The percentage of light transmittance can be plotted against the concentration; however, the relationship is not linear.

Absorbance is defined as the negative $\log_{10}$ of the transmittance of the light. Absorption based photometers are typically used for higher concentrations of process turbidity, in excess of 0.5 grams per liter.

### Light Scattering (Forward)

This technique uses suspended, finely dispersed particles which when exposed to infrared or visible light will scatter the light (Figure 3-84). The cloudier the process fluid (the higher its turbidity), the more scattering will occur, and therefore, less light will be transmitted through a

---

55. Image from Optek.
56. In general terms, photometers are used to measure the intensity of the light.

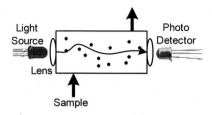

**Figure 3-83. Light Absorption Photometer**

sample. The amount of light scatter is measured via photometers and compared to the amount of scatter from known mixtures. Forward scatter is defined as less than 90° away from, or in the same general direction as the light source.

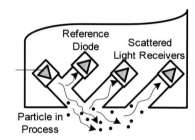

**Figure 3-84. Forward Light Scattering Turbidity Analyzer**

## 3.5.9 TOC (Total Organic Carbon) Analyzers

To arrive at the TOC content of a process sample, the TIC (total inorganic carbon) must be eliminated from the sample, then leftover carbon may be measured. Removing TIC is accomplished via various methods, all of which involve forming $CO_2$ and then driving the $CO_2$ out of the sample. The carbon that is left in the sample is totally organic.

Total Carbon = Inorganic Carbon + Organic Carbon. In the TOC process analyzers some methods will measure the organic carbon content directly after elimination of the inorganic carbon, and some methods will measure the inorganic carbon and subtract from the total carbon to arrive at the organic carbon content.

*Acidification:* This method (Figure 3-85) involves acidifying the sample to a pH level of 2 to 3, then purging with a carbon-free inert gas. This purging process drives off the $CO_2$ formed during acidification.

*Combustion/NDIR (Non-Dispersive Infrared):* In this technique, a sample is injected into a heated chamber that is packed with an oxidation catalyst. Within this heated chamber, the water in the sample is vaporized and the inorganic carbon is converted to $CO_2$. The $CO_2$ is carried with the gas stream from the heated chamber to the NDIR[57], where the $CO_2$ concentration is

---

57. With NDIR, a parabolic reflector is utilized to surround the IR source lamp to produce a parallel IR light source towards the detector.

---

measured by injecting this portion of the sample into a reaction chamber filled with a phosphoric acid solution. In this reaction chamber, all remaining inorganic carbon is converted to $CO_2$.

***UV Oxidation/Conductivity:*** In this technique the UV irradiation oxidizes the organic carbon within the sample to produce $CO_2$. The change in conductivity, as measured by a cell within the device, can be used to calculate the TOC concentration ($\Delta$ in conductivity before and after oxidation).[58]

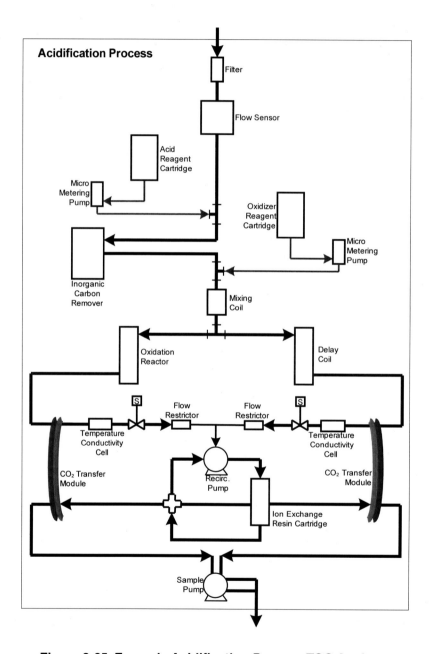

**Figure 3-85. Example Acidification Process TOC Analyzer**

---

58. This is a generalized acidification process, each TOC analyzer manufacturer utilizing the acidification process will be slightly different in construction.

### 3.5.10 Light Wavelength Type Analyzers

**Wavelengths per ISO 20473**

- Ultraviolet (UV) Wavelength Range: 100 to 400 nanometers

- Visible Light Wavelength Range: 400 to 780 nanometers

- Infrared (IR) Wavelength Range: 780 nanometers to 1 mm ($10^6$ nanometers)

  - Near Infrared (NIR): 780 to 3000 nanometers
  - Mid Infrared (MIR): 4000 to 50000 nanometers
  - Far Infrared (FIR): 50000 nanometers to 1 mm

#### *Ultraviolet (UV) Absorption Analyzers*

In this type of sensor, light from a UV lamp (e.g., mercury vapor) is passed through the process stream. The UV light is then passed through a UV filter and then to some form of photodetector. The purpose of the UV filter is to block all wavelengths except the desired UV wavelength. The resulting photocurrents induced in the photodetector are directly proportional to the remaining light intensity at this wavelength.

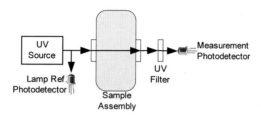

#### *Infrared (IR) Absorption Analyzers*

The infrared absorption spectrum of a substance is sometimes called its molecular fingerprint.

The principle of IR absorption analyzers is very similar to that of UV absorption analyzers except that the light source is different. Light from an IR lamp (e.g., incandescent) is passed through the process stream. The IR light is then passed through an IR filter and then to some form of photodetector. The purpose of the IR filter is to block all wavelengths except the desired IR wavelength.

The common technologies used to produce IR spectra are dispersive and Fourier Transform (FT).

*Dispersive* measures the intensity over a narrow range of wavelengths at a time.

*FTIR* simultaneously measures intensity over a wider range of wavelengths.

### 3.5.11 Chromatographs and Spectrometers

This book will provide only the high level principle of operation of these analytical devices as these devices have complete books written about their operation.

*Chromatographs*

Chromatography is a separation technique utilized in chemical analysis. Separation is achieved by distributing the components of a mixture between two phases: a stationary phase and a mobile phase. Chromatography may be used on either a liquid or a gas process. In general, gas chromatography is used for the separation of volatile materials and gases, and liquid chromatography is used for the separation of nonvolatile liquids.

**Gas Chromatographs**

It is estimated that 10%–20% of the known compounds may be analyzed by compound/molecule. To be suitable for gas chromatography analysis, a compound must have sufficient volatility and thermal stability. If all, or some, of a compound will be in the gas or vapor phase at 400–450°C or below, and the compound does not decompose at these temperatures, then the compound may be analyzed by gas chromatography (Figure 3-86).

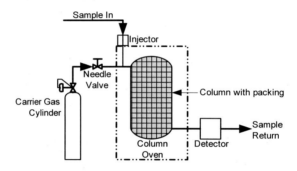

**Figure 3-86. Gas Chromatography (Schematic)**

One or more high purity gases is supplied to the gas chromatograph (mobile phase). One of the gases, which is the carrier gas (inert), flows into the injector, through the column (stationary phase) and then into the detector. The process sample is introduced into the injector via a sampling device. The injector is then heated to 150–250°C, which causes the volatile sample solutes to vaporize. The vaporized solutes are then transported into the column by the carrier gas. The column itself is maintained in a temperature controlled oven. The detector may use one of the following technologies:

*Thermal Conductivity Detector (TCD):* TCD is a technology that is based on the comparison of two gas streams, one stream containing the carrier gas, the other stream containing both the carrier gas and the compound. The carrier gas used should have a high thermal conductivity so as to maximize the temperature difference between two tungsten filaments. This temperature difference results in a difference in resistance in the filaments. The temperature difference between the reference cell filaments and the sample cell filaments is continuously monitored by a Wheatstone bridge circuit. This type of detector is considered non-destructive, because the sample can continue on further in the process, if necessary.

*Flame Ionization Detector (FID):* This type of detector is very sensitive toward organic molecules. The FID uses a hydrogen/air flame to ionize the sample gas and detect its concentration. Ionization occurs whenever electrons are ejected from the volatile organic compounds (VOCs) in the hot combustion flame. These ions are electrically charged, and produce a current that is measured by the sensor electrodes. This type of detector is considered destructive, because the sample is destroyed by the process.

*Other types* of gas chromatograph include the Photoionization Detector, Electron Capture Detector and Helium Ionization Detector.

## Liquid Chromatographs

Liquid chromatography (Figure 3-87) is useful for separating ions, or molecules that are dissolved in a solvent. If the sample solution is in contact with a second solid or liquid phase, the different solutes will interact with the other phase to differing degrees. This is due to differences in adsorption, ion-exchange, partitioning, or size. In standard liquid chromatography columns, the solvent is allowed to drip through the column under gravity. In high performance liquid chromatography (HPLC), the solvent is forced through the column by a pump. This forcing by a pump allows smaller particle size (1–10 μm) packing to be used in the column. There are four types of HPLC chromatography technologies:

- Partition (most popular)

- Adsorption

- Ion exchange

- Size exclusion or gel permeation (used for high molecular weight compounds)

The detector is frequently a UV absorbance or refractive index monitor.

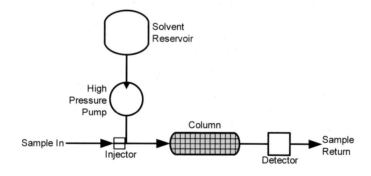

**Figure 3-87. Liquid Chromatography (Schematic)**

*Mass Spectrometers*

Mass spectrometers use the difference in mass-to-charge ratio of ionized atoms or molecules to separate them from each other. Mass spectrometry is useful in the quantization of atoms or molecules, and also for determining chemical and structural information about molecules. Molecules have distinctive fragmentation patterns that provide structural information to identify structural components. The general operation of a mass spectrometer is:

- Create gas-phase ions (positively charged) either by electron beam or by electrostatic field

- Separate the ions in space or time based on their mass-to-charge ratio (via electromagnet)

- Measure the quantity of ions of each mass-to-charge ratio

Mass spectrometers (Figure 3-88) consist of an ion source, a mass-selective analyzer and an ion detector, and they operate in a high-vacuum system ($10^{-5}$ torr).[59]

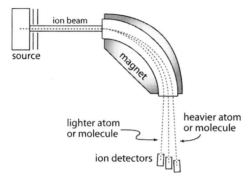

**Figure 3-88. Mass Spectrometer (Schematic)**

Many gas chromatograph instruments are coupled with a mass spectrometer. The gas chromatograph separates the compounds from each other, while the mass spectrometer helps to identify the compounds based on their fragmentation pattern.

### 3.5.12 Continuous Emission Monitoring Systems (CEMS)

Continuous emission monitoring systems (Figure 3-89) are typically used to analyze the off-gas from the stacks of incinerators, boilers, etc. to ensure compliance with EPA 40CFR Part 75 (acid rain). The compounds typically analyzed include the following (though different compounds may be required to be analyzed depending on the application):

- NOx (Nitrogen Oxide)

- CO (Carbon Monoxide)

- $SO_2$ (Sulfur Dioxide)

- $O_2$ (Oxygen)

Opacity may also have to be monitored.

*CEMS Sample Probe:* May contain a single port or multiple ports. The sample probe assembly (Figure 3-90) is typically heat traced to ensure that no condensation forms within the probe. The sample probe may additionally be equipped with an instrument air purge system to blow the probe clean.[60]

---

59. Image obtained from www.serc.carleton.edu.
60. Rosemount (Emerson) CEMS sample probe.

---

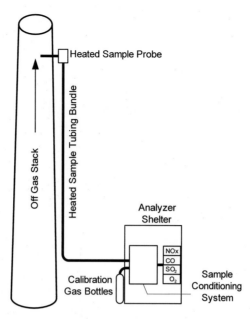

**Figure 3-89. Continuous Emissions Monitoring System**

**Figure 3-90. CEMS Sample Probe**

*CEMS Sample Tubing Bundle:* May contain a single tube or multiple tubes. The sample tube assembly (Figure 3-91) is typically heat traced to ensure that no condensation forms within the bundle.[61]

**Figure 3-91. CEMS Sample Tube Assembly**

*Sample Conditioning System:* Consists of a vacuum pump to draw the sample, sample and calibration valving, and flow regulation and moisture removal devices (Figure 3-92).

---

61. Thermon CEMS sample tube bundle.

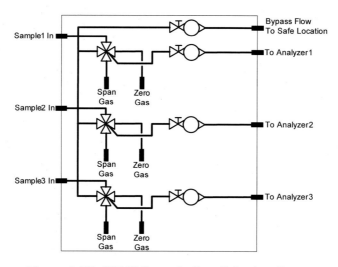

**Figure 3-92. CEMS Sample Conditioning System**

*Analyzer Shelter:* Most often constructed from either FRP (fiberglass reinforced plastic) or stainless steel. The shelter (Figure 3-93) typically contains the sample conditioning system, calibration gas cylinders, the analyzer electronic components and some sort of gas detection system for personnel protection.[62]

**Figure 3-93. CEMS Analyzer Shelter**

The following website contains useful information from the government concerning CEMS: http://www.epa.gov/ttn/emc/cem.html

### 3.5.13 Vibration Analysis

Vibration analysis is used to detect the early precursors to machine failure, allowing machinery to be repaired or replaced before an expensive failure occurs. All rotating equipment vibrates to some degree, but as older bearings and components reach the end of their product life, rotating equipment begins to vibrate more dramatically, and in distinct ways.

---

62. Image obtained from www.okbebco.com.

The two most important criteria of vibration to monitor are amplitude and frequency.

*Amplitude or Displacement* is the magnitude of the equipment vibration. The larger the amplitude (i.e., the larger the displacement), the greater the movement or stress that is experienced by the equipment.

- Velocity amplitude is the rate of change of the displacement (i.e., how fast something is vibrating back and forth). The velocity amplitude is the criteria that provides the best indication of the condition of the equipment being monitored. The unit associated with velocity amplitude is inches per second (in/sec).

- A velocity vibration transducer (velomitor) measures how fast the displacement is moving.

- An acceleration transducer (accelerometer) measures how fast the velocity is changing

- Vibration transducers (Figure 3-94) are typically mounted near the bearings of the equipment. The closer the transducer is mounted to the centerline of a bearing, the less likely the transducer will pick up distorted signals.

**Figure 3-94. Bently Nevada Vibration Transducer (Velomitor Shown)**

*Frequency* is the oscillation rate of the equipment vibration. The unit associated with frequency is Hz.

An operating piece of equipment oscillates at a certain frequency. The rate at which this piece of equipment oscillates is known as its frequency. The higher the vibration frequency, the faster the oscillations. This unit is important know so that the machine does run at or near its resonant (natural) frequency.

# 4. SIGNALS, TRANSMISSION AND NETWORKING

## 4.1 SIGNALS/TRANSMISSION

Signals and Transmissions are the terms used to describe the method(s) involved in information/data transfer whether it is from device to control system, or from control system to control system.

### 4.1.1 Copper Cabling

*Twisted Pair Cabling*

Twisting wires together decreases interference due to magnetic coupling (aka AC coupling) because the twisting causes the magnetic fields to cancel each other. The tighter the twist rate (the shorter the pitch), the more resistant the twisted pair cable is to interference. There are two designs of twisted pair cables: unshielded (UTP) and shielded (STP) (Table 4-1).

**Unshielded Twisted Pair:** UTP cable relies solely on the cancellation effect produced by the twisted wire pairs to limit signal degradation caused by electromagnetic interference (EMI) and radio frequency interference (RFI).

**Shielded Twisted Pair:** STP cable combines the techniques of shielding and cancellation by wire twisting. Multi-pair STP cables may be individually shielded or overall shielded or both. The shield may be a foil tape or a metallic braid. When foil tape is utilized it typically has a bare stranded drain wire in constant contact with the foil. Shields must be grounded to work. Care must be taken with grounding shields so as not to set up up-ground loop currents (i.e., ground at one end only).

## Table 4-1. Characteristics of STP/UTP Cables

| Characteristics of STP/UTP Cables | | |
|---|---|---|
| **Cable Type** | **Data Rate** | **Common Usage** |
| Category 1 | N/A | Voice Grade Analog |
| Category 2 | 4 Mbps | Digital Voice |
| Category 3 | 10 Mbps | 10BaseT |
| Category 4 | 16 Mbps | Token Ring |
| Category 5 | 100 Mbps* | 100BaseT |
| Category 5e | 1000 Mbps | 1000BaseT |
| Category 6 | 10 Gbps | |
| Category 6A | 10 Gbps | 10GBaseT |

* Indicates has been successfully used at 1000 Mbps
(though Cat 5e is the better choice)

*Category 1, 2 & 4 Cables:* These are no longer commonly used. They have been replaced by higher category cables.

*Category 3* cables are still in use today, though primarily in traditional phone systems.

*Category 5, 5e, 6 and 6A Cables:* These consist of four (4) twisted pairs in one overall cable. Reference Table 4-2 for the common wiring standards.

*RJ45 Connector:* This is the standard type of 8-pin connector for network cables. RJ45 connectors are most commonly seen with Ethernet cables and networks (Category 5, 5e, 6 and 6A cables).

### Table 4-2. Common Telecommunications Wiring Standards:

| TIA/EIA-568A Wiring | | | TIA/EIA-568B Wiring | | |
|---|---|---|---|---|---|
| Pair # | Insulation Color | RJ45 Pin # | Pair # | Insulation Color | RJ45 Pin # |
| | White/Blue | 5 | | White/Blue | 5 |
| | Blue/White | 4 | | Blue/White | 4 |
| 2 | White/Orange | 3 | 2 | White/Orange | 1 |
| | Orange/White | 6 | | Orange/White | 2 |
| 3 | White/Green | 1 | 3 | White/Green | 3 |
| | Green/White | 2 | | Green/White | 6 |
| | White/Brown | 7 | | White/Brown | 7 |
| | Brown/White | 8 | | Brown/White | 8 |

### Coaxial Cabling

A coaxial cable is one that consists of two conductors that share a common axis. The inner conductor is typically a straight wire, either solid or stranded, that is covered with flexible insulation. The outer conductor is typically a shield that might be braided or a foil. It can support up to 100 Mbps in data rate and may be run a greater distance than that of twisted pair cabling while maintaining transmission reliability.

**Coaxial cable types:** *Note: RG stands for Radio Guide.*

*50Ω impedance characteristic* coaxial cables are used in coaxial Ethernet networks and radio transmitter applications:

- RG-8: (Thicknet 10Base5) #12AWG inner conductor with foam insulation, braided and foil shield.

- RG-58: (Thinnet 10Base2) #19AWG inner conductor with foam insulation, braided shield.

RG8 and RG58 cables use BNC (Bayonet-Neill-Concelman) type connectors that employ a bayonet-mount mechanism for locking. They are good for frequencies up to 4 GHz.

*75Ω impedance characteristic* coaxial cables are used widely in video, audio and telecommunications applications.

- RG-6: #18AWG inner conductor with foam insulation, available with dual or quad shield.

- RG-11: #14AWG inner conductor. It can operate on frequencies as high as 3 GHz and has lower attenuation per foot than does RG-6 coaxial cable.

RG6 and RG11 cables use F type connectors.

## 4.1.2 Fiber Optic Cabling

Fiber optic cable is made of multiple long, thin strands of very pure glass, each about the diameter of a human hair. Each fiber in a fiber optic cable is made up of the following five parts:

- Core: The thin glass center of the fiber where the light travels

- Cladding: The outer optical material surrounding the core that reflects the light back into the core

- Buffer Coating: The plastic coating that protects the fiber from damage and moisture

- Strengthening Fibers: These are strands of aramid yarn. This is to prevent glass fracturing during the installation process.

- Cable Jacket: This serves a mechanical protection for the fiber core and cladding inside the cable.

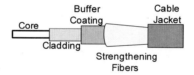

The light in a fiber optic cable travels through the core by constantly bouncing from the cladding, using a principle called total internal reflection. The cladding does not absorb any light from the core, so the light wave can travel great distances. Fiber optic cables function in reverse to that of copper cables. With copper cables, larger size means less resistance and therefore, more current flow. With fiber the opposite is true; the smaller the fiber, the less scattering of the light (intermodal dispersion) and therefore, more light transmission.

Optical fiber cores are made in two types, single-mode and multi-mode:

*Single-mode:*

- Single-mode has a smaller diameter core, 8–10 microns, which allows only one path for the rays of light to travel thru the fiber.

- A single-mode fiber yields a higher transmission rate and up to 50 times more distance than multi-mode fiber.

- The light source is typically a single-mode laser diode.

- The single-mode fibers can have a higher bandwidth than multi-mode fibers.

- The single-mode outer jacket color is typically yellow.

*Multi-mode:*

- This type has a larger diameter core, > 10 μm (typically 50 μm or 62.5 μm). this allows the rays of light to travel along several different angles between the core and cladding.

- It is used when higher power must be transmitted

- The light source is typically an LED.

- Due to the intermodal dispersion in the fiber, multi-mode fiber has higher pulse spreading rates than single mode fiber, limiting multi-mode fiber's information transmission capacity.

- It is usually less expensive than single-mode.

- The multi-mode outer jacket color is typically orange for 62.5 μm and aqua for 50 μm.

### *Fiber Optic Cable Connector Types[63]*

**SC** (Subscriber Connector) (Figure 4-1) is a connector with a push-pull latching snap in the connector. It is widely used for single-mode fibers.

**Figure 4-1. SC Connector**

**ST** (Straight Tip) (Figure 4-2) is a connector that uses a plug and socket which is locked in place with a half-twist bayonet lock. It is the most popular connector for multi-mode fibers.

**Figure 4-2. ST Connector**

---

63. Fiber optic connector photos from Lex-Tec.

**MU** (Figure 4-3) is a small form factor SC connector. It has the same push/pull style, but can fit two channels in the same footprint of a single SC.

**Figure 4-3. MU Connector**

**LC** (Lucent Connector) (Figure 4-4) is a small form factor connector that uses a 1.25 mm ferrule, half the size of the ST connector.

**Figure 4-4. Lucent Connector**

MT-RJ (Mechanical Transfer Registered Jack) (Figure 4-5) is a duplex connector used for multi-mode fiber. The MT-RJ has a single plastic locking clip similar to an RJ45 modular plug.

**Figure 4-5. MT-RJ Connector**

## 4.1.3 IEEE 802.11 & ISA 100.11a Wireless LAN Communication Protocols

802.11 is an evolving family of specifications for wireless local area networks (WLANs) developed by a working group of the IEEE. All the 802.11 substandards utilize the Ethernet protocol and CSMA/CA (Carrier Sense Multiple Access with Collision Avoidance) for path sharing.

With regards to CSMA/CA, as soon as a node on a network receives a packet that is to be sent, the node checks to be sure that no other node on the network is transmitting at the same time. If the channel is clear the node transmits the data. If the channel is not clear, the node waits for a pseudo-random period of time then checks again to see if the channel is clear. If the channel is still not clear, the node repeats the process over again until the channel is clear for transmission of data.

*802.11b:* 2.4 GHz, max theoretical speed 11 Mbit/s. It is subject to co-channel interference from other devices operating in this frequency range, such as microwave ovens, cordless telephones, Bluetooth, etc. It utilizes either frequency hopping spread spectrum (FHSS) or direct sequence spread spectrum (DSSS). All commercial 802.11b devices use DSSS.

*802.11a:* 5 GHz, max theoretical speed 54 Mbit/s. The 5 GHz band 802.11a currently has an advantage over 802.11b because the 802.11b 2.4 GHz band is very heavily used by other devices.

Because of the higher operating frequency range, 802.11a has a shorter range and less penetrating power than that of 802.11b. It utilizes an orthogonal frequency division multiplexing (OFDM) encoding scheme.

*802.11g:* 2.4 GHz, max theoretical speed 54 Mbit/s. An improvement on 802.11b to increase the speed. It utilizes orthogonal frequency division multiplexing (OFDM), the encoding scheme used in 802.11a, to obtain the higher data speed.

*802.11n:* 5 or 2.4 GHz, max speed 600 Mbit/s. Improves the range and coverage of the signals by allowing multiple antennas and splitting/combining signals in process known as multiple-input-multiple-output (MIMO). Some implementations of 802.11n provide backwards compatibility with previous 802.11 substandards, and can adapt between different substandards sequentially without serious impact.

*Orthogonal Frequency Division Multiplexing (OFDM):* The spread spectrum technique distributes the data over a large number of carriers, which are spaced apart at precise frequencies. This spacing provides the "orthogonality" in this technique, which prevents the demodulators from seeing frequencies other than their own frequency.

*Frequency Hopping Spread Spectrum (FHSS):* The frequency of the carrier signal is periodically modified (hopped) following a specific sequence of frequencies. The amount of time spent in each hop is called the dwell time (~100ms).

*Direct Sequence Spread Spectrum (DSSS):* For the duration of every message bit, the frequency carrier is modulated following a specific sequence of bits (chips). This results in the substitution of every message bit by the same sequence of the chips.

*ISA-100.11a:* This is a wireless sensor network and communications standard for a wide range of industrial applications. The standard comprises features that address robust and secure communications in harsh environments. It can be tailored specifically for devices with very low energy capability as well as devices that need to report control-specific information on a more repetitive basis. Various network topologies are supported including full mesh, hybrid mesh and star. The standard can support multiple application protocols simultaneously, which allows ISA-100.11a networks to serve as overall factory infrastructure networks for sensors and actuators. The standard includes various classes of devices including:

• Gateways that provide an application protocol interface enabling communications between various ISA100.11a network devices and user applications which may reside on the plant network.

• Wireless field devices that serve as the originator or terminus for information flowing through the network

- Security and Network Managers that control the detailed operation of the network, manage quality of service as well as access to the network.

- Routers that locally manage traffic flows.

*Wireless IEC 62591:* This is the wireless HART standard. The protocol utilizes a time synchronized, self-organizing, and self-healing mesh architecture. The protocol currently supports operation in the 2.4 GHz ISM Band using IEEE 802.15.4 standard radios. The radios employ direct-sequence spread spectrum technology and channel hopping for communication security

*Wireless HART:* *Wireless* HART is a wireless mesh network communications protocol for process automation applications. Each *Wireless* HART network includes three main elements:

- Wireless field devices that are connected to process or plant equipment. The device could be a device with WirelessHART built in or an existing installed HART-enabled device with a WirelessHART adapter attached to it.

- Gateways enable communication between these devices and host applications connected to a high-speed backbone or other existing plant communications network.

- The Network Manager is responsible for configuring the network, scheduling communications between devices, managing message routes, and monitoring network health. The Network Manager can be integrated into the gateway, host application, or process automation controller.

## 4.1.4 Transducers

Transducers convert one form of signal to another form of signal. Reference below for examples of transducers.

**Analog/Digital (A/D) Converter:** The typical range of an A/D converter is 0–5 V. The higher the bit output, the better the resolution and accuracy. For example, an 8-bit A/D converter would have 256 discrete values ($2^8$), a 16-bit A/D converter would have 65536 discrete values ($2^{16}$), and a 32-bit A/D converter would have $4.3 \times 10^9$ discrete values.

**Digital/Analog (D/A) Converter:** This type operates in reverse of an A/D converter and is used where a computer output must be interfaced with an analog device.

**Current/Pneumatic (I/P) Converter:** This type is frequently used on control valves where a 4–20 mA output from a control system is converted to the 3–15psig pneumatic signal used by a valve positioner.

## 4.1.5 Intrinsic Safety (I.S.)

Intrinsically safe transmission equipment is defined as "equipment and wiring which is incapable of releasing sufficient electrical or thermal energy under normal or abnormal conditions to cause ignition of a specific hazardous atmospheric mixture in its most easily ignited concentration"[64]. This section will discuss how I.S. barriers function while Section 8 will describe Hazardous Area Classifications. Reference Figure 4-6 for a General Intrinsic Safety Installation.

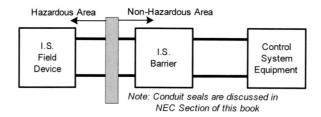

**Figure 4-6. General Intrinsic Safety Installation**

I.S. barriers generally fall into two classifications, active and passive:

*Passive:* Passive barriers are typically of the Zener diode design. If there should be a short circuit within the wiring or instrumentation in the hazardous area, there will be a corresponding drop in voltage going through the barrier. This short circuit will cause the fuse within the barrier to open and the Zener diode will conduct current to ground, thereby eliminating all possibility of any spark or thermal energy sufficient to ignite a flammable atmospheric condition. (Note the use of a ground in the generic schematic shown in Figure 4-7).

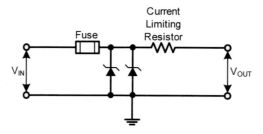

**Figure 4-7. Passive Barrier**

*Active:* Active barriers are typically of the galvanic isolator design. There is no physical connection between the input and output. A generic schematic of an active barrier is shown in Figure 4-8.

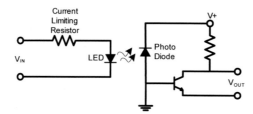

**Figure 4-8. Active Barrier**

---

64.  Reference ANSI/ISA-RP12.06.01-2003.

## 4.2 NETWORKING

### 4.2.1 OSI Model

OSI Model (Open Systems Interconnection) is an OSI standard (10731:1994) that defines the different stages that data must travel through in order to travel from one device to another over a network.

In its most basic form, the OSI model divides network architecture into seven layers (Figure 4-9).

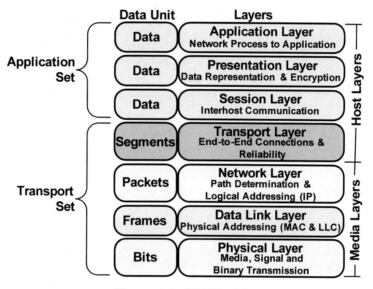

Figure 4-9. OSI Model

*Layer 7 (Application Layer):* This layer supports application and end-user processes as well as network access. Examples include HTTP, SNMP and SMTP.

*Layer 6 (Presentation Layer):* The presentation layer works to transform data into the form that the application layer can accept. It translates data into a format that can be read by many platforms. It also has support for security encryption and data compression.

*Layer 5 (Session Layer):* This layer establishes, manages and terminates connections between applications on a network.

*Layer 4 (Transport Layer):* This layer provides transparent transfer of data between end systems or hosts, and is responsible for end-to-end error recovery and flow control. It ensures complete data transfer.

*Layer 3 (Network Layer):* This layer provides switching and routing technologies, that is, it is responsible for establishing paths for data transfer through the network.

*Layer 2 (Data Link Layer):* This layer is responsible for communications between adjacent network nodes. Hubs and switches operate at the data link layer. The data link layer is divided into two sublayers: The *Media Access Control* (MAC) sublayer and the *Logical Link Control* (LLC) sublayer. The MAC sublayer controls how a computer on the network gains access to the data and permission to transmit it. The LLC sublayer controls frame synchronization, flow control and error checking.

*Layer 1 (Physical Layer):* This layer is responsible for bit-level transmission between network nodes over the transmission medium.

### 4.2.3 Protocol Stack

A protocol stack (Figure 4-10) is a group of protocols that all work together to allow software or hardware to perform a function. A common example of a protocol stack is TCP/IP.

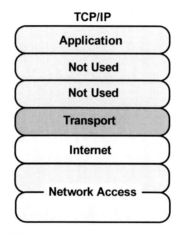

**Figure 4-10. Protocol Stack**

*Internet Protocol:* The format of an IP address is a 32 bit numeric address written as four numbers separated by periods. The range of each number can be from 0 to 255.

The following are the four IP address ranges reserved for private networks:

- 10.0.0.0 – 10.255.255.255

- 172.16.0.0 – 172.31.255.255

- 192.168.0.0 – 192.168.255.255

- 169.254.0.0 – 169.254.255.255

### 4.2.4 Network Hardware

*Gateway:* A gateway is network point that acts as an entrance to another network.

*Firewall:* A firewall examines all data traffic routed between two networks to ensure that the data meets predetermined criteria. When the data is cleared for passage, it is then routed between the networks, otherwise it is stopped. The firewall filters both inbound and outbound data traffic. A firewall is typically software that is loaded on hardware such as a gateway or router.

*Server:* A server is a computer that provides various shared resources to workstations and other servers on a computer network. A server also processes requests and delivers data to other computers over a network. A server may be either dedicated or shared.

**Dedicated:** Used for one particular application.

**Shared:** Used for multiple applications.

*Router:* A router joins networks by passing information from one network to the other. Routers operate at layer 3 of the OSI model. A router ensures that information doesn't go where it's not needed, as well as verifying that information does make it to its intended location.

*Switches:* Switches are used to connect computers, printers and servers within one local area network (LAN). A switch serves as a controller, enabling networked devices to talk to each other efficiently. Network switches are capable of inspecting data packets as they are received, determining the source and destination device of each packet, and forwarding them appropriately. Switches operate at Layer 2 of the OSI model.

**Managed Switch:** A type of switch that allows for control of the individual ports of the switch. It may also permit the ability to specify a particular MAC address to be allowed to connect via the switch.

**Unmanaged Switch:** A switch that has no configuration interface or options.

*Hub:* A hub is a form of a multi-port repeater. Unlike switches and routers, hubs do not manage data traffic. All data packets entering the input of the hub are replicated and sent out on the outputs of the hub. Hubs operate at Layer 1 of the OSI model.

*Disk Array:* Disk arrays (Figure 4-11) are storage systems that link multiple physical hard drives to appear as one large drive for advanced data control and security.

### RAID (Redundant Array of Independent Disks)

The purpose of RAID is for fault-tolerant systems (i.e., prevent the loss of data upon a failure of a component in the system).

**Disk Striping:** Disk striping is a process of dividing a chunk of data into data blocks across several partitions on the several hard disks that are used in a multiple hard disk system (RAID). Disk striping improves performance by using the hardware in all these drives in parallel so that data access is faster. By itself, disk striping does not provide fault tolerance; however, it is often used in conjunction with disk mirroring to provide both speed of data access and safety for the preservation of that data.

**Striping with Parity:** For example, a configuration utilizes three hard drives and stripes the data across them. The parity information is striped across the drives as well, so that if any one of the drives is lost, the information can be reconstructed from the parity information.

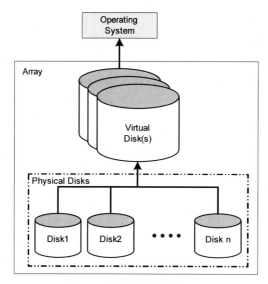

**Figure 4-11. Disk Array**

- The first stripe is data (on drive 1), data (on drive 2), parity (on drive 3)

- Then data (on drive 1), parity (on drive 2), data (on drive 3)

- Then parity (on drive 1), data (on drive 2), data (on drive 3)

This pattern continues so that if one drive fails, between the mix of parity and data on the remaining two drives, the data may be reconstructed.

**RAID 0:** This disk array (Figure 4-12) utilizes striping, but not with parity. There is no redundancy and the loss of a drive means loss of data.

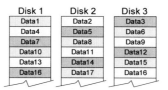

**Figure 4-12. RAID 0**

**RAID 1:** This disk array (Figure 4-13) utilizes complete duplication of data. This 100% data redundancy provides the best protection, but this type of protection comes at a cost and is the most expensive redundant array option.

**RAID 3:** This disk array (Figure 4-14) utilizes striping over three separate disks for the data and one dedicated disk for parity (redundancy).

*RAID 2, 3, and 4 are rarely used as many hardware controllers do not support these modes any longer.*

**Figure 4-13. RAID 1**

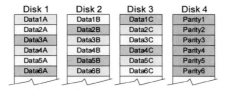

**Figure 4-14. RAID 3**

**RAID 5:** This array (Figure 4-15) is similar to RAID 3 except there is not a dedicated disk for the parity information. The parity information is distributed across all the disks within the array.

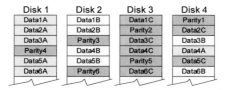

**Figure 4-15. RAID 5**

**RAID 6:** This array is similar to RAID 5 except there are two parity computations instead of one. This allows recovery of data from a two disk failure. It is not as popular as RAID 5.

### 4.2.5 Network Topology

*Bus*

Bus topology, such as Ethernet, uses a single communication backbone (Figure 4-16) for all devices. Bus topology utilizes a multi-drop transmission medium where all nodes on the network share a common bus and thus share communication. Failure of one of the stations does not affect any of the others; however, failure of the backbone takes down the entire network.

*Star*

This topology utilizes a central server to route data between clients (Figure 4-17). The central server switches data around the network. Data flow between the server and the nodes will, therefore, be relatively slow. Compared to the bus topology, a star network generally requires more cable, but a failure in a star network cable will only take down one computer's network access. However, if the central server fails, then the entire network also fails.

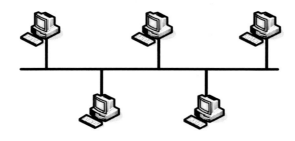

**Figure 4-16. Bus Network Topology**

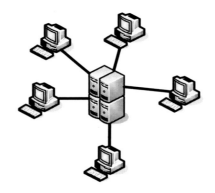

**Figure 4-17. Star Network Topology**

## Ring

This topology (Figure 4-18) connects computers in a circle of point-to-point connections with no central server, such as a series of desktop computers in an office. Each node handles its own applications and also shares resources over the entire network. In a "token ring" topology, the data is sent from one machine to the next and so on around the ring until it ends up back where it started. It also uses a token passing protocol, which means that a machine can only use the network when it has control of the token. This ensures that there are no collisions, because only one machine can use the network at any given time.

## Tree

Tree topologies (Figure 4-19) integrate multiple star topologies onto a bus.

## Mesh

Network topology (Figure 4-20) in which there are at least two pathways to each node. If one of the paths fails, the other is still available.

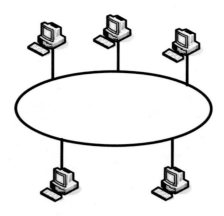

**Figure 4-18. Ring Network Topology**

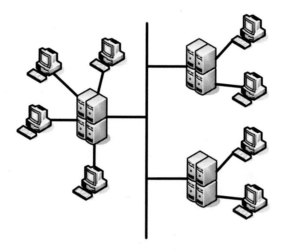

**Figure 4-19. Tree Network Topology**

## 4.2.6 Buses/Protocols

*Ethernet (IEEE 802.3)*

The original Ethernet was based upon the CSMA/CD LAN access method. CSMA/CD is an acronym for Carrier Sense Multiple Access/Collision Detection which is basically a set of rules for how networks respond when two devices attempt to use a data channel simultaneously. Later versions of Ethernet, 100Base-T or higher, may operate in full-duplex[65] mode, in which case the CSMA/CD function is switched off. Full duplex is similar to a phone, which allows the user to speak and listen simultaneously (whereas a walkie-talkie would be considered half-duplex). For the network level, full duplex means that a node may transmit and receive simultaneously with no possibility of contention.

---

65. Full duplex permits data to be sent and received at the same time. Half-duplex only permits data to move in one direction at a time.

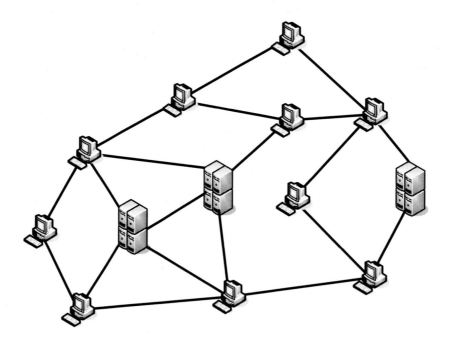

**Figure 4-20. Mesh Network Topology**

## 10 Mbps Ethernet (data rate $10^7$ bps)

**10Base-2:** This is a coaxial cable (RG-58) based Ethernet network, also referred to as "thinnet." The maximum segment length for 10Base-2 is 185 meters (~600 feet). One segment may contain NO more than 30 transceivers (nodes), including repeaters, with a maximum of 5 segments allowed. 10Base-2 cabling must be close-coupled to the transceiver, as the maximum distance between the medium and the transceiver is 4 inches. In addition, there must be a minimum of 0.5 meters (~1.6 feet) between nodes. The ends must be terminated with a 50 Ω terminating resistor. This is not a currently recommended network technology, but legacy systems may still be in use.

**10Base-5:** This is a coaxial (RG-8) based Ethernet network, also referred to as "thicknet." The maximum segment length for 10Base-5 is 500 meters (~1640 feet). The entire network scheme may not exceed 2500 meters (~8200 feet). One segment may contain NO more than 100 transceivers (nodes), including repeaters. Taps must be placed at a minimum of 2.5 meters (~8 feet) apart. The ends must be terminated with a 50 Ω terminating resistor with one end grounded. This is not a currently recommended network technology, but legacy systems may still be in use.

## 100 Mbps Ethernet (data rate $10^8$ bps, also known as "Fast" Ethernet)

**100Base-T:** The segment length for 100Base-T cabling is 100 meters (~328 feet). This means that the maximum distance between a workstation and a hub is 100 meters. Networks larger than 200 meters in length must therefore be logically connected together by store-and-forward type devices (routers, switches, etc.)

*100Base-T4* (four pairs of telephone twisted pair wire) is designed to operate on category 3 (Cat3) or above cables.

*100Base-TX* (two pairs of data grade twisted-pair wire) is designed to operate on category 5 (Cat5) or above cables.

**100Base-FX:** (a two-strand optical fiber cable) This standard extends the maximum allowed switch-to-switch distance from 100 meters (328 feet) for copper cabling to 2000 meters (~6400 feet) for fiber optic cabling.

**100Base Repeater Rules:** The cable distance and the number of repeaters that may be used in a 100Base system depend upon the delay as a result of the cable and delay as a result of the repeaters, as well as potential network interface controller (NIC) delays. The maximum roundtrip delay allowed is 5.12 μs. In other words, a frame must go from the transmitter (NIC) to the farthest end node and back again for collision detection within 5.12 μs. There are two types of repeaters within 100Base-T:

*Class I:* This type of repeater limits the quantity of repeaters in a physical domain to one. A class I repeater is allowed to have larger timing delays (< 0.70 μS) because it translates line signals on an incoming port to digital form. It then re-translates them to line signals when sending them out on the other ports. Only one Class I repeater may be within a single network segment.

*Class II:* This type of repeater is restricted to smaller timing delays (< 0.46 μs) because it immediately repeats the incoming signal to all other ports without a translation process. The maximum number of Class II repeaters within a single network segment is two.

### *1000 Mbps Ethernet (data rate $10^9$ bps, also known as "Gigabit" Ethernet)*

**1000Base-T:** (four pairs of data grade twisted-pair wire) This standard is designed to operate on category 5 (Cat5) or above cables. Just like 100Base-T, the maximum the distance between a workstation and a hub is 100 meters. (This standard is not to be confused with 1000Base-TX, which is a telecommunication industry association standard: TIA/EIA-854).

**1000Base-F:** (a two-strand optical fiber cable) This standard extends the maximum allowed switch-to-switch distance from 100 meters (328 feet) for copper cabling to 500 meter (~1600 feet) full duplex on multimode fiber cable or 2-3 km (6000–9800 feet) full duplex on single-mode fiber cable.

### *Profi®bus (<u>PRO</u>cess <u>FI</u>eld <u>BUS</u>):*

This standard supports two types of devices: Masters (Primary) and Slaves (Secondary).[66]

**Primary(M):** This is a device that controls the bus when it has the permission to access the bus. A Primary(M) device may transfer messages without the need for a remote request. This is an *active* type device.

**Secondary(S):** This is a device that may only acknowledge received messages, or transmit a message but only when requested to do so by the Primary(M). Secondaries (S) are typically peripheral devices such as sensors, transmitters, actuators, etc. This is a *passive* type device.

---

66. The county of Los Angeles has asked equipment manufacturers to not use the term master/slave as it may appear as offensive to some of its residents. This publication will utilize Primary(M)/Secondary(S) in lieu of Master/Slave.

Each segment may contain up to 32 nodes, including repeaters, which are laid out in a single node. Each node has one Primary(M) device and may have multiple Secondary(S) devices. Up to 126 nodes may be connected in up to five segments, which are separated by repeaters.

There are two versions of the Profibus standard:

*Profibus DP* (Primary(M)/Secondary(S)): DP stands for distributed peripheral. This protocol allows for the use of multiple Primary(M) devices, in which case each Secondary(S) device is assigned to only one Primary(M) device. In other words, multiple Primary(M) devices may read inputs from the Secondary(S) devices, but only one Primary(M) device has permission to write outputs to its own assigned Secondary(S) devices. Profibus DP normally operates using a cyclic transfer of data between Primary(M) and Secondary(S) on an EIA-485 network. An assigned Primary(M) device periodically requests (polls) each Secondary(S) device on the network. The bus cycle time is typically < 10ms.

*Profibus PA:* PA stands for process automation. This is essentially the same as Profibus DP, except that the voltage and current levels are reduced in order to comply with intrinsic safety for the process industry.

Data rate may vary between 9.6 kbps and 12 Mbps, dependent upon cable length (Table 4-3).

### Table 4-3. Profibus Data Rate and Segment Length

| Data Rate | Max. Segment Length |
|-----------|---------------------|
| 9.6 kbps | 1200 meters (~3900 feet) |
| 19.2 kbps | 1200 meters (~3900 feet) |
| 93.75 kbps | 1200 meters (~3900 feet) |
| 187.5 kbps | 600 meters (~1970 feet) |
| 500 kbps | 200 meters (~650 feet) |
| 1.5 Mbps | 200 meters (~650 feet) |
| 12 Mbps | 100 meters (328 feet) |

Profibus cabling is based on the EIA-485 bus and uses a non-powered 2-wire bus. The connection is half-duplex over a shielded, twisted-pair cable. The bus uses either a 9-pin D (DIN 19245) or a 4-pin M12 round B-coded connector. Data rates may be from 9600 to 12 Mbaud, with message lengths of 244 bytes. At 12 Mbps the maximum distance is 100 meters.

Profibus may also utilize optical link modules to permit the use of fiber optic cable to increase the maximum distance up to 3000 meters (~9800 feet) between modules.

### FOUNDATION *Fieldbus*™

FOUNDATION **Fieldbus (FF)** is an open, nonproprietary architecture that provides a communications protocol for control and instrumentation systems in which each device has its own "intelligence" and communicates via an all-digital, serial, two-way communication system. The architecture was developed and is administered by the Fieldbus Foundation.

Two related implementations (H1 & HSE) of Foundation Fieldbus have been introduced to meet different needs within the process automation environment. These two implementations use different physical media and communication speeds.

**H1:** This implementation is intended primarily for process control, field-level interface and device integration. Running at 31.25 kbps, the technology interconnects devices such as transmitters and actuators on a field network. H1 is designed to operate on existing twisted pair instrument cabling, with power and signal on the same wire. Fiber optic media is optional. H1 also supports intrinsically safe applications. In addition, H1 technology enables field instruments and other devices to execute control functions, thereby reducing the load on plant computers and workstations. Since the H1 network is digital, I/O conversion subsystems are eliminated. H1 can support up to 32 devices on one segment, though in reality, it is more like 4–15 because of power limitations. Minimum power requirement is 8 mA with a minimum device operating voltage of 9 V and a maximum bus voltage of 32 V.

**Segment Calculations:** When calculating how many devices can be accommodated on a FF segment, the factors to be taken into account are the maximum current requirement of each device and the resistance of the segment cable (because of the voltage drops along the cable length). The calculation is a simple Ohm's law problem, with the aim of showing that at least 9 V can be delivered at the farthest end of the segment, after taking into account all the voltage drops from the total segment current. Every segment must have a terminator at each end.

*Example:* Sourcing 16 devices at 20 mA each requires 320 mA, so if the segment is based on #18 AWG cable with 50 $\Omega$/km/loop and a 25 V power conditioner, the maximum cable length is 1000 m to guarantee 9 V at the end.

Voltage available for cable = 25 V (source) – 9 V = 16 V, therefore the allowable resistance is 16 V ÷ 0.320 A = 50 $\Omega$ and since #18 AWG cable has a resistivity of 50 $\Omega$/km, the maximum cable length is 1000 m. Note that you may also choose to allow for a safety margin on top of the 9 V minimum operating voltage to allow for unexpected current loads and for adding additional devices in the future.

**Recommended FF Spur Length** (Table 4-4)
A spur is an H1 branch line connecting to the trunk, which is a final circuit. A spur may vary in length from 1 meter (3.28 feet) to 120 meters (394 feet).

### Table 4-4. Recommended FF Spur Length

| Total Devices | (1) Device per Spur | (2) Devices per Spur | (3) Devices per Spur | (4) Devices per Spur |
|---|---|---|---|---|
| 25 – 32 | 1 meter (3.28 feet) | 1 meter (3.28 feet) | 1 meter (3.28 feet) | 1 meter (3.28 feet) |
| 19 – 24 | 30 meters (98 feet) | 1 meter (3.28 feet) | 1 meter (3.28 feet) | 1 meter (3.28 feet) |
| 15 – 18 | 60 meters (197 feet) | 30meters (98 feet) | 1 meter (3.28feet) | 1 meter (3.28 feet) |
| 13 – 14 | 90 meters (295 feet) | 60 meters (197 feet) | 30 meters (98 feet) | 1 meter (3.28 feet) |
| 1 – 12 | 120 meters (394 feet) | 90 meters (295 feet) | 60 meters (197 feet) | 30 meters (98 feet) |

## Recommended FF Segment Length

The maximum allowed length of a FF segment is 1900 meters (~6200 feet) except where repeaters are installed. The maximum segment length of 1900 meters will have to be reduced if another type of cable is selected or if cable types are mixed. Table 4-5 shows the maximum lengths of different types of cables. FF cable specification is FF-844.

**Table 4-5. Recommended FF Segment Length**

| Cable Type | Description | Cable Size | Maximum Length |
|---|---|---|---|
| A | Twisted Pair with shield | #18AWG | 1900 meters (~6200 feet) |
| B | Multi – Twisted Pair with overall shield | #22AWG | 1200 meters (~3900 feet) |
| C | Multi – Twisted Pair without shield | #26AWG | 400 meters (~1300 feet) |
| D | Multi – Conductor (no pairing) overall shield | #16AWG | 200 meters (~650 feet) |

The total segment length is calculated by:
Total Segment Length = Trunk Length + Length of All Spurs

**HSE** (High-speed Ethernet): This implementation works at 100 Mbps and is designed for device, subsystem and enterprise integration. HSE enhances access to H1 fieldbus technology via linking devices.

### DeviceNet

DeviceNet, which was developed by Allen Bradley (Rockwell), is a low-level, device-oriented network based on the CAN (Controller Area Network) protocol which was developed by Bosch GmbH for the automotive industry. The low level devices, such as sensors and actuators, connect to higher level controllers.

DeviceNet supports three data rates: 125, 250 and 500 kbps, with up to 64 devices on the bus. There is a distance tradeoff for the higher data rates. A four conductor cable provides both power and communication to the devices.

DeviceNet uses trunk and drop topology (similar to Bus Topology, reference Figure 4-20). The trunk is the main communication cable, and requires a 121 $\Omega$ resistor at both ends. The maximum length of the trunk depends on the communication rate and the cable type. Drops are branches off the trunk, and may be from zero to 20 feet in length. The cumulative drop lengths are dependent on the communication rate. Thick or flat cables are typically used for long distances because they are stronger and more resilient than thin cables. Thin cables are mainly used for the local drop lines connecting nodes to the trunk line.

**Thick Cable:** 0.48 inch diameter; data pair #18 AWG; power pair #15 AWG; overall braid shield and drain wire (Table 4-6; Figure 4-21).

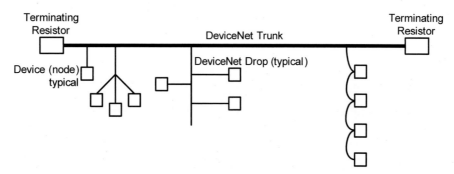

**Figure 4-20. DeviceNet Topology**

**Table 4-6. Thick Cable Lengths**

| Data Rate | 125 kbps | 250 kbps | 500 kbps |
|---|---|---|---|
| Trunk Distance | 500 meters (1640 feet) | 250 meters (630 feet) | 100 meters (328 feet) |
| Max. Drop Length | 20 feet | 20 feet | 20 feet |
| Cumulative Drops Length | 512 feet | 256 feet | 128 feet |
| Number of Nodes | 64 | 64 | 64 |

**Thin Cable:** 0.27 inch diameter, data pair #24 AWG; power pair #22 AWG; overall braid shield and drain wire (Table 4-7; Figure 4-21).

**Table 4-7. Thin Cable Lengths**

| Data Rate | 125 kbps | 250 kbps | 500 kbps |
|---|---|---|---|
| Trunk Distance | 100 meters (328 feet) | 100 meters (328 feet) | 100 meters (328 feet) |
| Max. Drop Length | 20 feet | 20 feet | 20 feet |
| Cumulative Drops Length | 512 feet | 256 feet | 128 feet |
| Number of Nodes | 64 | 64 | 64 |

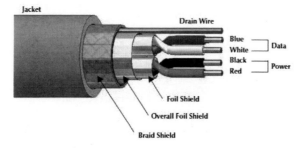

**Figure 4-21. Thick/Thin Cable**

**Flat Cable:** Highly flexible cable, 600 Volt, 8 Amp rating with a physical key molded into the overall jacket (Table 4-8; Figure 4-22).[67]

**Table 4-8. Flat Cable Lengths**

| Data Rate | 125 kbps | 250 kbps | 500 kbps |
|---|---|---|---|
| Trunk Distance | 420 meters (1640 feet) | 200 meters (630 feet) | 75 meters (246 feet) |
| Max. Drop Length | 20 feet | 20 feet | 20 feet |
| Cumulative Drops Length | 512 feet | 256 feet | 128 feet |
| Number of Nodes | 64 | 64 | 64 |

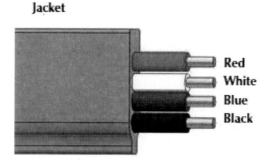

**Figure 4-22. Flat Cable**

## Data Highway

Data Highway Plus (DH+) and Data Highway 485 (DH-485) were both developed by Allen Bradley (Rockwell). The communication data rate for DH+ is 57.6 kbps. It utilizes peer-to-peer communications and allows up to 64 nodes. The interconnecting cable for the DH+ network link is called a "Blue Hose" due to the use of a shielded twin axial cable, with overall blue jacket insulation (Figure 4-23), and requires end termination resistors. The maximum trunk line cable distance is 10,000 feet, with a maximum dropline length of 100 feet.

The communication data rate for DH-485 is 19.2 kbps and DH-485 allows up to 32 nodes. The maximum interconnecting cable length for the DH-485 network link is 4000 feet.[68]

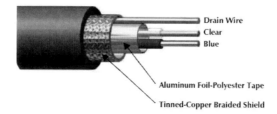

**Figure 4-23. "Blue Hose" DH+ Network Cable**

---

67. Images obtained from www.turck.com.
68. Image obtained from www.turck.com.

## HART (Highway Addressable Remote Transducer Protocol)

The protocol was developed by Rosemount (Emerson) as an open protocol. HART enables the use of existing 4–20 mA instrumentation cabling to simultaneously carry digital information superimposed on the analog signal. HART utilizes "frequency shift keying"[69] that uses two individual frequencies to represent the digits 1 (1200 Hz) and 0 (2200 Hz). These superimposed frequencies (Figure 4-24) cause NO interference to the 4–20 mA analog signal.[70]

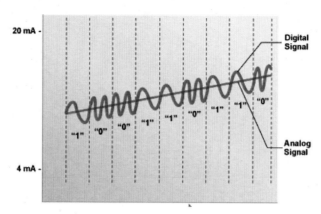

**Figure 4-24. HART Frequency Superimposition**

HART technology is a Primary(M)/Secondary(S) protocol, which means that a smart field (Secondary(S) device only speaks when spoken to by a Primary(M) device. The HART protocol can be used in various modes such as point-to-point or multidrop for communicating information to/from smart field instruments and central control or monitoring systems.

### ModBus®

ModBus (Table 4-9, 4-10) is a serial communications protocol published by Modicon (Schneider-Electric) that uses client/server architecture between devices connected to different types of buses (networks). The client/server architecture is basically in a request/response mode. The client makes a service request from another program and the server fulfills this service request.

Serial Modbus connections can use two basic transmission modes, **ASCII** or **RTU.**

*Modbus/ASCII:* the messages are in ASCII format (hexadecimal values represented with readable ASCII characters). This makes the message readable in monitoring format.

*Modbus/RTU:* This format uses binary coding, which makes the message unreadable when monitoring, but reduces the size of each message. This allows for more data exchange in the same time span.

---

69. Frequency shift keying is based upon Bell Systems 202 communication standard.
70. From www.hartcomm.org.

**Table 4-9. Protocol Structure**

| Address Field | Function Field | Data Field | Error Check Field |
|---------------|----------------|------------|-------------------|
| 1 byte | 1 byte | Variable | 2 bytes |

**Table 4-10. Modbus Addressing**

| Data Type | Absolute Address | Relative Address | Function Code Decimal (Hex) | Description |
|-----------|------------------|------------------|------------------------------|-------------|
| Coils | 00001 – 09999 | 0 – 9998 | 01 | Read coil status |
| Coils | 00001 – 09999 | 0 – 9998 | 05 | Force single coil |
| Coils | 00001 – 09999 | 0 – 9998 | 15 (0F) | Force multiple coils |
| Discrete Inputs | 10001 – 19999 | 0 – 9998 | 02 | Read input status |
| Input Registers | 30001 – 39999 | 0 – 9998 | 04 | Read input registers |
| Holding Registers | 40001 – 49999 | 0 – 9998 | 03 | Read holding registers |
| Holding Registers | 40001 – 49999 | 0 – 9998 | 06 | Preset single register |
| Holding Registers | 40001 – 49999 | 0 – 9998 | 16 (10) | Preset multiple registers |
| N/A | N/A | N/A | 07 | Read exception status |
| N/A | N/A | N/A | 08 | Loopback diagnostic test |

**Modbus Function Codes:** Due to the nature of Modbus serial communication, it is only possible to query one device at a time.

*Function 01: Read coil status:* This function allows the host to read the on/off status of one or more logic coils within a target device. The coil data bytes are packed with one bit for the status of each consecutive coil. If the number of coils read is less than 8, then the byte will be padded with leading zeroes (1 = ON; 0 = OFF).

*Function 02: Read input status:* This function allows the host to read one or more discrete inputs within a target device. Just like function 01, the discrete input data bytes are packed with one bit for the status of each consecutive discrete input. If the number of discrete inputs read is less than 8, then the byte will be padded with leading zeroes (1 = ON; 0 = OFF).

*Function 03: Read holding registers:* This function allows the host to read the contents of one or more holding registers within a target device. The data field of the requested frame consists of a relative address of the first holding register, followed by the number of holding registers to be read.

*Function 04: Read input registers:* This function allows the host to read the contents of one or more input registers within a target device. The data field of the requested frame consists of a relative address of the first input register, followed by the number of input registers to be read.

*Function 05: Force single coil:* This function allows the host to change the on/off status of a single logic coil within a target device. The data field of the requested frame consists of a relative address of the coil, followed by the desired status of that coil.

*Function 06: Preset single register:* This function allows the host to change the contents of single holding register within a target device. The data field of the requested frame consists of a relative address of the first holding register, followed by the desired value of that holding register.

*Function 07: Read exception status:* This function will provide the host the status of eight predefined digital points in the Secondary(S) device.

*Function 08: Loopback test:* This function provides a means to test the communication system without affecting any of the memory tables within the Secondary(S) device.

*Function 15 (0F): Force multiple coils:* This function allows the host to change the on/off status of a contiguous group of logic coils within a target device.

*Function 16 (10): Force multiple registers:* This function is similar to function 15 in operation.

### RS-232/EIA-232

**R**ecommended **S**tandard 232 (issued in 1969)

- It is now EIA-232

- The current version is EIA/TIA-232E (issued in 1991)

- This is the standard for serial binary data signals connecting between a *DTE* (Data Terminal Equipment, such as a computer or a printer)) and a *DCE* (Data Circuit-terminating Equipment, such as a modem). It is commonly used in computer serial ports.

- Maximum drivers =1; maximum receivers = 1

- Maximum data rate = 19.2 kbps @ 50 feet

The RS-232 standard defines the voltage levels that correspond to logical one and logical zero levels. Valid signals are as follows:

- Logic 1 = –3 to –12 V

- Logic 0 = +3 to +12 V

- Undefined logic level = –3 to +3 V

The range near zero volts is not a valid RS-232 level; logic one is defined as a negative voltage, and has the functional significance of OFF. Logic zero is positive, and has the function of ON.

The widely-used rule-of-thumb indicates that cables more than 50 feet (15 meters) long will have too much capacitance, unless special cables are used. Cable connectors may be either DB9 (9 pin) or DB25 (25 pin). In addition, the maximum data rate of 19.2 kbps is too slow for many applications. The RS-485 protocol is increasingly being used over RS-232 for instrumentation and control systems.

*RS-422/EIA-422*

Recommended Standard 422

- It is now EIA-422

- Maximum drivers =1; maximum receivers = 10

Unlike RS-232, which looks at the voltage *level* between the signal lines, RS-422 detects the voltage *difference* between the signal lines. This reduces the noise effect on the cable and allows for much higher speed data rates. In addition, RS-422 allows for connection to intelligent devices (sensors, etc.) rather than just as an interface between computer hardware. In addition, RS-422 allows a multi-drop network topology (similar to bus topology, reference Figure 4-25), rather than a multi-point network topology. In a multi-point network all nodes are considered equal and every node has the ability send and receive over the same cable. Whereas a multi-drop network allows one central control unit to send commands in parallel to up to ten secondary devices.

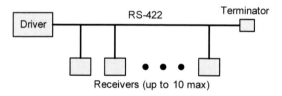

**Figure 4-25. RS-422 Multi-drop Topology**

*RS-485/EIA-485*

Recommended Standard 485

- It is now EIA-485

This is an extension of RS-422, but with the following enhanced features:

- Maximum drivers = 32; maximum receivers = 32

- Electrical specification of a two-wire, half-duplex, multipoint serial communications channel. Since it uses a differential balanced line over twisted pair (like EIA-422), it can span relatively large distances (up to 4,000 feet)

- It offers high data transmission speeds:

    - 10 Mbps at 1200 meters (~4000 feet)
    - 35 Mbps at 6 meters (~20 feet)

- Half-duplex operation can be made full-duplex by using four wires (duplex means the ability to both send and receive data).

### AS-i (Actuator Sensor Interface)

Actuator Sensor interface is an open system network developed by a consortium of manufacturers. AS-i is a bit-oriented communication link designed to connect binary sensors and actuators. AS-i operates as a Primary(M)/Secondary(S) system (reference standard IEC 62026-2).

AS-i uses two-wire, untwisted and unshielded cable (Figure 4-24) that serves as both the power supply and communication link (note that shielded cable may be used in high noise environments). The cable is trapezoidal shaped to ensure proper polarity. The AS-i network communication data rate is 167 kbps with a scan time of 5 ms. Network topology may be bus, ring, tree or star up to 100 meters (328 feet) in length, and may be increased up to 300 meters (~980 feet) with a repeater.[71]

**Figure 4-24. AS-I Cable**

AS-i can support up to 62 Secondary(S) devices. Each Secondary(S) device may have 4 byte input and 4 byte output, which equates to 496 inputs and 496 outputs.

### OPC

This is a compound document standard developed by Microsoft (the OPC specification was based on the OLE, COM, and DCOM technologies). It enables the creation of objects with one application and then linking (or embedding) them in a second application. Embedded objects retain their original format and the links to the application that created them. This effectively reduces the number of interfaces required to one per application.

- OPC = OLE for Process Control (OLE = Object Linking & Embedding)

- COM = Component Object Model

- DCOM = Distributed Component Object Model

**OPC Data Access (DA):** This standard is used to move real-time data from PLCs, DCSs, and other control devices to HMIs and other display clients. Each data point (SP, PV, etc.) contains a value, quality (reliability of reading), and timestamp. The OPC server provides the timestamp for protocols (e.g., Modbus) that do not transmit the timestamp information. (Note: OPC DA does not deal with historical data, only real-time data.) Reference OPC HDA for historical data.

**OPC Historical Data Access (HDA):** OPC HDA provides access to data already stored. Whether it be from a simple serial data logging system or a complex SCADA system, historical archives can be retrieved in a uniform manner with OPC HDA.

**OPC Alarms & Events (A&E):** OPC A&E is used to exchange process alarms and events (including operator actions). OPC A&E complements OPC DA & HDA, but is separate from them.

---

71. Images obtained from www.turck.com.

---

**OPC Batch:** OPC Batch is used to exchange information associated with batch processes: equipment capabilities, batch execution, recipe contents, and batch-specific events.

**OPC Data Exchange (DX):** OPC DX defines how OPC servers exchange data with other OPC servers.

**OPC Tunnel:** OPC Tunneling is connecting an OPC Server to an OPC Client over a network, using TCP/IP as the transport protocol. This avoids the problems associated with configuring DCOM to work over a network. Also, it eliminates the long DCOM timeouts that can effectively shut down an application.

**OPC Mirror:** This is software that enables two or more OPC DA servers to communicate with each other.

## 4.3 CIRCUIT CALCULATIONS

### 4.3.1 DC Circuits

*Units*

| | |
|---|---|
| Resistance | ohms ($\Omega$) |
| Voltage | volts (V) |
| Current | amps (I) |
| Capacitance | farad (F) |
| Inductance | Henry (H) |

*Resistive Circuits*

**Series Circuits**

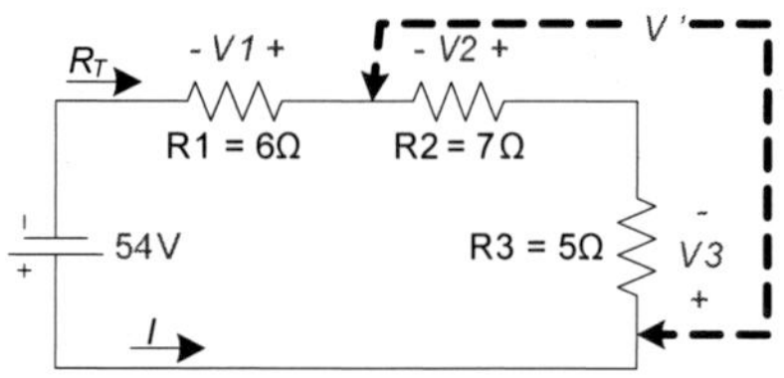

$$R_T = R_1 + R_2 + R_3 = 6\Omega + 7\Omega + 5\Omega = 18\Omega$$

$$I = \frac{E}{R_T} = \frac{54V}{18\Omega} = 3A \ \text{(Ohm's Law)}$$

$$\therefore V_1 = IR_1 = (3A)(6\Omega) = 18V$$

$$V_2 = IR_2 = (3A)(7\Omega) = 21V$$

$$V_3 = IR_3 = (3A)(5\Omega) = 15V$$

*Verify with Kirchhoff's Voltage Law:*

$$\sum V = +E - V_1 - V_2 - V_3 = 54 - 18 - 21 - 15 = 0 \ \textit{Checks}$$

Evaluating the voltage across any resistor or combination of series resistors in a series circuit can be reduced to one step by using the Voltage Divider Rule:.

$$V_X = \frac{R_X V}{R_T} \text{ For any one resistor } R_X$$

$$V'_T = \frac{R'_T V}{R_T} \text{ For two or more resistors in series having a total resistance equal to } R'_T$$

For the example circuit above, determine $V_2$ and $V'$ using the Voltage Divider Rule:

$$\text{Rule: } V_2 = \frac{R_2 V}{R_T} = \frac{(7\Omega)(54V)}{18\Omega} = 21V$$

$$V'_T = \frac{R'_T V}{R_T} = \frac{(R_2 + R_3)54V}{18\Omega} = \frac{(12\Omega)(54V)}{18\Omega} = 36V$$

## Parallel Circuits

The potential drops of each branch equal the potential rise of the source:

$$V_T = V_1 = V_2 = V_3 = ...V_N$$

The total current is equal to the sum of the currents in the branches:

$$I_T = I_1 + I_2 + I_3 = ...I_N$$

The inverse of the total resistance of the circuit (also called effective resistance) is equal to the sum of the inverses of the individual resistances:

$$\frac{1}{R_T} = \frac{1}{R_1} + \frac{1}{R_2} + \frac{1}{R_3} + ...\frac{1}{R_N}$$

For two parallel resistors: $R_T = \dfrac{R_1 R_2}{R_1 + R_2}$

For three parallel resistors: $R_T = \dfrac{R_1 R_2 R_3}{R_1 R_2 + R_1 R_3 + R_2 R_3}$

## Parallel Circuits Example:

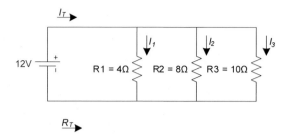

$$\frac{1}{R_T} = \frac{1}{R_1} + \frac{1}{R_2} + \frac{1}{R_3} = \frac{1}{4\Omega} + \frac{1}{8\Omega} + \frac{1}{10\Omega} = 0.475 \quad \therefore R_T \frac{1}{0.475} = 2.105\Omega$$

Using the formula for three parallel resistors:

$$R_T = \frac{R_1 R_2 R_3}{R_1 R_2 + R_1 R_3 + R_2 R_3} = \frac{(4\Omega)(8\Omega)(10\Omega)}{(4\Omega)(8\Omega)+(4\Omega)(10\Omega)+(8\Omega)(10\Omega)} = \frac{320}{(32+40+80)} = \frac{320}{152} = 2.105\Omega$$

$$I_1 = \frac{E}{R_1} = \frac{12V}{4\Omega} = 3A$$

$$I_2 = \frac{E}{R_2} = \frac{12V}{8\Omega} = 1.5A$$

$$I_3 = \frac{E}{R_3} = \frac{12V}{10\Omega} = 1.2A$$

$$I_T = \frac{E}{R_T} = \frac{12V}{2.105\Omega} = 5.7A \text{ OR } I_T = I_1 + I_2 + I_3 = 3A + 1.5A + 1.2A = 5.7A$$

Current Divider Rule for two parallel branch circuits:

$$I_1 = \frac{R_2 I_T}{R_1 + R_2} \text{ and similarly } I_2 = \frac{R_1 I_T}{R_1 + R_2}$$

$$I_1 = \frac{R_2 I_T}{R_1 + R_2} = \frac{(4\Omega)(12A)}{2\Omega + 4\Omega} = \frac{48}{6} = 8A \text{ AND } I_2 = \frac{R_1 I_T}{R_1 + R_2} = \frac{(2\Omega)(12A)}{2\Omega + 4\Omega} = \frac{24}{6} = 4A$$

*Verify with Kirchhoff's Current Law:*

$$I_T = I_1 + I_2 = 12A = 8A + 4A \quad Checks$$

## Series-Parallel Circuits

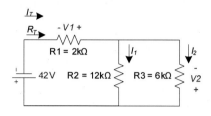

$$R_T = R_1 + \frac{R_2 R_3}{R_2 + R_3} = 2k\Omega + \frac{(12k\Omega)(6k\Omega)}{(12k\Omega + 6k\Omega)} = 2k\Omega + \frac{72k\Omega}{18k\Omega} = 2k\Omega + 4k\Omega = 6k\Omega$$

$$\therefore \, I_T = \frac{E}{R_T} = \frac{42V}{6k\Omega} = 7mA \quad \text{Using Current Divider} \quad I_1 = \frac{R_3 I}{R_2 + R_3} = \frac{(6k\Omega)(0.007A)}{12k\Omega + 6k\Omega} = \frac{42V}{18k\Omega} = 2.33mA$$

## AND

$$I_2 = \frac{R_2 I}{R_2 + R_3} = \frac{(12k\Omega)(0.007A)}{12k\Omega + 6k\Omega} = \frac{84V}{18k\Omega} = 4.67mA$$

*Verify with Kirchhoff's Current Law:*

$$I_T = I_1 + I_2 = 7mA = 2.33mA + 6.67mA \quad Checks$$

$$V_1 = \frac{R_1 V}{R_T} = \frac{(2k\Omega)(42V)}{6k\Omega} = 14V$$

## *R-L-C Circuits*

## Polar-Rectangular Conversions

*Rectangular Form:*

*Polar Form:*

*Exponential Form:*

Euler's equation:
$$\begin{aligned} a + jb &= ce^{j\theta} = c(\cos\theta + j\sin\theta) \\ a - jb &= ce^{j\theta} = c(\cos\theta - j\sin\theta) \end{aligned}$$
$\theta$ must be expressed in radians

Where c is the magnitude of the vector and $\theta$ is the angle expressed in radians.

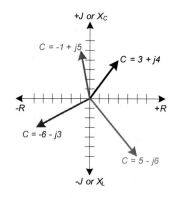

**Figure 4-26. Rectangular Form**

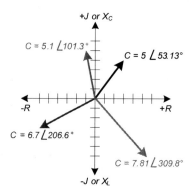

**Figure 4-27. Polar Form**

## Conversion Between Forms

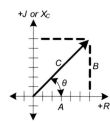

**Figure 4-28. Conversion Between Forms**

*Rectangular to polar:*

$$C = \sqrt{A^2 + B^2}$$

$$\theta = \tan^{-1}\frac{B}{A}$$

*Polar to rectangular:*

$$A = C\cos\theta$$

$$B = C\sin\theta$$

## Conversion Examples:

Convert $C = 3 + j4$ to polar:

$$C = \sqrt{3^2 + 4^2} = \sqrt{25} = 5$$

$$\theta = \tan^{-1}\frac{4}{3} = 53.13°$$

$$C = 5\underline{|53.13°}$$

Convert $10\underline{|230°}$ to rectangular:

$$A = C\cos\theta = 10\cos(230 - 180) = 10\cos 50 = 6.428$$

$$B = C\sin\theta = 10\sin 50 = 7.660$$

$$C = -6.428 - j7.660$$

### *Bridge Circuits*

The basic bridge circuit (Figure 4-29) consists of three known resistances ($R_1$, $R_2$, and $R_3$, which is variable) and an unknown variable resistor Rx, a voltage source and a sensitive ammeter. The bridge circuit has two arms ($R_1$ and $R_3$ constitute one arm here, and $R_2$ and $R_X$ constitute the other arm). Each arm is composed of two resistors in series, and you may want to think of each arm as a voltage divider. The output is the difference between the outputs of the two voltage dividers.

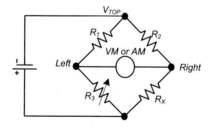

**Figure 4-29. Basic Bridge Circuit**

$$\frac{R_1}{R_3} = \frac{R_2}{R_X} \quad \therefore \quad R_X = \frac{R_2 R_3}{R_1}$$

$$V_{TOP} \times \frac{R_3}{R_1 + R_3} = \text{Voltage at Left Terminal}$$

$$V_{TOP} X \frac{R_X}{R_2 + R_X} = \text{Voltage at Right Terminal}$$

## 4.3.2 AC Circuits

Angular velocity is a vector quantity which specifies the angular speed of an object and the axis about which the object is rotating (as in an AC sine wave).

Angular Velocity $\omega$ = [distance (in degrees or radians) ÷ time (in seconds)]

$\omega = 2\pi f$ (in radians/second); in other words the higher the frequency of the generated sine wave, the higher the angular velocity.

*Phase Relations*

$2\pi$ radians = 360° and 1 radian ≈ 57.3°

30° = $\pi/6$ radians ≈ 0.524 rad

90° = $\pi/2$ radians ≈ 1.571 rad

180° = $\pi$ radians ≈ 3.141 rad

270° = $3\pi/2$ radians ≈ 4.712 rad

General format of a sine wave = $A_m \sin \omega t$ with $\omega t$ as the horizontal unit of measure (Figure 4-30).

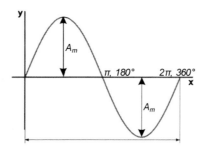

**Figure 4-30. General Format of a Sine Wave**

If the waveform is shifted to the right or left of 0°, the expression becomes:

$A_m (\sin \omega t \pm \theta)$

If the waveform passes through the horizontal axis with a positive going slope before 0° the expression is $A_m (\sin \omega t + \theta)$

If the waveform passes through the horizontal axia with a positive going slope after 0° the expression is **A$_m$ (sin ωt − θ)**

In Figure 4-31, the current waveform lags the voltage waveform. This is characteristic of an inductive circuit.

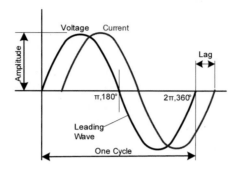

**Figure 4-31. Inductive Circuit**

*RMS Voltage*

$$V_{RMS} = \frac{V_{PEAK}}{\sqrt{2}} \therefore 0.707 x V_{PEAK} = V_{RMS} \qquad V_{RMS} = \sqrt{V_1 + V_2 + V_3 + ...V_N}$$

In general, electric power varies as voltage squared, ($P = V^2 \div R$) which is strictly non-negative. The average (mean) power can be computed using the mean squared voltage. Specifically, power is found using the square of the root mean squared voltage ($V_{RMS}$).

*Average Voltage*

$$V_{AVG} = \frac{2}{\pi} V_{PEAK} \therefore 0.636 x V_{PEAK} = V_{AVG}$$

As the name implies, $V_{avg}$ is calculated by taking the average of the voltage over an appropriately chosen interval. In the case of symmetrical waveforms like the sinewave, a quarter cycle faithfully represents all four quarter cycles of the waveform. Therefore, it is acceptable to choose the first quarter cycle, which goes from 0 radians (0°) through π/2 radians (90°).

*Capacitance*

The Farad (symbol: F) is the SI unit of capacitance. The term Farad is named after the English physicist Michael Faraday. A capacitor is a device that stores electrical energy in an electrostatic field. The energy is stored in such a way as to opposes any change in voltage. A farad is the charge in coulombs that a capacitor will accept for the potential across it to change 1 volt. A coulomb is 1 ampere second.

A simple capacitor (Figure 4-32) consists of two metal plates separated by an insulating material called a dielectric (Table 4-11).

## Table 4-11. Example Dielectric Constants

| Material | Constant | Material | Constant |
|---|---|---|---|
| Vacuum | 1.0000 | Rubber | 2.5 to 35 |
| Air | 1.0006 | Wood | 2.5 to 8 |
| Paraffin Paper | 3.5 | Glycerin (15°C) | 56 |
| Glass | 5 to 10 | Petroleum | 2 |
| Mica | 3 to 6 | Pure Water | 81 |

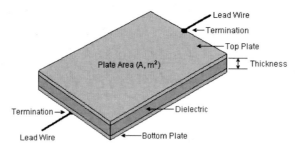

**Figure 4-32. Simple Capacitor**

$\varepsilon$ = dielectric constant    Q = Charge    W = Energy (work)

$$C = \frac{\varepsilon A}{d}$$

$$Q = (C)(V) \quad \frac{d_Q}{dt} = C\frac{dv}{dt} = I \quad W = \frac{CV_C^2}{2} = \frac{VQ}{2} \quad \textit{(Voltage cannot change instantaneously)}$$

Capacitance increases with area and decreases with separation.

**Example:** A 47 mA current causes the voltage across a capacitor to increase 1 Volt per second. It therefore has a capacitance of 47 mF. It has the base SI representation of $s^4 \cdot A^2 \cdot m^{-2} \cdot kg^{-1}$

$$\therefore F = \frac{(I)(t_{sec})}{E} \text{ AND } \frac{(I)(t_{sec})}{E} = \frac{C}{E} = \frac{s}{\Omega} \text{ Where C = Coulomb}$$

### Capacitive Reactance $X_C$

Unlike resistance, which has a fixed value (e.g. 100 $\Omega$, 1 k$\Omega$, 10 k$\Omega$ etc.), and obeys Ohm's Law, capacitive reactance varies with frequency (Figure 4-33). As the frequency applied to the capacitor increases, its reactance (measured in ohms) decreases. As the frequency decreases, its reactance increases. This is because electrons in the form of an electrical charge on the capacitor plates pass from one plate to the other more or less rapidly with respect to the varying frequency. As the frequency increases, the capacitor passes more charge across the plates in a given time. This results in a greater current flow through the capacitor, which appears as if the internal resistance of the capacitor has decreased. Therefore, a capacitor connected in an AC circuit can be said to be "frequency dependent" as indicated by the formula $Xc = 1 \div (2\pi fC)$.

In a purely capacitive circuit, the current $I_C$ "LEADS" the voltage by 90°.

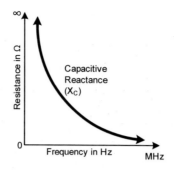

**Figure 4-33. Capacitive Reactance**

$$X_c = \frac{1}{2\pi fC}$$

Where:    $X_C$ = Capacitive Reactance (in Ohms)

f = Frequency (in Hertz)

C = Capacitance (in Farads)

## Inductance

The Henry (symbol: H) is the SI unit of induction. The term Henry is named after Joseph Henry, the American scientist who discovered electromagnetic induction. If the rate of change of current in a circuit is one ampere per second and the resulting electromotive force is one volt, then the inductance of the circuit is one Henry (unit is volt•second/ampere).

$$\therefore H = \frac{Wb}{A} = \frac{V \cdot S}{A} = \frac{J \cdot s^2}{C^2} = \Omega \cdot s \text{ Where C = Coulomb}$$

Inductance (symbol: L) is the property in an electrical circuit where a change in the electric current through that circuit induces an electromotive force (EMF) that opposes the change in current. Inductance does NOT oppose current, only a CHANGE in current. The unit for inductance is commonly abbreviated as Henry.

**Self-inductance** (Figure 4-34): The induction of a voltage in a current-carrying wire when the current in the wire is changing. In the case of self-inductance, the magnetic field created by a changing current in the circuit induces a voltage in the same circuit. Therefore, the voltage is self-induced.

**Mutual Inductance M:** When an EMF is produced in a coil because of the change in current in a magnetically coupled coil, the effect is called mutual inductance. The most common application of mutual inductance is the transformer.

**Lenz's Law:** An induced current is always in such a direction as to oppose the motion or change causing it. Lenz's law is named after Heinrich Lenz. This is the principle behind why an inductor opposes an instantaneous change in current (Figure 4-35).

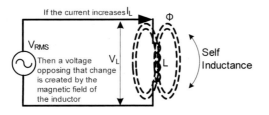

$$J = C \cdot V = N \cdot m = W \cdot s$$

**Figure 4-34. Self-inductance**

*Inductors Connected in Series:*

$$L_{TOTAL} = L_1 + L_2 + L_3 + ...L_N$$

*Inductors Connected in Parallel:*

$$\frac{1}{L_{TOTAL}} = \frac{1}{L_1} + \frac{1}{L_2} + \frac{1}{L_3} + ... \frac{1}{L_N}$$

## Inductive Reactance $X_L$

Just like capacitive reactance, inductive reactance varies with frequency. As the frequency applied to the inductor increases, its reactance (measured in ohms) increases. As the frequency decreases, its reactance decreases. This is because inductors drop voltage in proportion to the rate of current change. Therefore, they will drop less voltage for slower changing currents and more voltage for faster changing currents. Thus inductance reaction is directly proportional to the frequency change. When an alternating or AC voltage is applied across an inductor, the flow of current through it behaves very differently than with an applied DC voltage, by producing a phase difference between the voltage and the current waveforms. In an AC circuit, the opposition to current flow through a coil's windings not only depends upon the inductance of the coil, but also on the frequency of the AC waveform. The opposition to current flowing through a coil in an AC circuit is determined by the AC resistance of the circuit. But resistance is always associated with DC circuits, so to distinguish DC resistance from AC resistance, which is also known as Impedance, the term Reactance is used.

In a purely inductive AC circuit the current $I_L$ "LAGS" the applied voltage by 90°.

$$X_L = 2\pi fL$$

Where:     $X_L$ = Inductive Reactance (in Ohms)

f = Frequency (in Hertz)

L = Inductance (in Henries)

From the above equation for inductive reactance, it can be seen that if either the frequency or inductance was increased, the overall inductive reactance value would also increase. As the frequency approaches infinity, the inductor's reactance would also increase to infinity, acting like an open circuit. However, as the frequency approaches zero or DC, the inductor's reactance would decrease to zero, acting like a short circuit. This means that inductive reactance is proportional to frequency and is small at low frequencies and high at higher frequencies.

In the purely inductive circuit, the inductor is connected directly across the AC supply voltage. As the supply voltage increases and decreases with the frequency, the self-induced back EMF also increases and decreases in the coil with respect to this change. We know that this self-induced EMF is directly proportional to the rate of change of the current through the coil and is at its greatest as the supply voltage crosses over from its positive half cycle to its negative half cycle or vice versa at points 0° and 180° along the sine wave. Consequently, the smallest voltage change occurs when the AC sine wave crosses over at its maximum or minimum peak voltage level. At these positions in the cycle, the maximum or minimum currents are flowing through the inductive circuit.

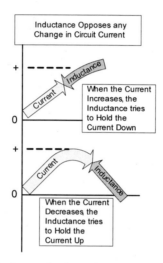

**Figure 4-35. Inductor's Opposition to Current Change**

## Table 4-12. Impedance of Some Common AC Circuits

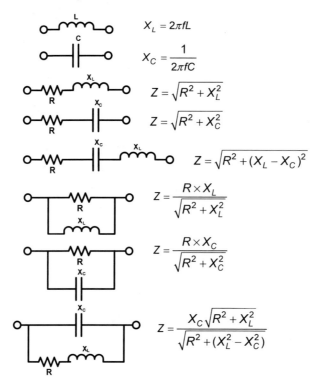

$$X_L = 2\pi f L$$

$$X_C = \frac{1}{2\pi f C}$$

$$Z = \sqrt{R^2 + X_L^2}$$

$$Z = \sqrt{R^2 + X_C^2}$$

$$Z = \sqrt{R^2 + (X_L - X_C)^2}$$

$$Z = \frac{R \times X_L}{\sqrt{R^2 + X_L^2}}$$

$$Z = \frac{R \times X_C}{\sqrt{R^2 + X_C^2}}$$

$$Z = \frac{X_C \sqrt{R^2 + X_L^2}}{\sqrt{R^2 + (X_L^2 - X_C^2)}}$$

### *RC Time Constants*

**RC Time Constant:** The time required to charge a capacitor to 63% (actually 63.2%) of full charge or to discharge it to 37% (actually 36.8%) of its initial voltage is known as the TIME CONSTANT (TC) of the circuit. The charge and discharge curves of a capacitor are shown in Figure 4-36.[72]

$$t(\text{seconds}) = \left[R('\Omega')\right] \times \left[C('F')\right]$$

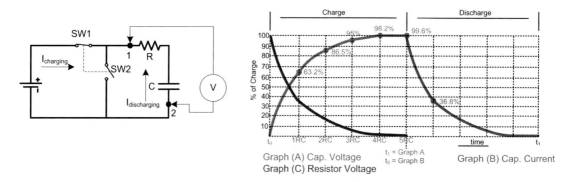

**Figure 4-36. RC Time Constant**

---

72.  Image obtained from www.interfacebus.com.

**Charge /Discharge Cycle (RC Circuit):** An RC series circuit is a circuit that includes only resistors and capacitors, along with a current or voltage source. In explaining the charge and discharge cycles of an RC series circuit, the time interval from time $t_0$ (time zero, when the switch is first closed) to time $t_1$ (time one, when the capacitor reaches full charge or discharge potential) will be used. (Note that switches SW1 and SW2 move at the same time and can never both be closed at the same time.)

When switch SW1 of the circuit in Figure 4-36 is closed at $t_0$, the source voltage ($E_S$) is instantly felt across the entire circuit. Graph (A) of the figure shows an instantaneous rise at time $t_0$ from zero to source voltage ($E_S$) (i.e., charging). The total voltage can be measured across the circuit between points 1 and 2. Graph (B) then represents the discharging current in the capacitor ($i_c$).

At time $t_0$, the charging current is at maximum. As time elapses toward time $t_1$, there is a continuous decrease in current flowing into the capacitor. The decreasing flow is caused by the voltage buildup across the capacitor.

At time $t_1$, current flow in the capacitor stops. At this time, the capacitor has reached full charge and has stored maximum energy in its electrostatic field. Graph (C) represents the voltage drop ($e_r$) across the resistor (R). The value of $e_r$ is determined by the amount of current flowing through the resistor on its way to the capacitor.

At time $t_0$ the current flowing to the capacitor is at maximum. Thus, the voltage drop across the resistor is at maximum ($E = IR$). As time progresses toward time $t_1$, the current flowing to the capacitor steadily decreases and causes the voltage developed across the resistor (R) to steadily decrease.

When time $t_1$ is reached, current flowing to the capacitor is stopped and the voltage developed across the resistor has decreased to zero.

### 4.3.3 Voltage Drop

*DC*

$$V_d = \left(\frac{2 \times L}{1000}\right) \times I \times R$$

*AC*

**Single Phase:**

$$V_d = \left(\frac{2 \times L}{1000}\right) \times I \times Z_e$$

$V_d$ = voltage drop; $Z_e$ = equivalent impedance; $L$ = conductor length; $I$ = current
(*Note: Ze with Power Factor = 100% is equal to DC resistance*)

**Three Phase:**

$$V_d = \left( \frac{\sqrt{3} \times L}{1000} \right) \times I \times Z_e \quad \textit{(Note: Ze with PF = 100\% is equal to DC resistance)}$$

## 4.3.4 Cable Sizing

**AC Single Phase:**

$$cm = \left( \frac{2 \times L \times I \times k}{V_d} \right) \quad cm = circular \; mil$$

1cm = area of a circle width a diameter of 1mil (0.001")

**AC Three Phase:**

$$cm = \left( \frac{\sqrt{3} \times L \times I \times k}{V_d} \right)$$

*k = specific resistance of copper = 12Ω•cmil per foot (for 75°C, most common conductor temperature rating))*

## 4.3.5 Electrical Formulas for Calculating Amps, HP, KW & KVA (Table 4-13)

### Table 4-13. Electrical Formulas for Calculating Amps, HP, KW and KVA

| To Find: | DC | AC | | |
|---|---|---|---|---|
| | | **1-Phase** | **2-Phase; 4-Wire** | **3-Phase** |
| **Amps** when **HP** known | $\dfrac{HP \times 746}{E \times Eff.}$ | $\dfrac{HP \times 746}{E \times Eff. \times PF}$ | $\dfrac{HP \times 746}{E \times Eff. \times PF \times 2}$ | $\dfrac{HP \times 746}{E \times Eff. \times PF \times \sqrt{3}}$ |
| **Amps** when **KW** known | $\dfrac{KW \times 1000}{E}$ | $\dfrac{KW \times 1000}{E \times PF}$ | $\dfrac{KW \times 1000}{E \times PF \times 2}$ | $\dfrac{KW \times 1000}{E \times PF \times \sqrt{3}}$ |
| **Amps** when **KVA** known | | $\dfrac{KVA \times 1000}{E}$ | $\dfrac{KVA \times 1000}{E \times 2}$ | $\dfrac{KVA \times 1000}{E \times \sqrt{3}}$ |
| KW | $\dfrac{E \times I}{1000}$ | $\dfrac{E \times I \times PF}{1000}$ | $\dfrac{E \times I \times PF \times 2}{1000}$ | $\dfrac{E \times I \times PF \times \sqrt{3}}{1000}$ |
| KVA | | $\dfrac{E \times I}{1000}$ | $\dfrac{E \times I \times 2}{1000}$ | $\dfrac{E \times I \times \sqrt{3}}{1000}$ |
| HP | $\dfrac{E \times I \times Eff.}{746}$ | $\dfrac{E \times I \times Eff. \times PF}{746}$ | $\dfrac{E \times I \times Eff. \times PF \times 2}{746}$ | $\dfrac{E \times I \times Eff. \times PF \times \sqrt{3}}{746}$ |

# 5. Final Control Elements

## 5.1 Control Valves

### 5.1.1 Selection Guide

**Table 5-1. Control Valve Selection Guide**

| Conveying Media | Nature of Media | Valve Function | Type of Plug/Disc |
|---|---|---|---|
| Liquid | Neutral (Water, Oil, etc.) | On / Off | Gate |
| | | | Rotary Ball |
| | | | Plug |
| | | | Diaphragm |
| | | | Butterfly |
| | | | Plug Gate |
| | | Modulating | Globe |
| | | | Butterfly |
| | | | Plug Gate |
| | | | Diaphragm |
| | | | Needle |
| | Corrosive (Acid, Alkaline, etc.) | On / Off | Gate |
| | | | Rotary Ball |
| | | | Plug |
| | | | Diaphragm |
| | | | Butterfly |
| | | | Plug Gate |
| | | Modulating | Globe |
| | | | Diaphragm |
| | | | Butterfly |
| | | | Plug Gate |
| | Hygenic (Food, Beverages, Drugs, etc.) | On / Off | Butterfly |
| | | | Diaphragm |
| | | Modulating | Butterfly |
| | | | Diaphragm |
| | | | Squeeze |
| | | | Pinch |
| | Slurry | On / Off | Rotary Ball |
| | | | Butterfly |
| | | | Diaphragm |
| | | | Plug |
| | | | Pinch |
| | | | Squeeze |
| | | Modulating | Butterfly |
| | | | Diaphragm |
| | | | Squeeze |
| | | | Pinch |
| | | | Gate |
| | Fibrous Suspensions | On / Off Modulating | Gate |
| | | | Diaphragm |
| | | | Squeeze |
| | | | Pinch |

| Conveying Media | Nature of Media | Valve Function | Type of Plug/Disc |
|---|---|---|---|
| Gas | Neutral (Air, Steam, etc.) | On / Off | Gate |
| | | | Globe |
| | | | Rotary Ball |
| | | | Plug |
| | | | Diaphragm |
| | | Modulating | Globe |
| | | | Needle |
| | | | Butterfly |
| | | | Diaphragm |
| | | | Gate |
| | Corrosive (Acid Vapors, Chlorine, etc.) | On / Off | Butterfly |
| | | | Rotary Ball |
| | | | Diaphragm |
| | | | Plug |
| | | Modulating | Butterfly |
| | | | Globe |
| | | | Needle |
| | | | Diaphragm |
| | Vacuum | On / Off | Gate |
| | | | Globe |
| | | | Rotary Ball |
| | | | Butterfly |
| Solids | Abrasive Powder (Silica, etc.) | On / Off Modulating | Pinch |
| | | | Squeeze |
| | Lubricating Powder (Graphite, Talcum, etc.) | On / Off Modulating | Pinch |
| | | | Gate |
| | | | Squeeze |

### 5.1.2 Control Valve Inherent Flow Characteristics

The inherent flow characteristic is the relationship between the valve flow coefficient $C_v$ and the valve travel. These flow characteristics not only apply to sliding stem (i.e., up and down movement) style control valves, but to rotary style control valves as well. The three major flow characteristics (Figure 5-1) are described below. Figure 5-2 shows the corresponding valve plug contours.

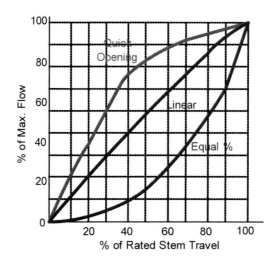

**Figure 5-1. Inherent Flow Characteristics**

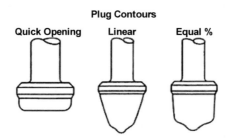

**Figure 5-2. Three Plug Contours**

*Quick Opening* flow characteristic provides a maximum change in flow rate at low plug travel, and small changes when the plug travel is near maximum. The curve is basically linear through the first 40% of plug travel, and then flattens out to indicate little increase in flow rate as plug travel approaches the wide open position.

Control valves with a quick opening flow characteristic are typically used for on/off type applications where a significant flow rate must be established quickly as the valve begins to open.

*Linear* flow characteristic shows that the flow rate is directly proportional to the valve travel (i.e., 50% of rated travel = 50% of maximum flow). This proportional relationship produces a characteristic with a constant slope so that with constant pressure drop, the steady state valve gain[73] will be constant.

Control valves with a linear flow characteristic are commonly applied for liquid level control and for certain flow control applications requiring constant valve gain over the full range of valve opening.

---

73. Valve gain is the slope of the valve characteristic curve.

*Equal %* flow characteristic: equal increments of valve travel produce equal % changes in the flow rate. This means that the change in the flow rate is exponential rather than linear.

Control valves with equal % flow characteristic are used on fast processes that require quick valve response (e.g., pressure and flow control) and on applications where a large % of the pressure drop is normally absorbed by the system itself with only a relatively small % of pressure drop available at the valve. It is the most commonly used flow characteristic.

### 5.1.3 Control Valve Shutoff (Seat Leakage) Classifications

As defined by ANSI/FCI 70-2 1976(R1982) (Table 5-2).

**Table 5-2. Valve Shutoff Classifications**

| Class I | No requirements |
|---------|-----------------|
| Class II | 0.05% of rated valve capacity |
| Class III | 0.1% of valve rated capacity |
| Class IV | 0.01% of valve rated capacity |
| Class V | $5 \times 10^{-4}$ ml/min of water per inch of seat diameter<br>This is the best expected of a metal seat valve |
| Class VI | The numbers range from:<br>  0.15 ml/min for a 1" valve to<br>  6.75 ml/min for an 8" valve<br>Soft Seat, which will have temperature and pressure limitations. |

### 5.1.4 Control Valve Choked Flow/Cavitation/Flashing

*Choked Flow:* With normal flow in liquids, lowering the downstream pressure will result in an increased flow rate through the valve, until such velocity is reached that partial vaporization of the liquid occurs (cavitation/flashing). When this partial vaporization pressure point is reached, the flow will no longer increase, regardless of how much more the downstream pressure is reduced (thus choked flow).

With regard to gas flow, the choked flow condition is reached when the outlet gas velocity reaches sonic velocity conditions.

*Cavitation:* If the speed through the valve is high enough, the pressure in the liquid may drop to a level where the liquid starts to bubble (Figure 5-3). The bubbles form as a result of a drop in fluid pressure to below the vapor pressure for that liquid, and the liquid actually begins to boil. When the pressure recovers sufficiently to be above the liquid's vapor pressure, the boiling effect will stop and the bubbles will collapse upon themselves.

When the bubbles collapse, they release a large amount of energy, which results in localized shock waves and liquid microjets within the valve body. These localized shock waves and microjets will cause significant pitting and erosion damage to the inner parts of the valve.

Cavitation is noisy, and in addition to the physical damage it causes to the valve (Figure 5-4), the noise may create a dangerous sound level condition for personnel nearby (it may sound like gravel flowing through the pipe).[74]

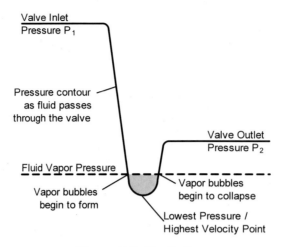

Valve Inlet
Pressure P$_1$

Pressure contour
as fluid passes
through the valve

Valve Outlet
Pressure P$_2$

Fluid Vapor Pressure

Vapor bubbles
begin to form

Vapor bubbles
begin to collapse

Lowest Pressure /
Highest Velocity Point

**Figure 5-3. Cavitation**

Cavitation Damaged Plug

**Figure 5-4. Cavitation Damage (1)**

*Flashing:* This is very similar to cavitation. The difference between flashing and cavitation is that with flashing the downstream pressure does not recover enough to be above the liquid's vapor pressure. The vapor bubbles do not collapse and they remain in the liquid as vapor (Figure 5-5). This results in 2-phase flow within the valve and downstream of the valve, which causes erosion damage (the "polishing effect") (Figure 5-6).

## 5.1.5 Control Valve Noise

Sources of control valve noise:

- Mechanical Noise: Mechanical vibration of the valve components

- The flow itself

    - Hydrodynamic Noise (Liquid Flow): Cavitation & Flashing
    - Aerodynamic Noise (Gas Flow): Generated as a by-product of a turbulent gas stream.

---

74. From Masoneilan (Dresser)

---

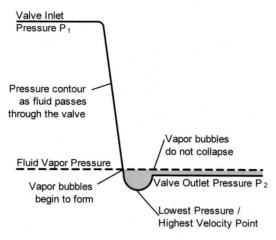

Figure 5-5. Flashin

Flashing Damaged Plug

Figure 5-6. Flashing Damage (1)

*OSHA Environmental Noise Classification*

### Table 5-3. Example OSHA Environmental Noise Classifications

| Noise Type | Sound Level Pressure |
|---|---|
| Weakest sound heard | 0 dB |
| Quiet whisper | 30 dB |
| Normal conversation at 3–5 ft | 60–70 dB |
| Vacuum cleaner | 60–80 dB |
| City traffic from inside car | 85 dB |
| Train whistle at 500 ft. or truck traffic | 90 dB |
| **Level at which exposure may result in hearing loss** | **90–95 dB** |
| Motorcycle (unmuffled) | 100 dB |
| Snowblower | 105 dB |
| Power lawn mower at 3 ft | 107 dB |
| Power saw at 3 ft. | 110 dB |
| Sandblasting | 115 dB |
| Thunder | 120 dB |
| **Pain begins** | **125 dB** |
| Chain saw | 125 dB |

| Short term exposure without hearing protection can cause permanent hearing damage | 140 dB |
|---|---|
| Jet taking off | 150 dB |
| Death of hearing tissue | 180 dB |
| Loudest sound possible | 194 dB |

OSHA has set a noise-exposure limit of a weighted 90-dBA maximum over 8 hours (Table 5-4). 85-dBA is the accepted maximum for control valve noise.

*OSHA Daily Permissible Noise Level Exposure*

### Table 5-4. OSHA Daily Permissible Noise Level Exposure

| Hours per Day | Sound Level Pressure |
|---|---|
| 8 | 90 dBA |
| 6 | 92 dBA |
| 4 | 95 dBA |
| 3 | 97 dBA |
| 2 | 100 dBA |
| 1.5 | 102 dBA |
| 1 | 105 dBA |
| ½ | 110 dBA |
| ≤ ¼ | 115 dBA |

Control valve noise may be remediated by source treatment, path treatment, or both.

- Source treatment reduces or attenuates the noise at its source and is the desired method. It typically uses some form of cage treatment (Figures 5-7 and 5-8).

- Path treatment does nothing to address the noise source, but rather attempts to attenuate the noise transmission from the source to the receiver (human ear). Path treatment methods include heavy wall pipe, extra acoustical insulation, and diffusers/silencers/mufflers.

The IEC noise standard 534-8-3[75] recognizes three categories of noise reducing trim for control valves:

---

75. IEC Standard 534-8-3 is used to calculate a sound pressure level for aerodynamic noise produced by a control valve.

- Single stage, multiple flow passage trim (tortuous path) (Figure 5-7[76]). Tortuous path indicates full of twists and turns so as to slow down the exit velocity of the fluid/gas.

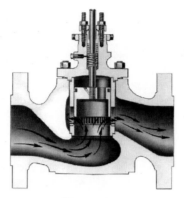

**Figure 5-7**

- Single flow path, multi-stage pressure reduction trim (Figure 5-8[77])

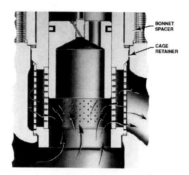

**Figure 5-8**

- Multiple flow passage, multi-stage trim (Figure 5-9[78])

**Figure 5-9**

---

76. Emerson (Fisher) Whisper Trim I.
77. Emerson (Fisher) Whisper Trim III.
78. Emerson (Fisher) WhisperFlo Trim.

"Single stage" means that the flowing fluid goes from the upstream pressure condition at the valve inlet to the downstream pressure condition at the valve outlet in one step or stage. This is the typical arrangement in most conventional control valves.

"Multiple Stage" (i.e., multiple flow passage) means that the flowing fluid, in going from the valve inlet to the valve outlet, passes through several flow openings rather than just one opening. The flow passages must be sufficiently separated in distance so there is no interaction between the microjets emanating from each flow opening.

### 5.1.6 Control Valve Plug Guiding

Control valve plug guiding is used in control valves to ensure that the control valve's plug and seat are always properly aligned with each other.

*Cage Guiding:* Cage guiding of the valve plug provides for stable control at high pressure drops. Cage guiding also reduces vibration and mechanical noise. The outside surface of the valve plug is close to the inside wall surface of the cylindrical cage throughout the travel range. This ensures correct valve plug/seat ring alignment when the valve closes. However, due to these close metal-to-metal tolerances, the most common maintenance problem with cage guiding is galling and sticking. The walls of the cage (Figure 5-10[79]) contain openings that can determine the flow characteristic of the control valve.

**Quick Opening**     **Linear**     **Equal %**

**Figure 5-10. Control Valve Cage Guides**

*Top Guiding:* The valve plug is aligned by a single guide in the bonnet or valve body or by packing arrangement. Top guiding minimizes the weight effect of the valve stem on the valve and increases resistance to trim vibration.

*Top-and-Bottom Guiding:* The valve plug is aligned by guides in the bonnet/body and the bottom flange. This is a modification of the top guided design in order to improve the valve's rigidity and its ability to withstand higher flow rates and larger pressure drops.

*Stem Guiding:* The valve plug is aligned with the seat ring by a guide in the bonnet that acts on the valve plug stem. This design is yet another variation on the top guided design.

*Port Guiding:* The valve plug is aligned by the valve body port. This construction is typical for valves that use small diameter plugs with fluted skirt projections to control low flow rates.

---

79. Emerson Process (Fisher) characterized cages for globe style control valves.

---

## 5.1.7 Control Valve Packing

The packing box (aka the stuffing box) is the chamber in a valve bonnet that surrounds the valve stem (Figure 5-11). It contains the packing that seals against leakage around the valve plug stem. The packing normally consists of deformable materials that include PTFE, graphite (Grafoil), Kalrez, glass fiber, etc. The material is commonly in the form of solid or split rings. Most control valves use packing boxes, with the packing retained and adjusted by a flange and stud bolts. Packing must be largely impermeable to the process fluid that it is required to seal.

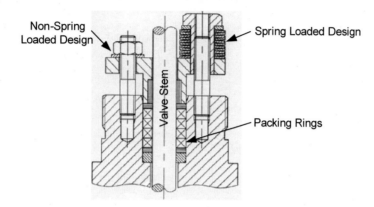

**Figure 5-11. Control Valve Packing**

**Live-Loaded Packing (internal and external):** This type uses a Belleville washer design (Figure 5-12[80]) to spring load the packing, which provides continuous pressure against the packing even when packing volume is lost by friction, extrusion, etc. and allows the packing to remain leak free without the need to periodically tighten the flange bolts.

**Figure 5-12. Belleville Washers**

*Packing Types*

For packing types available materials of construction reference packing temperature limitations section.

**V-Ring (Chevron):** These packings are molded V-shaped rings (Figure 5-13[81]) that are spring loaded and self-adjusting in the packing box. Packing lubrication is not required.

---

80. Image obtained from www.meadinfo.org.

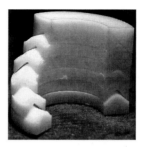

**Figure 5-13. V-Ring Packing**

**Fibers and Braids:** These packings (Figure 5-14[82]) provide leak free operation, high thermal conductivity, and long service life. *However,* this type of packing produces high stem friction and resultant hysteresis.[83] Packing lubrication is not required, but an extension bonnet or steel yoke should be used when packing temperature exceeds 800°F.

**Figure 5-14. Braided Fiber Packing**

*Packing Temperature Limitations*

**PTFE** (polytetrafluoroethylene aka Teflon[84]): This material should not be exposed to temperatures higher than 450°F (232°C). Note: PTFE is destroyed by radiation, therefore it should not be used in nuclear service.

**Graphite** (aka Grafoil[85]): Grafoil is a flexible material made from pure natural graphite flakes. It should not be exposed to temperatures higher than 750°F (232°C) when used in an oxidizing service application. When it is used in a non-oxidizing service application, the maximum temperature limit is 1200°F (649°C). If it is used in nuclear service, it must be 99.5% graphite at minimum.

**Kalrez** (perfluoroelastomer): Kalrez[86] is similar to PTFE, except that it can withstand higher temperatures and does not have the cold flow tendency that PTFE has. It should not be exposed to temperatures higher than 620°F (327°C).

---

81.  Image obtained from www.parksideind.com.
82.  Image obtained from www.supplierlist.com.
83.  Hysteresis is the tendency of the position of the valve to be dependent on the previous position of the valve when reacting to open/close movement.
84.  Teflon is a registered trademark of DuPont.
85.  Grafoil is a registered trademark of GrafTech International.
86.  Kalrez is a registered trademark of DuPont.

---

**Ceramic Fiber:** This is a replacement for asbestos in high temperature applications as it may be used up to 1800°F (982°C).

## 5.1.8 Control Valve Bonnets

The bonnet normally provides a means of mounting the actuator to the body and houses the packing box. Generally, rotary-style valves do not have bonnets.

A bolted flange bonnet (Figure 5-15[87]) is the most common type of bonnet. The bonnet may also be a screwed-in bonnet or a slip-on flange held in place with a split ring. Specialty bonnet extensions are available for use in either extremely high or extremely low temperature applications, and serve to protect the valve stem packing from extreme process temperatures. Bellows seal type bonnets should be used when no leakage along the stem can be tolerated (e.g., $< 10 \times 10^{-6}$ cc/sec of helium). They are often used when the process fluid is toxic, volatile, radioactive, or highly expensive.

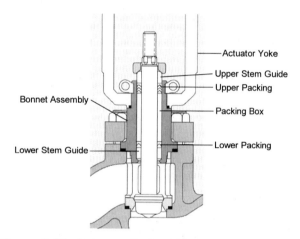

**Figure 5-15. Cutaway Bolted Flange Bonnet Valve**

## 5.1.9 Control Valve Body Styles

*Sliding Stem*

**Globe:** This is the most popular control valve body style in use today. The globe valve is available in both single-port and double-port design, with the single-port style accounting for approximately 75% of the globe valves in use. Single-port design have improved shutoff capabilities over those of the double-port design. However, the double-port design requires less actuator force to close open or close the valve. Normal flow direction is up through the seat ring, which has a self-flushing effect (it does not allow small particulates to accumulate in the bottom pocket of the valve body).

---

87.  Valtek (Flowserve) Mark One bonnet.

**Split Body Globe Valve:** A valve body that is made of two separate parts, with the seat ring sandwiched between the two body parts (Figure 5-16[88]). The advantage of this style over the single casting type body is compactness. This design also allows for easier inspection of the plug and seat ring. In addition, the outlet portion of the valve may be rotated in 90° intervals.

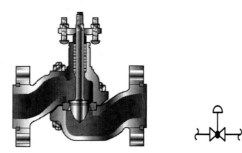

**Figure 5-16. Split Body Globe Valve with Symbol**

**Angle:** This style is typically reserved for severe service type applications such as erosive/dirty or high velocity flow conditions. The angle valve design (Figure 5-17[89]) also allows self-draining.

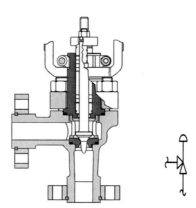

**Figure 5-17. Angle Valve with Symbol**

**3-Way:** This design (Figure 5-18[90]) is an extension of the double-ported globe valve. It is available in two configurations: diverting and mixing. Diverting is switching the flow from one outlet to the other. Mixing is bringing two separate flows together into one common output. The primary application for this style of globe valve is temperature control.

**Gate (Knife):** This style (Figure 5-19[91]) operates similarly to a guillotine. The valve opens by lifting a round or rectangular gate/wedge out of the path of the fluid. The distinct feature of a gate valve is that the sealing surfaces between the gate and seats are planar. Typically, gate valves are for on/off service.

---

88. Masoneilan (Dresser) split body globe valve.
89. Flowserve angle valve.
90. Emerson (Fisher) 3-way control valve.
91. DeZURIK knife gate valve.

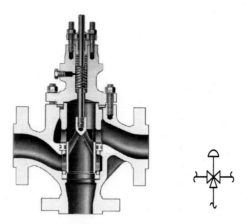

**Figure 5-18. 3-Way Valve with Symbol**

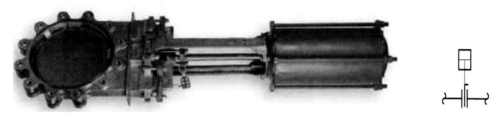

**Figure 5-19. Gate Valve with Symbol**

*Rotary Stem*

**Butterfly:** The butterfly valve design is basically a metal disk attached to a shaft that rotates 90°, which in turn positions the metal disk either parallel to the flow (open) or perpendicular the flow (closed). This style (Figure 5-20[92]) is usually not applied in pipelines below 3 inches, but is heavily used on larger diameter pipes because butterfly valves are cost friendly in larger sizes in comparison to other valve body styles. As mentioned, butterfly valves are mostly used in on/off type applications, though they may be used as control valves by using contoured disks. Butterfly valves should not be used for control-type applications where tight control is required. If a butterfly valve is required to shut off against high pressure, then care should be taken to over-size the actuator accordingly.

**Figure 5-20. Butterfly Valve with Symbol**

---

92.  Fisher butterfly valve.

**Ball:** The ball valve design utilizes a sphere (ball) (Figure 5-21[93]) attached to a shaft that rotates 90°, which in turn makes the hole in the ball either parallel to the flow (open) or perpendicular to the flow (closed). The ball has a hole drilled in the middle that may either have the same diameter as the pipeline internal diameter (full-bore) or a reduced diameter bore to allow for one of the pipelines (either upstream or downstream) to be one size smaller than the other. By far, the most common hole size is full-bore.

**Figure 5-21. Valve Ball with Ball Valve Symbol**

Because of the hole within the ball, if a ball valve is shut with fluid in the line, this fluid will become trapped within the hole space internal to the valve. As a result, care should be taken to avoid thermal expansion of this trapped fluid to eliminate the possibility of seal damage. Like butterfly valves, ball valves are used mainly in on/off type applications. If a ball valve is to be used in control applications, then the better design is the segmented ball valve.

**Segmented Ball:** This style (Figure 5-22) is similar to a ball valve but the v-notch allows for better control capabilities and produces an equal % flow characteristic. It has better shutoff capability than that of a globe valve. This design also does not trap media within the ball body when the valve is closed, as the traditional full sphere ball valve design does.

**Figure 5-22: Segmented Ball Valve with Symbol**

**Plug:** This style is similar in operation to ball valve except that in lieu of a sphere-shaped plug the design uses a cylindrical or conical-shaped plug. The plug has a port in it that allows fluid to flow when the port is parallel with the flow path. The plug valve (Figure 5-23[94]) is typically used for on/off service.

**Eccentric Plug:** This style (Figure 5-24[95]) is similar in operation to a plug valve. The difference between a standard plug valve and an eccentric plug valve is that the plug is not cylindrical in shape and the rotational operation is not concentric but rather eccentric. Whereas the cylindrical plug valve is primarily used in on/off applications, the eccentric plug may be used in control

---

93. Image obtained from www.dwvalve.com.
94. Durco (Flowserve) plug valve.
95. Emerson (Fisher) eccentric plug valve.

**Figure 5-23. Plug Valve with Symbol**

applications as well. Eccentric plugs also offer better shutoff capability than that of butterfly valves.

**Figure 5-24. Eccentric Plug Valve with Symbol**

**3-Way Ball:** This style is similar in operation to a 2-way ball valve (which was described above), except that the ball will have either an "L" or "T" shaped hole in the ball (Figure 5-25). The "L" shaped ball is used more for diverting applications, and the "T" shaped ball is used more for mixing/diverting applications.

**Figure 5-25. 3-Way Ball Valve Variations with 3-Way Valve Symbol**

*Special Purpose Valves*

**Needle Valve:** Needle valves are similar in design and operation to globe valves. Instead of a globe-shaped valve plug, the needle valve has a long tapered point at the end of the valve stem and a relatively small orifice seat (Figure 5-26[96]). Typically, this is a manually operated valve, and is used for fine tuning (precise) control.

**Pinch Valve:** This style employs a flexible tube liner within the valve body (Figure 5-27). When a compressed gas is applied externally to the tube, the pressure causes the tube to pinch off and close. When the pressure is released, the flexible liner returns to its original shape. Another style

---

96. Swagelok needle valve.

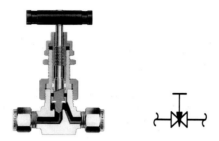

**Figure 5-26. Needle Valve with Symbol**

(shown in Figure 5-27[97]) uses a pneumatic actuator to pinch off the tube liner. Pinch valves are typically used in on/off solids-containing process fluid applications.

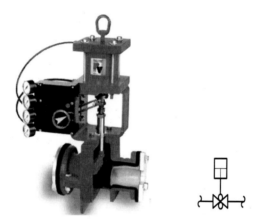

**Figure 5-27. Pinch Valve with Symbol**

**Diaphragm Valve:** This style consists of a flexible diaphragm and a valve body with either a straight-way or weir design (Figure 5-28[98]). To close the valve, the diaphragm is pressed against the valve body or weir to pinch off the flow. The valve body may have two or more ports and is available in multiple configurations. Their application is generally as on/off valves in process systems within the food and beverage, pharmaceutical and biotech industries.

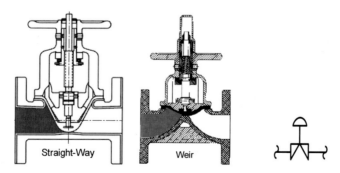

**Figure 5-28. Two Types of Diaphragm Valve with Diaphragm Valve Symbol**

---

97. Red Valve pinch valve.
98. ITT diaphragm valve.

**Mixproof Valve:** This type provides the functionality of a double block and bleed[99] within a single unit. Mixproof valves are generally used in process systems within the food and beverage, pharmaceutical, and biotech industries. Each valve portion has two separate flow paths, separated by a pair of seats (Figure 5-29[100]). This allows the process fluid and CIP (clean in place) or SIP (steam in place) fluid to flow through the same valve assembly.

**Figure 5-29. Mixproof Valve**

## 5.1.10 Common Valve Trim Material Temperature Limits

**Table 5-5. Common Valve Trim Material Temperature Limits**

| Material | Lower | | Upper | |
|---|---|---|---|---|
| | °F | °C | °F | °C |
| 304 SS | −450 | −268 | 600 | 316 |
| 316 SS | −450 | −268 | 600 | 316 |
| 317 SS | −450 | −268 | 600 | 316 |
| 416 SS | −20 | −29 | 800 | 427 |
| 440 SS | −20 | −29 | 800 | 427 |
| 17-4 PH (SS) | −80 | −62 | 800 | 427 |
| Monel | −325 | −198 | 800 | 427 |

99.   Equivalent double block and bleed:

100. Tuchenhagen mixproof valve.

| Hastelloy B and C | −325 | −198 | 800 | 427 |
|---|---|---|---|---|
| Titanium | −75 | −59 | 600 | 316 |
| Nickel | −325 | −198 | 600 | 316 |
| Alloy 20 | −325 | −198 | 600 | 316 |

### 5.1.11 Control Valve Installation

- The process flow direction must match the flow arrow on the control valve body.

- Most control valves are installed with the actuator facing vertically up. If the installation requires the actuator to be in a horizontal plane, then sufficient vertical support is required for the actuator. Valves should not be installed with the actuator facing down.

- The piping fittings must be properly aligned with the valve end fittings to ensure that there is no piping stress on the valve.

- The pipeline should be checked for the presence of foreign material before the installation of a control valve, because foreign material may obstruct the flow, prevent tight shutoff, and damage the valve seat.

## 5.2 ACTUATORS

An actuator is a device that uses some form of energy to operate a valve. The most common actuator movement types are linear and rotary. The most common forms of energy used to power valve actuators are compressed air, hydraulic fluid, and electricity. An actuator may be equipped with limit switches for end-of-travel feedback, or a linear position feedback device.

### 5.2.1 Failure State

Actuators may be configured to have one of the following failure states:

*Fail Open:* This is an actuator style where the valve is a normally-open valve and an external energy source is required to close the valve. When the external energy source is removed, the valve will return to the open position (Figure 5-30).

OR

**Figure 5-30. Fail Open Actuator Symbols**

*Fail Closed:* This is an actuator style where the valve is a normally-closed valve and an external energy source is required to open the valve. When the external energy source is removed, the valve will return to the closed position (Figure 5-31).

**Figure 5-31. Fail Closed Actuator Symbols**

*Fail Last Position:* With this actuator style, the valve will stay in its current position when the external energy source is removed (Figure 5-32).

**Figure 5-32. Fail Last Position Actuator Symbols**

## 5.2.2 Action

Actuators may be single-acting or double-acting:

*Single-Acting:* A single-acting actuator requires an external energy source to either open or close the valve, but the valve will return on its own to its appropriate fail position (usually by spring return) (Figure 5-33).

**Figure 5-33. Single-acting Actuator Symbol**

*Double-Acting:* A double-acting actuator requires an external energy source to move the valve to the open position, as well as back to the closed position. This design is typically used on fail-last-position valves (Figure 5-34).

**Figure 5-34. Double-acting Actuator Symbol**

The difference between direct and reverse acting:

*Direct Acting:* An external energy source is required to <u>extend</u> the valve stem (Figure 5-35). This arrangement is typically associated with "Fail Open" valves. However, whether direct or reverse acting is dependent upon the design of the valve body and how the stem plug and seat interact with each other (whether push up or down to open or close).

*Reverse Acting:* The external energy source is required to <u>retract</u> the valve stem (Figure 5-35[101]). This arrangement is typically associated with "Fail Closed" valves. However, whether direct or reverse acting is dependent upon the design of the valve body and how the stem plug and seat interact with each other (whether push up or down to open or close).

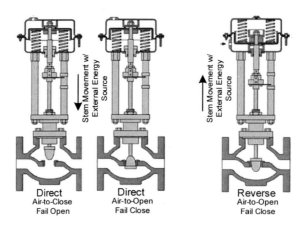

**Figure 5-35. Direct-acting and Reverse-acting Actuators**

## 5.2.3 Valve Positioner

A valve positioner senses the valve opening through a position feedback link connected to the valve stem (note: an I/P converter does not have this feedback link). Based on the feedback, the positioner adjusts the output to the actuator, if required. Positioners are used on valves where the following is desired:

- Precise positioning

- Limiting the effects of valve hysteresis

- Faster speed of response to changes in process conditions

- Split ranging (reference Section 6.2.6 for a description of split ranging)

## 5.2.4 Actuator Types

*Spring & Diaphragm:* This type consists of a flexible diaphragm with pre-compressed springs within a pressurized casing (Figure 5-36[102]). When compressed air is applied to the diaphragm, the diaphragm will extend, causing the valve stem to extend or retract. Whether the valve stem extends or retracts is dependent upon which side of the diaphragm the compressed air is applied to. When the compressed air is applied to the top of the diaphragm, the valve stem will extend, and when it is applied to the bottom of the diaphragm, the valve stem will retract. When the compressed air is removed, the spring will drive the valve stem to its fail position. This is the type of actuator that is most widely used today.

---

101. From Spirax Sarco.
102. Emerson (Fisher) spring & diaphragm actuators.

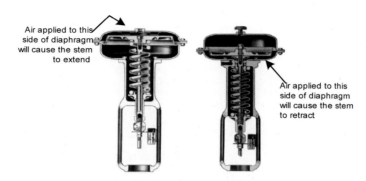

**Figure 5-36. Spring & Diaphragm Actuator**

Air applied to this side of diaphragm will cause the stem to extend

Air applied to this side of diaphragm will cause the stem to retract

*Piston:* This type (Figure 5-37[103]) is similar in operation to a spring and diaphragm style actuator. This design incorporates a sliding piston within a cylindrical seal, as opposed to a flexible diaphragm. It is better suited for higher actuating pressures than a spring and diaphragm type actuator.

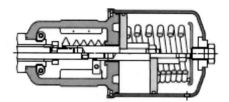

**Figure 5-37. Piston Actuator**

*Rotary Vane:* This type consists of a chamber with a movable vane connected to a drive shaft (Figure 5-38[104]). Within the chamber there are two ports, one on either side of the vane. When compressed air is applied to one of the ports the vane rotates in one direction, and the drive shaft rotates the valve stem accordingly. When compressed air is applied to the other port, the vane rotates back to its original position, as does the valve stem. This type of actuator is typically used in fail-last-position type valves. To make a rotary vane actuator fail-safe usually requires the use of an external compressed air accumulator tank mounted to the actuator.

Drive Shaft

Vane

**Figure 5-38. Rotary Vane Actuator**

---

103. Flowserve (Valtek) piston actuator.
104. Emerson (Xomox) vane actuator with top cover removed.

---

*Rack & Pinion:* This is a design that uses a rack-and-pinion gear assembly (Figure 5-39) to convert linear motion to rotary motion. A "Scotch-Yoke" actuator (Figure 5-40) is very similar to a rack-and-pinion actuator in that it converts linear motion into a rotational motion. The Scotch-Yoke utilizes a piston that is directly coupled to a sliding yoke which in turn engages with a rotating part.[105]

Rack-Pinion Assembly

**Figure 5-39. Rack and Pinion Actuator**

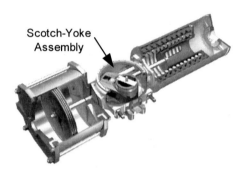

Scotch-Yoke Assembly

**Figure 5-40. Scotch-yoke Actuator**

*Electric:* This design is well suited to remote areas when compressed air is not available. It consists of an electric motor drive that provides the torque to turn the valve stem (Figure 5-41[106]). In addition to their use in remote areas, electric actuators are typically used on multi-turn type valves. Electric actuators inherently fail in the last position, unless some form of backup power is available to make them fail safe.

## 5.2.5 Actuator Selection

The following items should be taken into consideration when selecting an actuator:

- Is the actuator compatible with the valve selected (e.g., sliding stem, rotary, mounting arrangement compatibility)?

- Force or torque requirements

---

105. Flowserve (Automax) rack and pinion actuator and scotch-yoke actuator.
106. Flowserve (Limitorque) electric actuator.

**Figure 5-41. Electric Actuator**

- What will the external energy source be?

- What fail position is required for the valve?

- What type of control signal is required for the actuator?

- In what environmental conditions will the actuator be mounted?

- Initial purchase cost vs overall lifecycle costs.

- What types of accessories are required for the actuator?

    - Solenoid valve(s)
    - Local position indicators
    - Limit switches
    - Positioner
    - Emergency shutdown override
    - Override handwheel

## 5.3 CONTROL VALVE SIZING

The first step in sizing a control valve is to determine the valve's required flow coefficient 'C$_V$'. Reference ANSI/ISA-75.01.01 (IEC 60534-2-1 Mod)-2007 standard.

The flow coefficient (C$_V$) of a control valve is expressed as the flow rate of water in GPM for a pressure drop of 1 psi across a flow passage.

*Liquid (Basic):*

$$Q = C_V \sqrt{\frac{(P_1 - P_2)}{G_f}} \quad or \quad C_V = Q \sqrt{\frac{G_f}{(P_1 - P_2)}}$$

*Liquid (with Correction Factors):*

**For volumetric flow rate:**

$$C_V = \frac{Q}{N_1 F_P} \sqrt{\frac{G_f}{P_1 - P_2}} \quad \text{OR} \quad Q = N_1 F_P C_V \sqrt{\frac{P_1 - P_2}{G_f}}$$

**For mass flow rate:**

$$C_v = \frac{w}{N_6 F_P \sqrt{(P_1 - P_2)\gamma_1}} \quad \text{OR} \quad w = N_6 F_P C_V \sqrt{(P_1 - P_2)\gamma_1}$$

Q = Volumetric Flow Rate
w = Weight or Mass Flow Rate
$G_f$ = Liquid Specific Gravity
$P_1$ = Inlet Pressure in psia
$P_2$ = Outlet Pressure in psia
N = Numerical Constants of Units of Measure Used (Ref. Table 5-6)
$\gamma_1$ = Specific Weight (upstream conditions)
d = Nominal Valve Size
D = Pipe ID

$F_P$ = Piping Geometry Factor $\quad F_P = \sqrt{\left( \frac{\Sigma K \times C_V{}^2}{N_2 d^4} + 1 \right)}$

When the valve is mounted near pipe reducers there is a decrease in the actual flow capacity of the valve.

Inlet reducer only: $K_1 = 0.5\left(1 - \frac{d^2}{D^2}\right)^2$     Outlet reducer only: $K_2 = 1.0\left(1 - \frac{d^2}{D^2}\right)^2$

Between identical reducers: $K_1 + K_2 = 1.5\left(1 - \frac{d^2}{D^2}\right)^2$

*Numerical Constants N for Liquid Flow (from Fisher Control Valve Handbook)*

**Table 5-6. Numerical Constants N for Liquid Flow**

| Constant | | Units Used in Equations | | | | | |
|---|---|---|---|---|---|---|---|
| **N** | | w | Q | $P_1 \Delta P$ | d, D | $\gamma_1$ | v |
| $N_1$ | 0.0865 | | $m^3/h$ | kPa | | | |
| | 0.865 | | $m^3/h$ | Bar | | | |
| | 1.00 | | gpm | psia | | | |
| $N_2$ | 0.00214 | | | | mm | | |
| | 890 | | | | in | | |

| | | | | | | |
|---|---|---|---|---|---|---|
| **N₄** | 76000 | | m³/h | | mm | Centistokes* |
| | 17300 | | gpm | | in | Centistokes* |
| **N₆** | 2.73 | kg/h | | kPa | kg/m³ | |
| | 2.73 | kg/h | | Bar | kg/m³ | |
| | 63.3 | lb/h | | psia | lb/ft³ | |

* To convert $m^2/s$ to centistokes multiply by $10^6$;
to convert centipoise to centistokes divide by $G_f$

### Choked Flow, Cavitation and Noise in Liquid Flow:

- Valves in <u>flashing</u> service can be recognized using the comparison below:

When $P_2 < P_V$ and $\Delta P_{(choked)} < \Delta P_{(actual)}$ = Flashing Service

$P_V$ = Vapor pressure of fluid at the inlet temperature.

- Valves in <u>cavitation</u> service can be recognized using the comparison below:

When $P_2 > P_V$ and $\Delta P_{(choked)} < \Delta P_{(actual)}$ = Cavitation Service

### Gas and Vapor (Compressible Fluids)

**For volumetric flow rates:**

**For mass flow rates:**

When the specific gravity $G_g$ is known:

When the specific weight $\gamma_1$ is known:

$$C_v = \frac{Q}{N_7 F_P P_1 Y \sqrt{\dfrac{x}{C_g T_1 Z}}}$$

$$C_v = \frac{w}{N_6 F_P Y \sqrt{x P_1 \gamma_1}}$$

When molecular weight M is known:

When molecular weight M is known:

$$C_v = \frac{Q}{N_9 F_P P_1 Y \sqrt{\dfrac{x}{M T_1 Z}}}$$

$$C_v = \frac{w}{N_8 F_P P_1 Y \sqrt{\dfrac{xM}{T_1 Z}}}$$

### Aerodynamic Noise Prediction

$$C_g = 40 \times C_v \sqrt{X_T}$$

Q = Volumetric Flow Rate
w = Weight or Mass Flow Rate
M = Molecular Weight (MW of air = 29)
$C_g$ = Specific Gravity of Gas or vapor ($C_g$ = MW ÷ 29)
$P_1$ = Inlet Pressure in psia
$T_1$ = Inlet Temperature in °R
N = Numerical Constants of Units of Measure Used (Ref. Table 5-7)
$y_1$ = Specific Weight (upstream conditions)

$F_K$ = Ratio of Specific Heats (use 1.0 if unknown)

Z = Compressibility Factor (1.0 for pressures less than 100 psia – ideal gas)

d = Nominal Valve Size

D = Pipe ID

Y = Expansion Factor (accounts for the change in density of the fluid as it travels through the valve)

X = Pressure Drop Ratio

$$X = \frac{\Delta P}{P_1}$$

$X_T$ = Pressure Drop Ratio Factor, this is developed experimentally by the valve manufacturer Quick rule of thumb factors: $= 0.85 F_L^2$ ($F_L$ depends on valve style: globe = 0.85; ball = 0.060)

$F_P$ = Piping Geometry Factor $\quad F_P = \sqrt{\left( \frac{\Sigma K \times C_v^2}{N_2 d^4} + 1 \right)}$

Inlet reducer only: $\quad K_1 = 0.5 \left( 1 - \frac{d^2}{D^2} \right)^2 \qquad$ Outlet reducer only: $\quad K_2 = 1.0 \left( 1 - \frac{d^2}{D^2} \right)^2$

*Numerical Constants N for Gas and Vapor Flow (from Fisher Control Valve Handbook)*

### Table 5-7. Numerical Constants N for Gas and Vapor Flow

| Constant | | Units Used in Equations | | | | | |
|---|---|---|---|---|---|---|---|
| **N** | | **w** | **Q** | **P₁ΔP** | **d,D** | **Υ₁** | **T₁** |
| N₅ | 0.00241 | | | | mm | | |
| | 1000 | | | | in | | |
| N₂ | 2.73 | kg/h | | kPa | | kg/m³ | |
| | 27.3 | kg/h | | bar | | kg/m³ | |
| | 63.3 | lb/h | | psia | | lb/ft³ | |
| N₇ | 4.17 | | m³/h | kPa | | | °K |
| | 417 | | m³/h | bar | | | °K |
| | 1360 | | SCFH | psia | | | °R |
| N₈ | 0.948 | kg/h | | kPa | | | °K |
| | 94.8 | kg/h | | bar | | | °K |
| | 19.3 | lb/h | | psia | | | °R |
| N₉ | 22.5 | | m³/h | kPa | | | °K |
| | 2250 | | m³/h | bar | | | °K |
| | 7320 | | SCFH | psia | | | °R |

## 5.4 Pressure Regulators

A pressure regulator uses the downstream pressure to balance the force exerted by a spring that causes the valve portion of the device to open and close. Rising downstream pressure exerts force on a diaphragm, which overcomes the spring force and closes the valve. Downstream pressure is adjusted by adjusting spring tension. Within the broad categories of direct-operated and pilot-operated regulators fall virtually all of the general regulator designs, including:

- Pressure reducing regulators

- Backpressure regulators

- Vacuum regulators and breakers

### 5.4.1 Pressure Reducing Regulator

A pressure reducing regulator automatically maintains a desired reduced outlet pressure while providing the required fluid flow to satisfy the downstream demand. All pressure reducing regulators fit into one of the following two categories: direct-operated or pilot-operated.

*Direct-Operated* (also called Self-Operated): This design consists of three major components:

- Flow restricting element such as a disk or plug

- Measuring element, usually a diaphragm

- Loading element, usually a spring

In operation, a direct-operated pressure reducing regulator (Figure 5-42[107]) allows flow via the restricting element and senses the downstream pressure through the measuring element (diaphragm). The measuring element is usually attached to the flow restricting element. As the diaphragm moves, so does the restricting element. The loading element (spring) counterbalances the downstream pressure. The amount of unbalance between the spring and diaphragm is the resulting position of the flow restricting plug. When there is an increasing demand P2 will drop which allows the restricting element to open further from its current position. When there is a decreasing demand P2 will rise which allows the restricting element to close further from its current position. When P2 = Setpoint, the restricting element is closed.

*Pilot-Operated:* A pilot is also known as a pressure amplifier, and provides an amplified pressure signal (referred to as regulator gain) to the regulator's main valve operating mechanism. The pilot's main purpose is to increase the regulator's sensitivity. Under a constant flow demand, the pilot valve and main regulator valve remain relatively motionless.

This style of regulator is preferred for high flow rates or where precise pressure control is required. In addition, lower regulator droop (defined below) is associated with a pilot-operated regulator ($\leq 3\%$). Pilot-operated regulators are available two different designs: loading and unloading.

---

107. Emerson (Fisher) type 95H regulator.

---

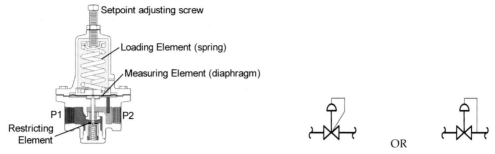

**Figure 5-42. Direct-operated Pressure Reducing Regulator with Symbols**

**Loading:** This design (Figure 5-43[108]) is also known as two-path control. This style of pilot-operated regulator provides loading pressure onto the main regulator's measuring element. The initial change in downstream pressure is first sensed by the main regulating valve's sensing element. The pilot, in turn, provides supplemental loading pressure at a precise level to the moving action of the main regulator valve to attain better output pressure control.

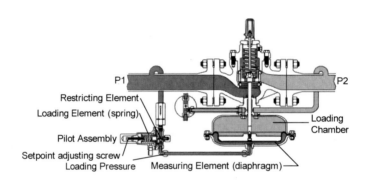

**Figure 5-43. Loading Style Diaphragm Pressure Regulator**

**Unloading:** This style of pilot-operated regulator (Figure 5-44[109]) unloads the pressure on the main regulator's measuring element. The adjustment of the main regulator valve is accomplished by exhausting (unloading) the pressure from the main regulator's measurement chamber more quickly than the regulator's restricting element can fill the loading chamber.

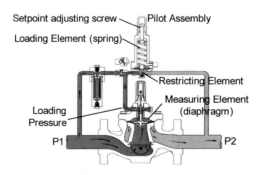

**Figure 5-44. Unloading Style Diaphragm Pressure Regulator**

---

108. Emerson (Fisher) type 1098-EGR regulator.
109. Emerson (Fisher) type EZR regulator.

## 5.4.2 Back Pressure Regulator

A back pressure regulator (Figure 5-45[110]) is installed at the end of a piping system to provide an obstruction to flow and thereby regulate upstream (back) pressure. The back pressure regulator is called upon to provide pressure in order to draw process fluid off the system so that all the equipment prior to the back pressure regulator will have sufficient pressure. When the pressure becomes too high, the back pressure regulator opens more and removes additional process fluid from the system.

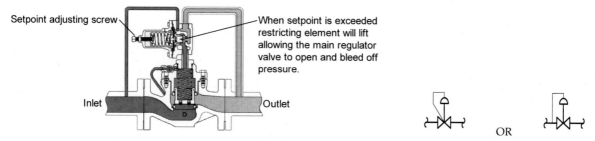

**Figure 5-45. Back Pressure Regulator with Symbols**

## 5.4.3 Vacuum Regulators & Breakers

Vacuum regulators (Figure 5-46) and vacuum breakers (Figure 5-47) are devices used to control vacuum. A vacuum regulator maintains a constant vacuum at the regulator inlet. The vacuum regulator will remain closed until a vacuum decrease exceeds the regulator setpoint and opens the valve disk.

A vacuum breaker operates to prevent the vacuum level from exceeding a setpoint. The vacuum breaker will remain closed until an increase in vacuum exceeds the setpoint and opens the valve disk.

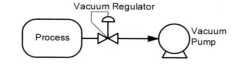

**Figure 5-46. Vacuum Regulator Installation**

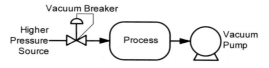

**Figure 5-47. Vacuum Breaker Installation**

---

110. Emerson (Fisher) type 63EG regulator.

## 5.4.4 Regulator Droop

Droop is the reduction of outlet pressure experienced by pressure-reducing regulators as the flow rate increases, and indicates the difference between the outlet pressure setting made at low flow rates and the actual outlet pressure at the published maximum flow rate. The extent of droop is determined by three factors: diaphragm area, spring rate, and valve stroke length.

## 5.4.5 Regulator Hunting

Hunting is the rapid opening and closing of the regulator, which occurs when the regulator tries to respond to cyclic fluctuations caused by pulsations in the system (a very strong indication that the regulator is over-sized). To minimize hunting, the smallest orifice size or regulator that will allow adequate flow at full opening should be used.

## 5.4.6 Pressure Regulator Sizing

*Liquids*

**For volumetric flow rates:**          **For mass flow rates:**

$$C_V \frac{Q}{N_1 F_P \sqrt{\dfrac{P_1 - P_2}{G_f}}} \qquad C_V = \frac{w}{N_6 F_P \sqrt{(P_1 - P_2)\gamma}}$$

Where:
Q = volumetric flow rate
w = mass flow rate
$P_1$ = upstream absolute pressure
$P_2$ = downstream absolute pressure
$G_f$ = Liquid specific gravity
$\gamma$ = Specific weight at inlet conditions (standard is 60°F at 14.7 psia)
$N_1$ = Numerical constant from table below
$N_6$ = Numerical constant from table below
        *(reference Emerson (Fisher) technical document D351798x012_02.pdf)*

$F_P$ = Piping geometry factor

$$F_P = \left[ 1 + \frac{\Sigma K}{N_2} \left( \frac{C_V}{d^2} \right)^2 \right]^{-1/2}$$

d = Nominal valve size
D = Internal diameter of piping
$\Sigma K = K_1 + K_2 + K_{B1} + K_{B2}$
$K_1$ = Resistance coefficient of upstream fittings

For an inlet reducer $K_1 = 0.5 \left( \dfrac{d^2}{D^2} \right)^2$

$K_2$ = Resistance coefficient of downstream fittings

For an outlet reducer $K_1 = 1.0 \left( \dfrac{d^2}{D^2} \right)^2$

$K_{B1}$ = Inlet Bernoulli coefficient
$K_{B2}$ = Outlet Bernoulli coefficient

$K_{B1}$ and $K_{B2}$ are only used if the inlet piping is a different size than the outlet piping.

$$K_{B1} \text{ or } K_{B2} = 1 - \left( \dfrac{d}{D} \right)^4$$

### Table 5-8. Numerical Constants N for Liquid Flow

|       | N       | w     | Q        | P (abs) | $\gamma$   | d, D |
|-------|---------|-------|----------|---------|------------|------|
| $N_1$ | 0.0865  | -     | Nm³/h    | kPa     | -          |      |
|       | 0.865   | -     | Nm³/h    | bar     | -          |      |
|       | 1.00    | -     | GPM      | psia    | -          |      |
| $N_2$ | 0.00214 | -     | -        | -       | -          | mm   |
|       | 890     | -     | -        | -       | -          | inch |
| $N_6$ | 2.73    | kg/hr | -        | kPa     | kg/m³      |      |
|       | 27.3    | kg/hr | -        | Bar     | kg/m³      |      |
|       | 63.3    | lb/hr | -        | psia    | lb/ft³     |      |

*Gas and Steam*

**For volumetric flow rates:**

**For mass flow rates:**

When specific gravity $G_g$ is known:

When the specific weight $\gamma_1$ is known:

$$C_V \dfrac{Q}{N_7 F_P P_1 Y \sqrt{\dfrac{x}{G_g T_1 Z}}}$$

$$C_V \dfrac{w}{N_6 F_P Y \sqrt{x P_1 \gamma_1}}$$

When molecular weight M is known:

When molecular weight M is known:

$$C_V \dfrac{Q}{N_9 F_P P_1 Y \sqrt{\dfrac{x}{M T_1 Z}}}$$

$$C_V \dfrac{w}{N_8 F_P P_1 Y \sqrt{\dfrac{x M}{T_1 Z}}}$$

Where:
Q = Volumetric flow rate
w = Mass flow rate
$P_1$ = Upstream absolute pressure
$P_2$ = Downstream absolute pressure

$G_g$ = Gas specific gravity
$T_1$ = Absolute upstream temperature
$\gamma$ = Specific weight at inlet conditions (standard is 60°F at 14.7 psia)
M = Molecular weight of gas
Z = Compressibility factor
$N_5$ = Numerical constant from table 5-9
$N_6$ = Numerical constant from table 5-8
$N_7$ = Numerical constant from table 5-9
       *(reference Emerson (Fisher) technical document D351798x012_02.pdf)*
$F_P$ = Piping geometry factor
x = Pressure drop ratio $\Delta P / P_1$
Y = Expansion factor (can never be < 0.667)

$$Y = 1 - \frac{x}{3F_k x_T} \text{ or } 1 - \frac{x}{3F_k x_{TP}}$$

$F_k$ = k / 1.4 ratio of specific heats factor
k = Ratio of specific heats
$x_T$ = Pressure drop ratio factor for valves installed without attached fittings
$x_{TP}$ = Pressure drop ratio factor for valves installed with attached fittings

$$x_{TP} = \frac{x_T}{F_P{}^2}\left[1 + \frac{x_T K_i}{N_5}\left(\frac{C_V}{d^2}\right)^2\right]^{-1} \quad \text{Where } K_i = K_1 + K_{B1}$$

## Table 5-9. Numerical Constants N for Gas or Vapor Flow

| | | N | w | Q | P (abs) | $\gamma$ | T | d, D |
|---|---|---|---|---|---|---|---|---|
| $N_5$ | - | 0.00241 | - | - | - | - | - | mm |
| | - | 1000 | - | - | - | - | - | inch |
| $N_7$ | Normal Conditions $T_N$ = 0ºC | 3.94 | - | Nm³/h | kPa | kg/m³ | ºK | - |
| | | 394 | - | Nm³/h | bar | | ºK | - |
| | Standard Conditions $T_S$ = 16ºC | 4.17 | - | Nm³/h | kPa | kg/m³ | ºK | - |
| | | 417 | - | Nm³/h | bar | | ºK | - |
| | Standard Conditions $T_S$ = 60ºF | 1360 | - | SCFH | psia | - | ºR | - |
| $N_8$ | - | 0.948 | kg/hr | - | kPa | - | ºK | - |
| | - | 94.8 | kg/hr | - | bar | - | ºK | - |
| | - | 19.3 | lb/hr | - | psia | - | ºR | - |
| $N_9$ | Normal Conditions $T_N$ = 0ºC | 21.2 | - | Nm³/h | kPa | kg/m³ | ºK | - |
| | | 2120 | - | Nm³/h | bar | | ºK | - |
| | Standard Conditions $T_S$ = 16ºC | 22.4 | - | Nm³/h | kPa | kg/m³ | ºK | - |
| | | 2240 | - | Nm³/h | bar | | ºK | - |
| | Standard Conditions $T_S$ = 60ºF | 7320 | - | SCFH | psia | - | ºR | - |

# 5.5 MOTORS

## 5.5.1 Types of Motors

### Table 5-10. Types of Motors

A variant of the wound field DC motor is the universal motor

*DC Motors*

DC motors are seldom used in ordinary industrial applications because electrical utility companies furnish alternating current. However, for special applications such as in steel mills, mines and electric trains, it is sometimes advantageous to convert AC to DC in order to use a DC motor. The reason is that the torque-speed characteristics of DC motors can be varied over a wide range while retaining high efficiency. As the torque-speed characteristics of VFDs (variable frequency drives) improve, AC motors are starting to enter applications that would normally require a DC motor.

The main DC motor components are:

- **Armature**: The rotating part of a DC generator, consisting of a commutator, an iron core and a set of coils. The iron core is composed of slotted iron laminations that are stacked to form a solid cylindrical form. The laminations are individually coated with an insulating film so that they do not come into contact with each other, this in turn reduces eddy-current losses.

- **Commutator**: Rotary electrical switch in certain types of electric motors or electrical generators that periodically reverses the current direction between the rotor and the external circuit. As a switch, it has exceptionally long life, considering the number of circuit makes and breaks that occur in normal operation. A commutator is a common feature of direct current rotating machines. By reversing the current direction in the moving coil of a motor's armature, a steady rotating force (torque) is produced. Similarly, in a generator, reversing of the coil's connection to the external circuit provides unidirectional direct current to the external circuit. In its simplest form it is composed of a slip ring that is cut in half, with each segment insulated from the other, as well as from the shaft. One segment is connected to one end of the coil and the other half connected to the other end of the coil. The commutator revolves with coil and the voltage between the segments is picked up by two stationary brushes. The voltage between the brushes pulsates but never changes polarity, thus the AC voltage in the coil is rectified by the commutator.

- **Brushes**: A DC motor has brushes fixed diametrically opposite to each other. They slide on the commutator and ensure good electrical contact between the revolving armature and the stationary external load. The brushes are typically made from carbon because carbon has good electrical conductivity and its softness does not score the commutator. To improve conductivity a small amount of copper is sometimes mixed with the carbon.

- **Field Coil**: Electro-magnetic winding used for speed control of the DC motor.

**Series-wound:** In series-wound motors, the field coils and armature are connected in series and the armature current also flows through the field coils (Figure 5-48). This allows the motor to develop high torque at low speeds.

The series field is composed of a few turns of wire having a cross section sufficiently large to carry the armature current. If the motor operates at less than full load, the armature current and the flux per pole are smaller than at full load, the weaker field causes the speed to rise (e.g., if the current drops by half, then the flux will diminish by half, causing the speed to double). Therefore, if the load is extremely small, the motor can accelerate to speeds that could cause self-destruction. A large series-wound motor must NEVER be permitted to operate at no-load conditions.

Series motors are used on equipment requiring high starting torque. The major disadvantage of a series DC motor is that the speed varies widely between no-load and full-load conditions, and they should not be used where a relatively constant speed is required.

**Figure 5-48. Series Wound Motor (Schematic)**

**Shunt-wound:** The field coils and the armature in a shunt-wound motor are connected in parallel, (also known as shunt) formation, causing the field current to be proportional to the load on the motor (Figure 5-49).

The speed of a DC shunt motor stays relatively constant from no-load to full-load, but it does not develop the same high torque at low speed that the series motor does. In small shunt-wound motors the speed drops by about 10%–15% when full load is applied. In large motors the speed will drop even less when full load is applied.

When full voltage is applied to a stationary shunt motor, the starting current will be very high, and you run the risk of:

- Burning out the armature

- Damaging the commutator and brushes due to heavy sparking

- Overloading the motor feeder

- Fracturing the shaft due to mechanical shock

As a result all DC motors, not just shunt type, must be provided with a means to limit the starting current to a reasonable limit (usually between 1 1/2 and 2 times the full-load current). One traditional solution is to connect a stepping rheostat in series with the armature. The resistance is gradually decreased as the motor accelerates. Today's solutions are usually electronic.

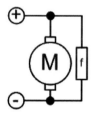

**Figure 5-49. Shunt Wound Motor (Schematic)**

**Compound-wound:** A compound-wound motor is hybrid of both the shunt-wound and series-wound types and features both configurations (Figure 5-50).

In a *cumulative* compound motor, the MMF (magnetomotive force)[111] of the two fields adds, with the shunt field being the stronger of the two. When the motor runs at low speed, the MMF in the series field is negligible; however the shunt field is fully excited so the motor acts like a shunt motor. As the load increases, the EMF of the series field increases but the MMF of the shunt field remains constant. Therefore, the total MMF is greater under load than at no load. The motor speed falls with increasing load and the speed drop from no-load to full-load is generally 10%–30%.

---

111. Magnetomotive force is the driving force that produces magnetic flux, it is the magnetic equivalent of EMF (electromotive force – voltage).

If the series field is connected so that it opposes the shunt field, the result is a *differential* compound motor. In such a motor, the total MMF decreases with increasing load, thus the speed rises as the load increases. This may lead to instability, so as a result the differential compound motor has very few applications.

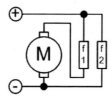

**Figure 5-50. Compound wound Motor (Schematic)**

### Types of DC Motor Braking Methods

**Dynamic Braking:** To activate dynamic braking on any type DC motor, the armature is placed in parallel with a current-limiting resistor while the field coils remain energized (Figure 5-51). As a result, the armature current and torque reverse direction and the motor tries to reverse its rotation. The speed in the forward direction rapidly decreases, as does the voltage generated within the armature. At the point of reversal or zero speed, the generated voltage is zero and the motor stops at this point since the current cannot flow and no more reversing torque is developed. The value of the shunt current-limiting resistor controls the braking rate. A small resistance allows large current flow, and since torque is proportional to armature current, the load and current flow stop in a short time.

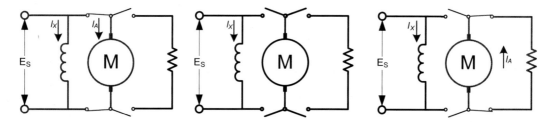

**Figure 5-51. Dynamic Braking Current Flow**

**Plugging:** This technique consists of suddenly reversing the armature current by reversing the terminals of the source (Figure 5-52). However, to limit the catastrophic effects of the resultant high current a resistor MUST be used in series with the reversing circuit. At the instant the motor is plugged, the armature impressed EMF and its CEMF (counter EMF) are nearly equal and in the same direction.

### AC Motors

Reference Table 5-11 for voltage levels typically used for single-phase motors and various three phase motors.

**Single-Phase:** As the name suggests, this type of motor has only one stator winding (main winding) and operates with a single-phase power supply. In all single-phase induction motors,

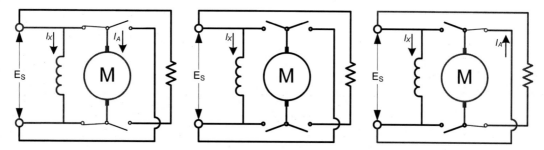

**Figure 5-52. Plugging Current Flow**

the rotor is the squirrel cage type (named squirrel cage due to its resemblance to the exercise wheel inside a squirrel's cage, reference Figure 5-59a).

A single-phase induction motor is not self-starting. When the motor is connected to a single-phase power supply, the main winding carries an alternating current. This current produces a pulsating magnetic field. Due to induction, the rotor is energized. However, though the main magnetic field is pulsating, the torque necessary for motor rotation is not generated. The rotor will vibrate, but not rotate. Hence, a single-phase induction motor is required to have a starting mechanism that can provide the starting "kick" and cause the motor to rotate.

The starting mechanism of the single-phase induction motor is most often an additional stator winding (start/auxiliary winding). The start winding can have a series capacitor and a centrifugal switch. When the supply voltage is applied, current in the main winding lags the supply voltage due to the main winding impedance. At the same time, current in the start winding leads/lags the supply voltage depending on the starting mechanism impedance. Interaction between magnetic fields generated by the main winding and the starting mechanism generates a resultant magnetic field rotating in one direction. The motor starts rotating in the direction of the resultant magnetic field.

Once the motor reaches about 75% of its rated speed, a centrifugal switch disconnects the start winding. From this point on, the single-phase motor can maintain sufficient torque to operate on its own. Except for special capacitor start/capacitor run types, single-phase motors are generally used for applications ≤ 3/4 HP only.

**Split Phase:** This type uses both a starting and a running winding (Figure 5-53). The starting winding is displaced 90 electrical degrees from the running winding. The running winding has many turns of large diameter wire wound in the bottom of the stator slots to get high reactance. Therefore, the current in the starting winding leads the current in the running winding, causing a rotating field.

During startup, both windings are connected to the line. As the motor comes up to speed (at about 25% of full-load speed), a centrifugal switch actuated by the rotor, or an electronic switch, disconnects the starting winding. The starting torque is low, typically 100% to 175% of the rated torque. The motor draws high starting current, approximately 700% to 1,000% of the rated current. The maximum generated torque ranges from 250% to 350% of the rated torque. Good applications for split-phase motors include small grinders, small fans and blowers and other low starting torque applications with power needs from 1/20 to 1/3 hp. This type of motor should not be used in any applications requiring high on/off cycle rates or high torque.

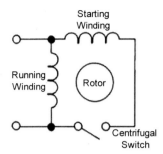

**Figure 5-53. Split Phase Motor (Schematic)**

**Capacitor Start:** This is a modified split-phase motor with a capacitor in series with the start winding to provide a start "boost" (Figure 5-54). The capacitor start motor has a centrifugal switch that disconnects both the start winding and the capacitor when the motor reaches about 75% of the rated speed. Since the capacitor is in series with the start circuit, it creates more starting torque than the split-phase motor, typically 200% to 400% of the rated torque, and the starting current (usually 450% to 575% of the rated current) is much lower than that of the split-phase due to the larger wire in the start circuit.

A modified version of the capacitor start motor is the *resistance start* motor. In this motor type, the starting capacitor is replaced by a resistor. The resistance start motor is used in applications where the starting torque requirement is less than that provided by the capacitor start motor. Apart from the cost, this motor does not offer any major advantage over the capacitor start motor. Capacitor start and resistance start motors are used in a wide range of belt-drive applications like small conveyors, large blowers, and pumps, as well as in many direct-drive or geared applications.

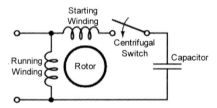

**Figure 5-54. Capacitor Start Motor**

**Permanent Split Capacitor (PSC) Run:** Similar in construction to a capacitor start motor except that there is no centrifugal switch (i.e., the PSC motor has a run capacitor permanently connected in series with the start winding). This makes the start winding an auxiliary winding once the motor reaches its running speed. Since the run capacitor must be designed for continuous use, it cannot provide the starting boost of a starting capacitor. The typical starting torque of the PSC motor is low, from 30% to 150% of the rated torque. PSC motors have low starting current, usually less than 200% of the rated current, making them excellent for applications with high on/off cycle rates.

PSC motors have several advantages. The motor design can easily be altered for use with speed controllers. They can also be designed for optimum efficiency and high Power Factor (PF) at the rated load. They're considered to be the most reliable of the single-phase motors, mainly because no centrifugal starting switch is required.

Permanent split capacitor motors have a wide variety of applications depending on the design. They include fans and blowers with low starting torque needs, and intermittent cycling uses such as adjusting mechanisms, gate operators and garage door openers.

**Capacitor Start/Capacitor Run:** This type of motor has a start capacitor in series with the auxiliary winding, like the capacitor start motor, for high starting torque (Figure 5-55). Like a PSC motor, it also has a run capacitor that remains in series with the auxiliary winding after the start capacitor is switched out of the circuit. Doing this allows high overload torque. This type of motor can be designed for lower full-load currents and higher efficiency than a simple capacitor start motor. It is costly due to having both start and run capacitors and a centrifugal switch. It is able to handle applications too demanding for any other type of single-phase motor. These include woodworking machinery, air compressors, high-pressure water pumps, vacuum pumps and other high torque applications requiring 1 to 10 hp.

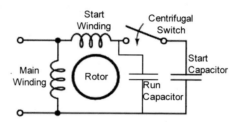

**Figure 5-55. Capacitor Start/Capacitor Run Motor (Schematic)**

**Shaded Pole:** Shaded-pole motors have only one main winding and no start winding. Starting is by means of a design that includes a continuous copper loop around a small portion of each of the motor poles. This "shades" that portion of the pole, causing the magnetic field in the shaded area to lag behind the field in the unshaded area. The reaction of the two fields gets the shaft rotating (Figure 5-56).

Because the shaded-pole motor lacks a start winding, starting switch or capacitor, it is electrically simple and inexpensive. In addition, the speed can be controlled by merely varying voltage, or through a multi-tap winding. Mechanically, the shaded-pole motor construction allows high-volume production. In fact, these are usually considered as "disposable" motors, meaning they are much cheaper to replace than to repair.

The shaded-pole motor has many positive features, but it also has several disadvantages. Its starting torque is low, typically only 25% to 75% of the rated torque. It is a high slip motor with a running speed 7% to 10% below the synchronous speed. Generally, the efficiency of this motor type is very low (below 20%). The low initial cost lends shaded-pole motors to low horsepower or light duty applications. Perhaps their largest use is in multi-speed fans for household use. However, the low torque, low efficiency and less sturdy mechanical features make shaded-pole motors impractical for most industrial or commercial use, where higher cycle rates or long-term continuous duty are the norm.

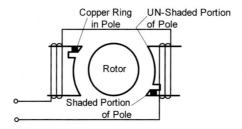

**Figure 5-56. Shaded Pole Motor (Schematic)**

**Repulsion-induction:** A repulsion motor is based upon the principle of two magnetic field repelling each other. In the case of a repulsion motor, when current flows through the armature a magnetic field of the same polarity is produced within the armature. This magnetic field is repelled by the stator's magnetic field that is the opposite polarity causing the armature to rotate. However the repulsion-induction motor has a single phase stator winding but it has two separate windings on the rotor, in common slots. The inner winding is a squirrel cage winding with the rotor bars permanently short-circuited. Placed over the squirrel cage winding is a repulsion winding similar to a DC motor armature winding. The repulsion winding is connected to a commutator on which ride short circuited brushes (Figure 5-57). When the motor starts, due to its high reactance the squirrel cage winding has no effect and the motor starts as a repulsion motor, giving high starting torque. As the motor picks up speed, the squirrel cage winding comes into action.

The shifting from repulsion to induction motor characteristics is thus done without any switching arrangement. Both of the armature windings are active during the entire period of operation. The starting torque is about 225% - 300% the full load torque, the lower being for larger motors, and the starting current is 300% - 400% the full load current.

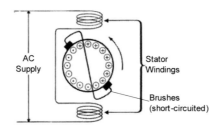

**Figure 5-57. Repulsion-induction Motor (Schematic)**

**Universal:** A single-phase universal motor (Figure 5-58a & b[112]) is very similar to a DC series motor. A universal motor can operate on either DC or AC. When on DC the motor operates as would a DC series motor. However, when the motor is connected to AC, the current flows through the armature and the series field. The field produces an AC flux $\phi$ that reacts with the current flowing in the armature to produce torque. Because the armature current and the flux reverse simultaneously, the torque always acts in the same direction. The main advantage of fractional HP universal motors is their high speed and high starting torque, which makes them very popular in portable tools such as drills and saws.

---

112. Image obtained from www.globe-usa.com.

---

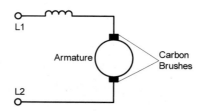

**Figure 5-58a. Universal Motor (Schematic)**

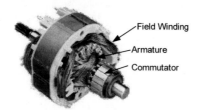

**Figure 5-58b. Universal Motor (Cutaway)**

**Hysteresis:** This is a form of synchronous motor, and consists of a wound stator and a solid rotor of uniformly high magnetic permeability material. With a smooth rotor of homogenous material, the noise and vibration produced by the motor are inherently low. Since there are no salient (protruding) pole faces, the magnetic path is of constant permeability, thus eliminating the magnetic pulsations that are the major cause of noise in the salient pole type. The torque of the motor is uniform throughout each revolution due to the homogenous rotor and constant permeability. The availability of new materials and designs has increased the torque and horsepower values of the hysteresis motor into the one horsepower range. A hysteresis motor is a very popular choice in analog electric clock use because of its synchronous motor design and small packaging.

**Poly-phase:** Three-phase AC induction motors are widely used in industrial and commercial applications. They are classified either as squirrel cage or wound-rotor motors. These motors are self-starting and use no capacitor, start winding, centrifugal switch or other starting device.

They produce medium to high starting torque. The power capabilities and efficiency of these motors range from medium to high compared to their single-phase counterparts. Popular applications include grinders, lathes, drill presses, pumps, compressors, conveyors, printing equipment, farm equipment, electronic cooling and other mechanical equipment duty applications.

**Squirrel Cage:** Almost 90% of three-phase AC Induction motors are of this type. Here, the rotor is of the squirrel cage type and it works as explained in the single-phase section. The power ratings range from one-third to several hundred horsepower in three-phase motors. Motors of this type rated one horsepower or larger cost less and can start heavier loads than their single-phase counterparts.

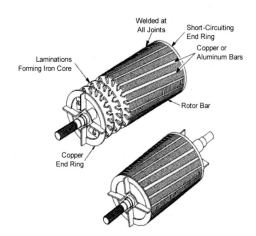

**Figure 5-59a. Squirrel Cage Rotor**

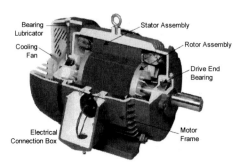

**Figure 5-59b. Cutaway Squirrel Cage Motor**

Note how the slots are skewed in the squirrel cage rotor assembly (Figure 5-59a[113] & b[114]). The rotor slots are skewed for the following reasons:

- Aids in the reduction of magnetic hum, thus helping to keep the motor quiet

- Aids in avoiding "cogging." Cogging is the tendency of rotor teeth (poles) to remain under the stator teeth due to the direct magnetic attraction between the two. If cogging were to occur then the motor would not develop any counter electromotive force. This in turn would cause the current draw developed by the motor to rise to very high levels.

- Provides an increase in the effective ratio of transformation[115] between stator and rotor.

**Wound Rotor (Slip Ring):** The slip-ring or wound-rotor motor is a variation of the squirrel cage induction motor. While the stator is the same as that of the squirrel cage motor, it has a set of windings on the rotor that are not short-circuited, but are terminated to a set of slip rings. These are helpful in adding external resistors and contactors (Figure 5-60 a & b[116]).

---

113. Image obtained from www.siemens.com.
114. Image obtained from www.reliance.com.
115. Just as in a transformer, there is a transformation of voltage from the motor winding to the motor cage assembly.
116. Image obtained from www.electricalengineeringbasics.com.

---

**FINAL CONTROL ELEMENTS – MOTORS**

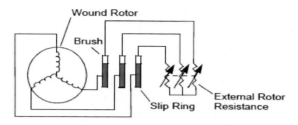

**Figure 5-60a. Wound Rotor (Schematic)**

NOTE: No Shorting rings

Slip Rings

**Figure 5-60b. Stator with Slip Rings**

The slip necessary to generate the maximum torque (pull-out torque)[117] is directly proportional to the rotor resistance. In the slip-ring motor, the effective rotor resistance is increased by adding external resistance through the slip rings. Thus, it is possible to get higher slip and hence, higher pull-out torque at a lower speed.

A particularly high resistance can result in the pull-out torque occurring at almost zero speed, providing a very high pull-out torque at a low starting current. As the motor accelerates, the value of the resistance can be reduced, altering the motor characteristic to suit the load requirement. Once the motor reaches its nameplate speed, the external resistors are disconnected from the rotor. This means that the motor is now working as a standard induction motor.

This motor type is ideal for very high inertia loads, where it must generate pull-out torque at almost zero speed and accelerate to full speed in the minimum time with minimum current draw.

The downside of the slip ring motor is that slip rings and brush assemblies need regular maintenance, which is a cost not applicable to the standard squirrel cage motor. If the rotor windings are shorted and a start is attempted (i.e., the motor is converted to a standard induction motor), it will exhibit extremely high locked rotor current—as high as 1400%—and very low locked rotor torque, perhaps as low as 60%. In most applications, this is not an option.

By modifying the speed torque curve by altering the rotor resistors, the speed at which the motor will drive a particular load can be altered. At full load, it is possible to reduce the speed effectively to about 50% of the motor synchronous speed, particularly when it is driving variable torque/variable speed loads. Reducing the speed below 50% results in very low efficiency due to higher power dissipation in the rotor resistances. This type of motor is used in applications for driving variable torque/variable speed loads such as in printing presses, compressors, conveyor belts, hoists and elevators.

---

117. Pull-out torque is the largest torque under which a motor can operate without sharply losing speed.

---

*Synchronous Motors*

Synchronous motors have the following characteristics:

- A three-phase stator similar to that of an induction motor.

- A wound rotor (rotating field) that has the same number of poles as the stator, and is supplied by an external source of direct current (DC). Both brush-type and brushless exciters are used to supply the DC field current to the rotor. The rotor current establishes a north/south magnetic pole relationship in the rotor poles, enabling the rotor to "lock-in-step" with the rotating stator flux.

- It starts as an induction motor. The synchronous motor rotor also has a squirrel-cage winding, known as an Amortisseur winding (Figure 5-61[118]), which produces torque for motor starting.

Synchronous motors may be used to provide power factor correction. Over-exciting the synchronous motor makes the current lead the voltage, similar to what occurs in a capacitive circuit. Power Factor correction on large current consumers is usually required by the utility company. The majority of loads in an industrial facility are usually highly inductive in nature (e.g., motors), and this type of load causes the power factor to drop below 100% due to the fact that an inductive type current lags the voltage by 90°. Reference the high level description of power factor to understand the power factor concept.

Synchronous motors are typically used for very large compressors, pumps and industrial grinders.

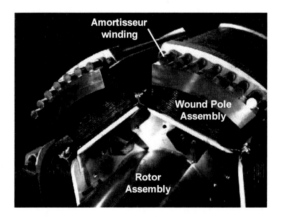

**Figure 5-61. Synchronous Motor (Rotor Cutaway)**

**Power Factor:** Power factor is the ratio of active power (watts) to total RMS 'VA'. PF can be an important aspect to consider in an AC circuit, because any PF less than 1 means that the circuit's wiring has to carry more current than what would be necessary with zero reactance in the circuit to deliver the same amount of (true) power to the resistive load. It is for this reason that many utility companies substantially penalize customers for low power factors (usually less than 0.90),

---

118. TECO Westinghouse synchronous rotor assembly.

therefore PF correction shall be needed to avoid this penalty. If a PF is too low (e.g., < 0.85), a utility company may refuse to provide service to the customer.

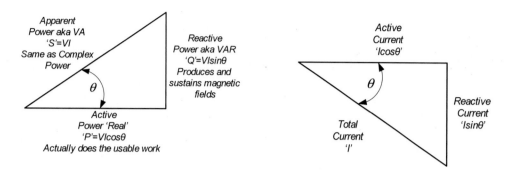

'Active' power is usually less than apparent power due to the following reasons:

- The current wave is usually out of phase with the voltage wave at the fundamental frequency of the power system (except for purely resistive loads).

- The current wave could be distorted from a pure sinusoidal waveform.

The current that will flow in a power system is the vectorial resultant of the active component and the reactive component of the circuit.

- If much of the reactive current is furnished locally (i.e., not at the utility), then the utility and the plant's distribution systems could be more effectively utilized, since the current that they would have to carry could consist mainly of the power components. The transformers, cables, etc. could then be sized to carry a smaller magnitude of current.

$$total\ PF = \frac{active\ power}{apparent\ power} = \frac{kW}{kVA} \therefore kW = kVA \times PF = kVA \times PF$$

for 3-phase circuits: $\dfrac{\sqrt{3} \times V_{L-L_{RMS}} \times I_1 \cos\theta}{\sqrt{3} \times V_{L-L_{RMS}} \times I_L} = \dfrac{I_1 \cos\theta}{I_L}$

Where :
$\cos\theta$ = displacement PF
$I_1$ = The fundamental frequency value of the line current
$I_L$ = The RMS value of line current including harmonics
$\dfrac{I_1}{I_L}$ = The distortion PF

$$total\ current\ I = \sqrt{(active\ current)^2 + (reactive\ current)^2} =$$
$$\therefore I = \sqrt{(I\cos\theta)^2 + (I\sin\theta)^2}$$

$$apparent\ power\ 'in\ VA' = \sqrt{(active\ power)^2 + (reactive\ power)^2} =$$
$$|S| = |VI| = \sqrt{(VI\cos\theta)^2 + (VI\sin\theta)^2}$$
$$\therefore S = VI = P + jQ = VI\cos\theta + j\sin\theta$$

## 5.5.2 Motor Enclosure Types

*Drip Proof (ODP)(IP12):* The frame protects the windings against liquid drops and solid particles that fall at any angle between 0° and 15° from vertical (Figure 5-62). These motors are cooled by means of a fan directly coupled to the rotor.

*Splash Proof (IP54):* The frame protects the windings against liquid drops and solid particles that fall at any angle between 0° and 100° from vertical (Figure 5-62[119]). These motors are cooled by means of a fan directly coupled to the rotor.

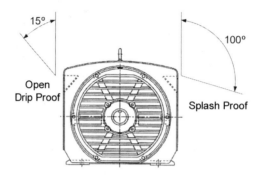

**Figure 5-62. Drip Proof, Splash Proof**

*Weather Protected (WPI and WPII):*

**WPI:** Open machine (not totally enclosed) with its ventilating passages constructed so as to minimize the entrance of rain, snow and airborne particles to the motor's internal electrical parts and having its ventilation openings constructed so as to prevent the passage of a cylindrical rod 3/4 inch in diameter.

**WPII:** In addition to the construction requirements of the WPI described above, the WPII ventilating passages at both the intake and discharge are arranged so that high-velocity air and airborne particles blown into the machine by storms or high winds can be discharged without entering the internal passages leading directly to the electric parts (Figure 5-63[120]).

*Totally Enclosed Non-Ventilated (TENV):* The frame prevents the free exchange of air between the inside and outside of the case. It is designed for very wet and dusty locations. The motor losses are dissipated by natural convection and radiation from the frame. Most are rated < 10kW because it is difficult to get rid of the heat of larger motors via just natural convection and radiation.

*Totally Enclosed Fan Cooled (TEFC):* Medium to high-power motors that are totally cooled by an external blast of air. An external fan, directly coupled to the shaft, blows air over the ribbed motor frame. A concentric outer shield prevents physical contact with the fan and serves to channel the air downstream.

---

119. Image obtained from www.eemsco.com.
120. Toshiba weather protected type II medium voltage motor.

**Figure 5-63. WPII Motor**

***Totally Enclosed Forced Ventilated (TEFV):*** A totally enclosed motor, except that inlet and outlet openings are provided for connection to forced ventilation cooling air systems.

***Explosion Proof:*** Used in highly flammable or explosive applications. These motors are totally enclosed, but are not airtight, and the frames are built to withstand the enormous pressure that may build up inside the motor due to an internal explosion. In addition, the flanges on the end bells[121] are made extra long in order to cool any escaping gases generated by an explosion.

### 5.5.3 Nameplate Voltage Ratings of Standard Induction Motors

**Table 5-11. Nameplate Voltage Ratings of Standard Induction Motors**

| Nominal System Voltage | Nameplate Voltage |
|---|---|
| Single-Phase Motors | |
| 120 | 115 |
| 240 | 230 |
| Three-Phase Motors | |
| 208 | 200 |
| 240 | 230 |
| 480 | 460 |
| 600 | 575 |
| 2400 | 2300 |
| 4160 | 4000 |
| 4800 | 4600 |
| 6900 | 6600 |
| 13800 | 13200 |

### 5.5.4 Motor Speed

Motor speed is typically defined in revolutions per minute and is dependent on the number of poles within the motor (except for variable frequency drives, described in section 5.5.2).

---

121. The motor end bells are the end portions of the motor frame that contain the bearing assemblies.

---

$$S_{Synchronous} = \frac{120 \times f}{\# \, poles}$$  # of poles is poles per phase.

### Table 5-12. Motor Speeds

| # of Poles per Phase | Synchronous RPM |
|---|---|
| 2 | 3600 |
| 4 | 1800 |
| 6 | 1200 |
| 8 | 900 |
| 10 | 720 |
| 12 | 600 |

**Slip:** The interaction of currents flowing in the rotor bars and the stator's rotating magnetic field generates torque. In actual operation, rotor speed always lags the magnetic field's speed, allowing the rotor bars to cut magnetic lines of force and produce useful torque. This speed difference is called slip speed.

$$\text{Slip: } S_L = \frac{(S_S - S_R)}{S_R} \quad S_L = \text{Slip} \quad S_S = \text{Synchronous Speed} \quad S_R = \text{Rotor Speed}$$

$$f_R = S_L \times f \quad f_R = \text{Rotor Frequency, } f = \text{Stator power line frequency}$$

The required speed of a motor is determined by the speed of the machine that it has to drive. However, for low speed machines, it is often preferable to use a high speed motor and a gearbox instead of a direct coupling. There are advantages to using a gearbox:

- For a given power output, the size and cost of a high speed motor (lesser # of poles), is less than that of a lower speed motor, and its efficiency and power factor are higher.

- The locked-rotor torque of a higher speed motor is always greater (as a % of full-load torque) than that of a similar low speed motor of equal power.

A gearbox is mandatory when the equipment has to operate at very low speeds (< 100rpm) and also when the equipment must run above 3600rpm. Reference Table 5-13 for an example cost differential.

### Table 5-13. Comparison Between Two Motors of the Same HP but Different Speeds

| Power | | Synchronous Speed | P.F. | Efficiency | LockedRotor Torque | Weight | ~ Price |
|---|---|---|---|---|---|---|---|
| HP | kW | RPM | % | % | % | lbs | $ |
| 10 | 7.5 | 3600 | 89 | 90 | 150 | 110 | 815 |
| 10 | 7.5 | 900 | 75 | 85 | 125 | 375 | 2600 |

*Two Speed Motors*

**Two Speed, Two Winding:** A two-winding motor is made in such a manner that it is really two motors that share one stator. One winding, when energized, gives one of the speeds. When the second winding is energized and the first is de-energized, the motor takes on the speed that is determined by the second winding. The two speed, two winding motor design can be used to get virtually any combination of normal motor speeds, and the two speeds need not be related to each other by a 2:1 speed factor. Thus, a two speed motor requiring 1750 RPM and 1140 RPM would, of necessity, have to be a two winding motor.

**Two Speed, One Winding:** The second type of two speed motor is the two speed, one winding motor. In this type of motor, a 2:1 relationship between the low and high speed must exist. Two speed, one winding motors are of a design that is called "consequent pole." These motors are wound for one speed but when the winding is reconnected, the number of magnetic poles within the stator is doubled and the motor speed is reduced to one-half of the original speed. The two speed, one winding motor is, by nature, more economical to manufacture than the two speed, two winding motor. This is because the same winding is used for both speeds and the slots in which the conductors are placed within the motor do not have to be nearly as large as they would have to be to accommodate two separate windings that work independently. Thus, the frame size on the two speed, one winding motor can usually be smaller than on an equivalent two winding motor.

## 5.5.5 Motor NEMA Designations

$$HP = \frac{T \times RPM}{5250} \quad \text{AND} \quad T = \frac{HP \times 5250}{RPM} \text{ (See Figure 5-64)}$$

**Design A** motors have a higher breakdown torque[122] than Design B motors and are usually designed for a specific use. Slip at full load is 5%, or less.

**Design B** motors account for most of the induction motors sold. They are often referred to as general purpose motors. Slip at full load is 5% or less. The per-unit locked-rotor torque decreases as the size of the motor increases. The corresponding locked-rotor current should not exceed 6.4 times the full-load current.

**Design C** motors have high starting torque with normal starting current and low slip. This design is normally used where breakaway loads are high at starting, but these motors normally run at rated full load and are not subject to high overload demands after running speed has been reached. Slip at full load is 5% or less. They are suitable for equipment with high inertia starts (e.g., positive displacement [PD] pumps and piston type compressors). The locked-rotor torque is 200% of the full-load torque. The corresponding locked-rotor current should not exceed 6.4 times the full-load current. In general these motors are equipped with a double-cage rotor.

**Design D** motors exhibit high slip (5% to 13%), very high starting torque, low starting current, and low full load speed. Because of high slip, the speed can change when fluctuating loads are encountered. This design is subdivided into several groups that vary according to slip or the shape of the speed-torque curve. These motors are usually available only on a special order basis and are suitable for equipment with very high inertia starts (e.g., cranes, hoists).

---

122. Breakdown torque is the torque point in which the motor begins to lose speed and stall.

*Speed vs. Torque Characteristics of the Four NEMA Designs*

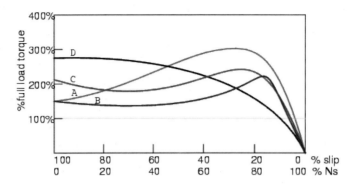

**Figure 5-64. Speed vs. Torque Characteristics**

## 5.5.6 Life Expectancy of Electric Motors

The unbreakable rule is the HIGHER the temperature, the SHORTER the electrical equipment's life. Tests have shown that the service life of piece of electrical equipment diminishes by half every time the temperature increases by 10°C. This means if a motor has a normal life expectancy of eight years at a temperature of 105°C, it will have a service life of only four years at 125°C, and only one year at 135°C.

The primary cause of temperature-related equipment failure is insulation deterioration. The factors that contribute the most to deterioration of the insulating materials used in electrical equipment are:

- Heat (temperature)

- Humidity

- Vibration

- Acidity

- Oxidation

- Time

Because of these factors, the physical state of the insulating material changes gradually. At normal temperatures it slowly begins to crystallize, but the transformation takes place much more rapidly as the temperature rises. When the organic insulating material crystallizes, it becomes hard and brittle. Eventually, the slightest shock or vibration will cause the insulation to break. Humidity can causes arcing within the motor components which will degrade the insulation. Vibration can cause misalignment in the motor which can lead to bearing failure and an unexpected heat source. Acidic and oxidation type environments can chemically react with the motor's insulation material causing degradation. Time – components within a motor age with regards to time as does everything else in the world.

*Thermal Classification of Motor Insulation*

Ratings are based upon a normal life expectancy of 20,000–40,000 hours of insulation life at the stated temperature (Table 5-14).

**Table 5-14. Thermal Classification of Insulation**

| Class | Temperature | Description |
|-------|-------------|-------------|
| A | 105ºC | Materials or combinations of materials such as cotton, silk and paper that are suitable impregnated or coated when immersed in a dielectric liquid such as oil. |
| B | 130ºC | Materials or combinations of materials such as mica, glass fiber, asbestos, etc., with suitable bonding substances. |
| F | 155ºC | Materials or combinations of materials such as mica, glass fiber, asbestos, etc., with suitable bonding substances. |
| H | 180ºC | Materials or combinations of materials such as silicone elastomer, mica, glass fiber, asbestos, etc., with suitable bonding substances such as appropriate silicone resins. |
| N | 200ºC | |
| R | 220ºC | |
| S | 240ºC | |
| C | > 240ºC | |

**Hot Spot Temperature:** Standards organizations have established a maximum *ambient* temperature, which is usually 40ºC. The temperature of a machine varies from point to point, but there are places where the temperature is warmer than anywhere else. This *hot spot* temperature must not exceed the maximum allowable temperature of the particular insulation class used.

### 5.5.7 Motor Positioning

Motor positioning involves driving a motor to certain exact rotational position.and receiving feedback that the motor is in the desired position. An example application for motor positioning is in the field of robotics.

*Servo Motor:* A servo motor has a positionable shaft that can be arranged in a number of angled positions via a coded signal (a pulse of variable width) (Figure 5-65). The angle is determined by the duration of the pulse that is applied (a technique called PWM, pulse width modulation). The servo expects to see a pulse every 20ms. The length of the pulse determines how far the motor shaft turns with each pulse (Figure 5-66).

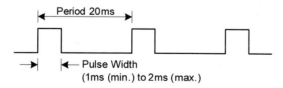

**Figure 5-65. Pulse Width**

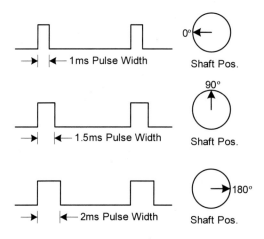

**Figure 5-66. Servo Motor Shaft Position**

*Stepper Motor:* Similar to the DC motors discussed previously in this section, a permanent magnet style stepper motor consists of a stator and a rotor. The rotor contains a set of permanent magnets, and the stator contains the coils. There more fixed on the stator the more precise positioning (stepping). Unlike a DC motor, a stepper motor will only rotate when it is energized by a pulse train, with each pulse causing the motor to rotate an incremental amount.

*Shaft Encoder:* A shaft encoder is an electrical-mechanical device mounted to a shaft that converts linear (rotary distance the shaft has travelled) or rotary displacement (angle of shaft travel) into digital or pulse signals. An incremental encoder generates a pulse for each incremental step in its rotation which in turn is processed by the controller to calculate position.

*Selsyn* (abbreviation for Self Synchronous): A selsyn is a type of rotary electrical transformer that is used for measuring and remotely indicating the angle of a rotating shaft. A typical system consists of a transmitter and a receiver, connected by three wires (no mechanical connection exists between the units). The transmitter and receiver may be identical units. Rotation at one location (the shaft) will create the same rotation in the other (the indicator).

## 5.5.8 Example Motor Elementary Diagrams

Italicized text is for informational purposes only and should not be included in the elementary wiring diagram drawing.

## Simple Stop-Start Pushbutton

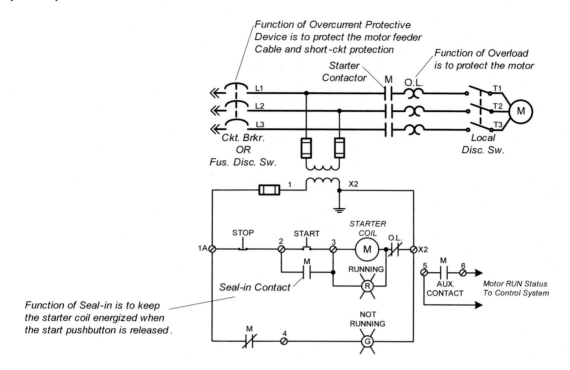

**Figure 5-67. Stop-Start Pushbutton**

## Simple Reversing Motor

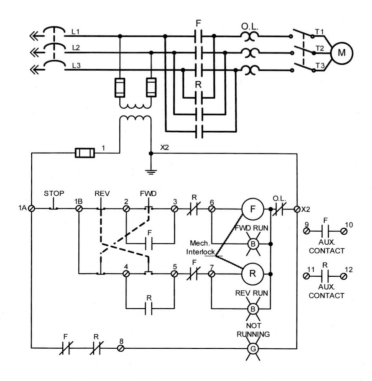

**Figure 5.68: Simple Reversing Motor**

## Simple Hand-Off-Auto (HOA) Selector Switch

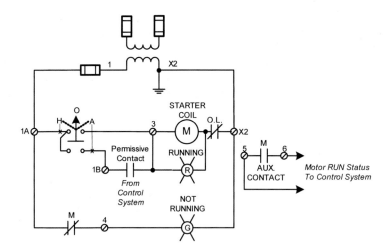

**Figure 5-69. H-O-A Selector Switch**

## Simple Two-Speed One-Winding Motor

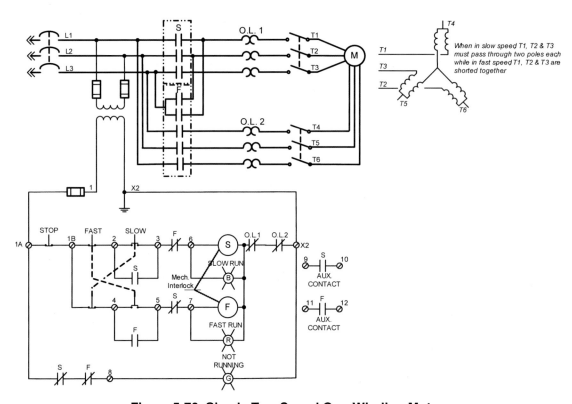

**Figure 5-70. Simple Two-Speed One-Winding Motor**

## Simple Two-Speed Two-Winding Motor

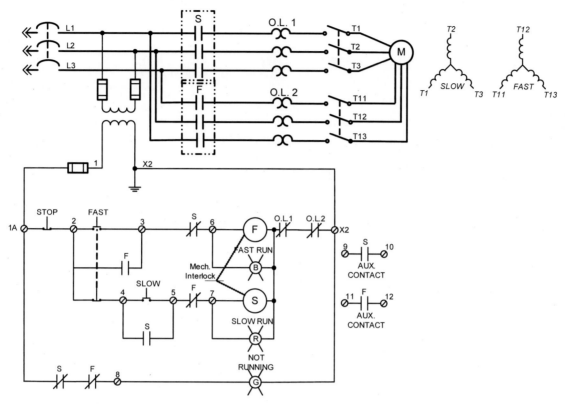

**Figure 5-71. Simple Two-Speed Two-Winding Motor**

## Simple Synchronous Motor

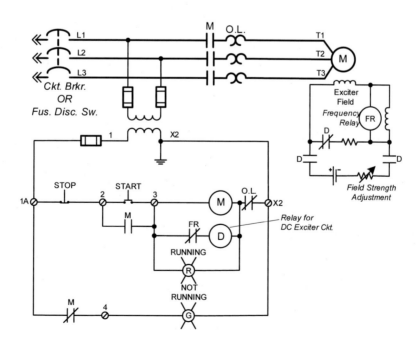

**Figure 5-72. Simple Synchronous Motor**

# 5.5.9 Motor Feeder Sizing Table

## Table 5-15. Motor Feeder Sizing Table

| **** 460VAC, 3Phase, MOTOR STARTERS, FEEDERS AND SIZES **** (Based on NEC Tables 310.16 and 430.250) | | | | | | | | | | | | |
|---|---|---|---|---|---|---|---|---|---|---|---|---|
| Motor HP | Starter Size (1) | Space Factor (2) | FLA | HMCP Rating (Adj Setting) | HMCP CAM Setting | Feeder Wire Size (7) | Max. Fdr. Ampacity (3) | Fdr Cable XLP-HP OD / Area | Conduit Size Reqd. (5) | Ohms / K Feet (6) | Max Length (in feet) 5% V. D. (24V) | 3% V. D. (14.4V) |
| 0.5 | 1 | 1 | 1.1 | 3A (0.69-2.5) | C | 4/C #12AWG 7/C #12AWG | 16 | 0.475" 0.1772sq in | 3/4" | 2.00 | 6915 | 3983 |
| 0.75 | 1 | 1 | 1.6 | 3A (0.69-2.5) | E | 4/C #12AWG 7/C #12AWG | 16 | 0.475" 0.1772sq in | 3/4" | 2.00 | 4754 | 2739 |
| 1 | 1 | 1 | 2.1 | 3A (0.69-2.5) | G | 4/C #12AWG 7/C #12AWG | 16 | 0.475" 0.1772sq in | 3/4" | 2.00 | 3622 | 2087 |
| 1.5 | 1 | 1 | 3.0 | 7A (1.5-5.7) | C | 4/C #12AWG 7/C #12AWG | 16 | 0.475" 0.1772sq in | 3/4" | 2.00 | 2536 | 1461 |
| 2 | 1 | 1 | 3.4 | 7A (1.5-5.7) | D | 4/C #12AWG 7/C #12AWG | 16 | 0.475" 0.1772sq in | 3/4" | 2.00 | 2237 | 1289 |
| 3 | 1 | 1 | 4.8 | 15A (3.4-12.6) | B | 4/C #12AWG 7/C #12AWG | 16 | 0.475" 0.1772sq in | 3/4" | 2.00 | 1585 | 913 |
| 5 | 1 | 1 | 7.6 | 15A (3.4-12.6) | D | 4/C #12AWG 7/C #12AWG | 16 | 0.475" 0.1772sq in | 3/4" | 2.00 | 1001 | 577 |
| 7.5 | 1 | 1 | 11 | 30A (6.9-25.2) | B | 4/C #12AWG 7/C #12AWG | 16 | 0.475" 0.1772sq in | 3/4" | 2.00 | 692 | 399 |
| 10 | 1 | 1 | 14 | 30A (6.9-25.2) | D | 4/C #12AWG 7/C #12AWG | 16 | 0.475" 0.1772sq in | 3/4" | 2.00 | 544 | 313 |
| 15 | 2 | 1 | 21 | 50A (11.5-42.1) | C | 4/C #10AWG | 24 | 0.56" 0.2463sq in | 1" 1.5" | 1.20 | 604 | 348 |
| 20 | 2 | 1 | 27 | 50A (11.5-42.1) | E | 3/C #8AWG w/GRD | 44 | 0.66" 0.3421sq in | 1" 1.5 | 0.78 | 723 | 417 |
| 25 | 2 | 1 | 34 | 50A (11.5-42.1) | F | 3/C #8AWG w/GRD | 44 | 0.66" 0.3421sq in | 1" 1.5 | 0.78 | 574 | 331 |
| 30 | 3 | 1.5 | 40 | 100A (23-84.5) | C | 3/C #6AWG w/GRD | 60 | 0.745" 0.4359sq in | 1.5" | 0.49 | 777 | 448 |
| 40 | 3 | 1.5 | 52 | 100A (23-84.5) | D | 3/C #6AWG w/GRD | 60 | 0.745" 0.4359sq in | 1.5" | 0.49 | 597 | 344 |
| 50 | 3 | 1.5 | 65 | 100A (23-84.5) | F | 3/C #4AWG w/GRD | 76 | 0.89" 0.6221sq in | 1.5" | 0.31 | 755 | 435 |
| 60 | 4 | 2 | 77 | 150A (34.6-126.7) | D | 3/C #2AWG w/GRD | 104 | 1.02" 0.8171sq in | 1.5" | 0.20 | 988 | 569 |
| 75 | 4 | 2 | 96 | 150A (34.6-126.7) | F | 3/C #1/0AWG w/GRD | 136 | 1.24" 1.2076sq in | 2" | 0.12 | 1321 | 761 |
| 100 (4) | 4 | 2 | 124 | 150A (57-130.7) | D | 3/C #2/0AWG w/GRD | 152 | 1.34" 1.4103sq in | 3" | 0.10 | 1227 | 707 |
| 125 (4) | 5 | 3.5 | 156 | 250A (86.6-183.6) | G | 3/C #4/0AWG w/GRD | 208 | 1.578" 1.9557sq in | 3" | 0.063 | 1548 | 892 |
| 150 (4) | 5 | 3.5 | 180 | 250A (96.2-204) | H | 3/C #4/0AWG w/GRD | 208 | 1.578" 1.9557sq in | 3" | 0.06 | 1342 | 773 |

(1) Starter size based upon FVNR Starter Unit with Ckt. Brkr.
(2) Space factor is based upon Allen Bradley "Centerline" model MCC combination starter.
(3) The above wire sizes are based on 90 deg. C conductor termination rating, derated 80% for continuous load.
(4) Soft start style starters shall be considered for these HP levels.
(5) NEC Chapter 9 Table 4 (Art. 344 Rigid Metal Conduit) Based upon 40% fill over 2 wires.
(6) NEC Chapter 9 Table 9 AC resistance for uncoated copper wires in steel conduit.
(7) Top size shown is for motor feeder only; bottom size shown is for motor feeder and control wires in same cable.

Continuous Load Definition: A load where the maximum current is expected to continue for 3 hours or more. *Conductors that get too hot for too long can lead to a loss of temper of the metal components and a potential softening of the insulation material.*

Conductors that supply a single motor used in a continuous duty application shall have an ampacity of not less than 125% of the motor's full load current (or stated another way, the conductor may only be loaded to 80% of its capacity due to a continuous load).

## 5.6 VARIABLE FREQUENCY DRIVES (VFDS)

### 5.6.1 Types of Variable Frequency AC Drives

The three main types of VFD are IGBT, BJT, and GTO.

*IGBT:* Insulated Gate Bipolar Transistor (three-terminal power semiconductor device, noted for high efficiency and fast switching).

*BJT:* Bipolar Junction Transistor (three-terminal electronic device constructed of doped semiconductor material that may be used in amplifying or switching applications).

*GTO:* Gate Turn-Off Thyristor (special thyristor that can be turned on by a positive gate signal and turned off by a negative gate signal).

All VFDs use their output devices (IGBTs, transistors, thyristors) only as switches to approximate a sine wave output (Figure 5-73). Due to the heating effect from the approximated (not pure) sine wave output, <u>motors must be rated for inverter duty</u> and motor feeder cable lengths should be kept as short as possible (Table 5-16).

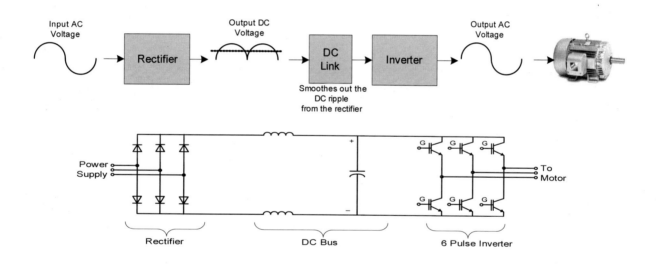

**Figure 5-73. Physical Layout and Schematic of VFD Installation**

**Table 5-16. Maximum Cable Length vs Drive Type for 480 VAC Motor Insulation**

| Drive Type | Semiconductor $t_{rise}$ range (µs) | Motor Allowable Line-to-Line Surge Voltage | | |
|---|---|---|---|---|
| | | 1000 $V_{PK}$ | 1200 $V_{PK}$ | 1600 $V_{PK}$ |
| IGBT | 0.05 – 0.4 | 30 – 1000 feet | 50 – 200 feet | Unlimited |
| BJT | 0.5 – 2.0 | 25 – 600 feet | 500 – 1000 feet | Unlimited |
| GTO | 2.0 – 4.0 | 600 – 1000 feet | 1000 – 2000 feet | Unlimited |

In addition, steps should be taken to reduce the "reflected wave" phenomenon[123] associated with VFDs by the addition of output filters if the motor feeder cable length is above 150 feet or utilizing 12-pulse or 18-pulse VFD units.

## 5.6.2 VFD Applications

**Constant Torque:** Requires the motor to produce full load torque at zero speed.

- Crane

- Elevator

- Conveyor

- P.D. Pumps

**Variable Torque:** A type of load that requires high torque at low speeds and low torque at high speeds. Horsepower remains constant, because speed and torque are inversely proportional.

- Centrifugal and axial pumps

- Fans and blowers

- Mixers and agitators

**Constant HP:** A type of load that has decreasing torque requirements at higher speeds.

- Grinders or Lathes

- Winding machines

**Regenerative:** Used instead of mechanical friction brake shoes for high speed loads that must be stopped more rapidly than coast-down speed.

- Centrifuges

CAUTION: Care should be taken not to operate a motor at too low a frequency output from a VFD (typically $\leq 5$ Hz) because the motor's self-cooling capability is lost at very low speed.

## 5.6.3 Harmonics Associated with VFDs

Harmonics are deviations from the ideal sinusoidal AC line voltage and current waveforms.

- These distortions are typically low in magnitude but can have an impact on the performance of connected devices.

---

123. The reflected wave phenomenon is caused by the electronic switching devices utilized within a VFD which can result in shaft currents (currents induced within the motor shaft) which results in premature bearing failure. The reflected wave occurs when cable impedance, or electrical resistance, does not match motor surge impedance.

- With the advent of power electronics, non-linear loads, and power switching devices such as Variable Frequency Drives, there is concern for a potential increase in the level of harmonic distortion conducted back onto utility power lines.

### Solutions for VFD-induced Harmonics

Several techniques can be used to mitigate harmonic distortion associated with VFDs, including line reactors, DC chokes, and multi-pulse drives (12 pulse and 18 pulse).

There are applications in existence where the VFD may output a frequency > 60 Hz, as many motors are tested up to 90 Hz. When this feature is to be utilized, the equipment should be evaluated thoroughly to ensure that it can safely handle a frequency in excess of 60 Hz.

## 5.7 Pressure Safety Devices

Pressure safety devices are used to protect equipment and personnel from the hazards of excessive pressure in process vessels. They include pressure safety valves, tank vents, and rupture discs.

### 5.7.1 Terminology

*Pressure Relief Device:* A device designed to prevent the internal pressure in a pressure vessel that is exposed to abnormal or emergency conditions from rising above a pre-determined maximum pressure.

### Relief Device Terms

**Accumulation:** The pressure increase above the Maximum Allowable Working Pressure (MAWP) of a vessel during discharge through a pressure relief device.

**Back Pressure:** The pressure that exists at the outlet of a pressure relief device as a result of the pressure in the discharge system. It is the sum of the built-up and superimposed back pressure.

- *Built-Up Back Pressure:* The increase in pressure in the discharge header that develops as a result of flow after the relief device opens.

- *Superimposed Back Pressure:* The static pressure that exists at the outlet of a pressure relief device at the time the device is required to operate.

**Blowdown:** The difference between the set pressure and the closing pressure of a pressure relief valve.

**CCF (Combination Capacity Factor):** The calculated value that is derived from data obtained during certified capacity testing of a stand-alone relief valve and a relief valve/rupture disk combination. CCF = Flow $_{\text{Combination Capacity}}$/Flow $_{\text{Stand-Alone Relief Valve Capacity}}$

**Coincident Temperature:** A value normally used with rupture discs. It is the temperature of the flowing fluid after a rupture disc has burst.

**Design Pressure:** The pressure at which the vessel will normally operate.

**Gag:** A device that can hold a pressure relief valve in the closed position. A gag is used to allow equipment pressure testing without removing or blinding (blocking off) the pressure relief valve.

**Lifting Lever:** A device for manually opening a relief valve. Section VIII of the ASME Code covering pressure vessels requires pressure relief valves to have lifting levers on air, steam, and hot water (over 140°F) service.

**MAWP:** The maximum allowable working pressure. This is what the fabricated vessel is rated for.

- Maximum vessel pressure with one working relief valve is (110% x MAWP)

- Maximum vessel pressure with two working relief valves is (116% x MAWP)

### 5.7.2 Types of Pressure Relief/Safety Valves

*Spring Loaded Design:* The valve consists of an inlet valve or nozzle mounted on the pressurized system, a disc held against the nozzle to prevent flow under normal operating conditions, a spring to hold the disc closed, and a body/bonnet to contain the operating elements (Figure 5-74). The spring loading is adjustable to vary the pressure at which the valve will open.[124]

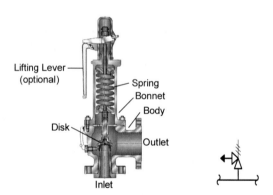

**Figure 5-74. Spring Loaded Pressure Relief Valve with Symbol**

*Pilot:* This type of relief valve consists of a main valve with a piston or diaphragm operated disc, and a pilot (Figure 5-75[125]). Under normal operating conditions, the pilot allows system pressure into the piston chamber. Since the piston area is greater than the disc seat area, the disc is held closed. When the set pressure is reached, the pilot actuates to shut off system fluid to the piston chamber and simultaneously vents the piston chamber. This causes the disc to open.

*Pressure Safety Valve:* A pressure safety valve (PSV) is actuated by static inlet pressure, and characterized by rapid opening or "pop" action. It is typically used for steam and air service.

---

124. Image obtained from www.tycoflowcontrol.com.
125. Image obtained from www.tycoflowcontrol.com.

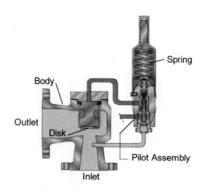

**Figure 5.75: Pilot Operated Relief Valve**

## 5.7.3 Tank Venting

*Emergency Relief Vent* (Typically used on storage tanks–API rated): Emergency relief vents (Figure 5-76[126]) are designed to provide emergency relief capacity beyond what is furnished by the operating vent on tanks. Under normal operating conditions, the emergency vent pallet assembly is closed, providing an effective vapor seal. In the event of an emergency (such as fire involvement of the tank), the pallet lifts in response to the increased pressure in the tank's vapor space. Vapor is expelled, protecting the tank from dangerous over-pressurization. The pallet assembly automatically closes and reseals when the pressure is reduced. Emergency vents do not provide vacuum relief. Vacuum relief must be supplied by independently operating conservation vents.

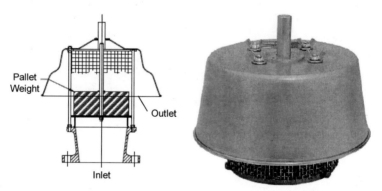

**Figure 5-76. Emergency Tank Relief Vent (Cutaway and Exterior View)**

*Conservation Vent:* A device intended for use where both pressure, and vacuum relief are required. The pallets in the vent housing allow intake of air and escape of vapors as the tank normally breathes in and out (Figure 5-77[127]). The pallets open and close to permit only the intake or outlet relief necessary to maintain the tank within permissible working pressures and avoid damage to the tank.

*Flame Arrester:* A flame arrester is installed where it is not necessary to conserve vapors, but where low flash point liquids must be protected against fire and explosion from exterior sources

---

126. Protectoseal pressure relief.
127. Protectoseal conservation vent.

---

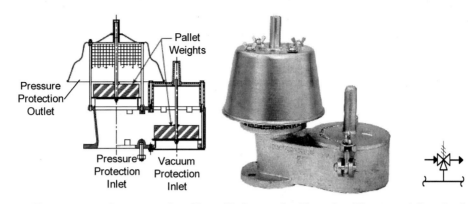

**Figure 5-77. Conservation Vent (Schematic, Exterior View, and Symbol)**

of ignition. The tightly spaced circular flame arrester grid plates are integral with the vent housing (Figure 5-78[128]). The grid plates are mounted on the end of a vent pipe from the tank (the flame arrester may also be integral to a conservation vent). The vapors are allowed to escape into the atmosphere, and air can be drawn into the tank through the specially designed flame arrester grid assembly. If an ignition source outside the tank ("unconfined deflagration") is encountered, the flame arrester provides protection for the tank's vapor space.

**Figure 5-78. Flame Arrester (Schematic, Exterior View, and Symbol)**

## 5.7.4 Types of Rupture Disks

A rupture disk is a type of pressure relief device that protects a vessel or system from over- or under-pressurization. A rupture disk is a type of sacrificial part, because it has a one-time-use membrane that fails at a predetermined pressure, either positive or a vacuum. The membrane is usually a thin metal foil, but nearly any material can be used to suit a particular application. A rupture disk provides fast response to an increase in system pressure, but once the membrane has failed it will not reseal.

*Reverse Buckling:* When a reverse buckling disk (Figure 5-79) is loaded in compression, it is able to resist operating pressures up to 100% of minimum burst pressure, even under pressure cycling or pulsating conditions. It is designed for non-fragmentation upon activation. Reverse buckling disks are recommended for use in combination with pressure relief valves to isolate them from normal process conditions, ensuring excellent leak tightness. Forward acting rupture

---

128. Protectoseal flame arrester.

disks are not recommended for use in combination with a relief valve because its design nature could allow pieces of the ruptured disk to enter the relief valve.

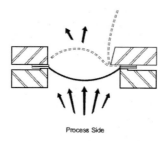

Process Side

**Figure 5-79. Reverse Buckling Rupture Disk**

*Forward Acting (Tension):* Because pressure-induced loading is applied to the concave side, the disk is subjected to tensile forces (Figure 5-80). Forward acting disks establish burst pressure by the tensile strength of the material.

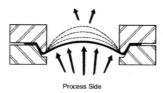

Process Side

**Figure 5-80. Forward Acting Rupture Disk**

## 5.7.5 Rupture Disk Accessories

*Holder (Safety Head):* A rupture disk holder (Figure 5-81[129]) allows the rupture disk to be pre-assembled. This ensures that it is properly seated before installation into the system.

**Figure 5-81. Rupture Disk Holder**

*Tell-Tale Connection:* The ASME code requires that the space between a rupture disk (bursting disk) device and a pressure relief valve be provided with a pressure gauge, tri-cock valve[130], free vent (vent port on disk holder contains no plug assembly), or other suitable telltale indicator (Figure 5-82[131]).

---

129. BS&B holder.
130. Tri-cock valve is opened to check if pressure is above or below the level of the valve.
131. Oseco holder.

**Figure 5-82. Rupture Disk Telltale Indicators**

*Burst Indicator:* A burst indicator is designed to operate in a "normally closed" electrical circuit. A membrane is used to support an electrical conducting circuit. When the pressure event (disk rupture) occurs, the flow of fluid places the membrane in tension, which leads to a break of the electrical conductor. This permanently changes the electrical status of the sensor to "open."

## 5.7.6 Rupture Disk Performance

**Tolerances:** The burst pressure tolerance at the specified rupture disk temperature may not exceed ±2 psi for marked burst pressures up to and including 40 psi and ±5 psi for marked burst pressures above 40 psi.

**Operating Ratios:** Defined as the relationship between operating pressure and the stamped burst pressure, and usually expressed as a % (i.e., $P_o$, $P_b$). In general, good service life can be expected when operating pressures do not exceed the following:

- 70% of stamped burst pressure for conventional pre-bulged rupture disks (domed style).

- 80% of stamped burst pressure for composite design rupture disks (pressure differential style that contain multiple pre-drilled holes).

- 80%–90% of stamped burst pressure for forward acting rupture disks.

- Up to 90% of stamped burst pressure for reverse acting (reverse buckling) rupture disks.

## 5.7.7 Pressure Relief Device Sizing Contingencies

Reference Table 5-17 for contingencies to be considered when sizing a pressure relief device.

## Table 5-17. Pressure Relief Device Sizing Contingencies

| Pressure Relief Contingencies | | | WORST CREDIBLE SCENARIO |
|---|---|---|---|
| PRV Designer: | Plant Area: | Proj# | |
| Process Knowledgeable Person: | P & ID No. with latest Rev. and Rev. Date: | | |
| Date: | Relief Device Tag No.: | | |
| Equipment to be protected: | | | |
| Possible Pressure or Vacuum Contingencies | Relief Flow Rate Lb/Hr | Orifice Area Sq In | |
| 1 Exposure to External Fire | | | |
| 2 Incorrect Valve Position | | | |
| 3 Overfilling a Vessel | | | |
| 4 Emptying a Vessel | | | |
| 5 Thermal Expansion | | | |
| 6 Coolant Failure | | | |
| 7 Loss of Heat | | | |
| 8 Power Failure | | | |
| 9 Instrument Failure | | | |
| 10 Utility Failure | | | |
| Air | | | |
| Water | | | |
| Nitrogen | | | |
| Steam | | | |
| Refrigeration | | | |
| Other (Specify) | | | |
| 11 Abnormal Heat Input | | | |
| 12 Abnormal Flow Input | | | |
| 13 Exchanger Tube Failure | | | |
| 14 Chemical Reaction | | | |
| 15 Accumulation of Non-Condensables | | | |
| 16 Mechanical Failures | | | |
| 17 Pump Overpressure | | | |
| 18 Sudden Mixing of Hot & Highly Volatile Materials | | | |
| 19 Transient Pressure Surges | | | |
| 20 Absorbent Flow Failure | | | |
| 21 Plugging/Fouling | | | |
| 22 Other (Specify) | | | |
| Comments: | | | |

## 5.7.8  Pressure Levels as a % of MAWP

*Reference API 521*

### Table 5-18. Pressure Levels as a % of MAWP

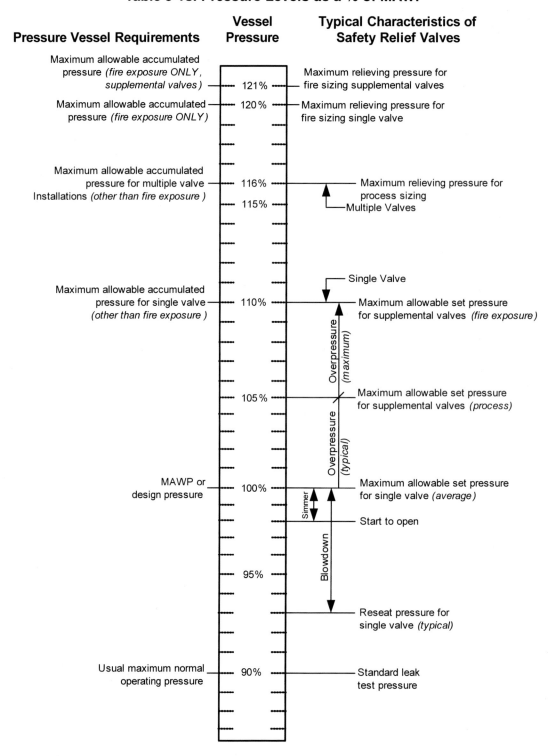

## 5.7.9 Pressure Relief Valve Sizing (per ASME/API RP520)

Liquid Service                Gas & Vapor Service                Steam

$$A = \frac{W_{gpm}}{38K_dK_vK_w(CCF)}\sqrt{\frac{G}{P_R - P_B}} \qquad A = \frac{W_{scfm}\sqrt{TZG}}{1.175C_gK_dP_RK_b(CCF)} \qquad A = \frac{W_{lb/hr}}{51.5C_gP_RK_dK_{SH}(CCF)}$$

CCF = Combination De-rating Factor (1 if not combination, otherwise 0.9)
A = Minimum required orifice area, in$^2$
W = Required relieving rate, [gpm (liquids); scfm (gas/vapor); lb/hr (steam)]
T = Relieving temperature (°R)
TR = Actual relieving temperature (°K)
G = Specific gravity
k = Ratio of specific heats
$K_d$ = Coefficient of discharge (by manufacturer, ~0.975)
$K_b$ = Back pressure correction factor (use 1.0 for atmospheric): ratio of the capacity with backpressure, $C_1$, to the capacity when discharging to atmosphere, $C_2$

$$K_b = \frac{C_1}{C_2}$$

$K_v$ = Correction factor due to viscosity at flowing conditions (reference Table 5-19)
$K_W$ = Back pressure correction factor for liquids (bellows balanced valves only)
$K_{SH}$ = Superheat correction factor (for saturated use 1.0; otherwise reference Table 5-20)
$P_R$ = Upstream relieving pressure (psia)
$P_B$ = Absolute back pressure (psia)
M = Molar mass (kg/kmol)
$R_u$ = Universal gas constant (8.314 Nm/kmol K)
$v$ = Specific volume of the gas at the actual relieving pressure and temperature (m/kg)

    $C_g$ = Nozzle gas constant

$$C_g = 520\sqrt{k\left(\frac{2}{(k+1)}\right)^{\left(\frac{k+1}{k-1}\right)}} \quad \text{for } k > 1$$

$$C_g = 315 \quad \text{for } k = 1$$

Z = Compressibility factor

$$Z = \frac{10^5 P_R M v}{R_u T_R}$$

## Table 5-19. $K_v$ = Correction Factor Due to Viscosity at Flowing Conditions

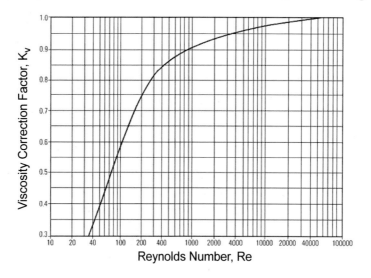

## Table 5-20. Superheat Correction Factor ($K_{SH}$)

| Set Pressure psig | Temperature °F | | | | | | | | | |
|---|---|---|---|---|---|---|---|---|---|---|
| | 300 | 400 | 500 | 600 | 700 | 800 | 900 | 1000 | 1100 | 1200 |
| 15 | 1.00 | 0.98 | 0.93 | 0.88 | 0.84 | 0.80 | 0.77 | 0.74 | 0.72 | 0.70 |
| 20 | 1.00 | 0.98 | 0.93 | 0.88 | 0.84 | 0.80 | 0.77 | 0.74 | 0.72 | 0.70 |
| 40 | 1.00 | 0.99 | 0.93 | 0.88 | 0.84 | 0.81 | 0.77 | 0.74 | 0.72 | 0.70 |
| 60 | 1.00 | 0.99 | 0.93 | 0.88 | 0.84 | 0.81 | 0.77 | 0.75 | 0.72 | 0.70 |
| 80 | 1.00 | 0.99 | 0.93 | 0.88 | 0.84 | 0.81 | 0.77 | 0.75 | 0.72 | 0.70 |
| 100 | 1.00 | 0.99 | 0.94 | 0.89 | 0.84 | 0.81 | 0.77 | 0.75 | 0.72 | 0.70 |
| 120 | 1.00 | 0.99 | 0.94 | 0.89 | 0.84 | 0.81 | 0.78 | 0.75 | 0.72 | 0.70 |
| 140 | 1.00 | 0.99 | 0.94 | 0.89 | 0.85 | 0.81 | 0.78 | 0.75 | 0.72 | 0.70 |
| 160 | 1.00 | 0.99 | 0.94 | 0.89 | 0.85 | 0.81 | 0.78 | 0.75 | 0.72 | 0.70 |
| 180 | 1.00 | 0.99 | 0.94 | 0.89 | 0.85 | 0.81 | 0.78 | 0.75 | 0.72 | 0.70 |
| 200 | 1.00 | 0.99 | 0.95 | 0.89 | 0.85 | 0.81 | 0.78 | 0.75 | 0.72 | 0.70 |
| 220 | 1.00 | 0.99 | 0.95 | 0.89 | 0.85 | 0.81 | 0.78 | 0.75 | 0.72 | 0.70 |
| 240 | - | 1.00 | 0.95 | 0.90 | 0.85 | 0.81 | 0.78 | 0.75 | 0.72 | 0.70 |
| 260 | - | 1.00 | 0.95 | 0.90 | 0.85 | 0.81 | 0.78 | 0.75 | 0.72 | 0.70 |
| 280 | - | 1.00 | 0.96 | 0.90 | 0.85 | 0.81 | 0.78 | 0.75 | 0.72 | 0.70 |
| 300 | - | 1.00 | 0.96 | 0.90 | 0.85 | 0.81 | 0.78 | 0.75 | 0.72 | 0.70 |
| 350 | - | 1.00 | 0.96 | 0.90 | 0.86 | 0.82 | 0.78 | 0.75 | 0.72 | 0.70 |
| 400 | - | 1.00 | 0.96 | 0.91 | 0.86 | 0.82 | 0.78 | 0.75 | 0.72 | 0.70 |
| 500 | - | 1.00 | 0.96 | 0.92 | 0.86 | 0.82 | 0.78 | 0.75 | 0.73 | 0.70 |
| 600 | - | 1.00 | 0.97 | 0.92 | 0.87 | 0.82 | 0.79 | 0.75 | 0.73 | 0.70 |
| 800 | - | - | 1.00 | 0.95 | 0.88 | 0.83 | 0.79 | 0.76 | 0.73 | 0.70 |
| 1000 | - | - | 1.00 | 0.96 | 0.89 | 0.84 | 0.78 | 0.76 | 0.73 | 0.71 |
| 1250 | - | - | 1.00 | 0.97 | 0.91 | 0.85 | 0.80 | 0.77 | 0.74 | 0.71 |
| 1500 | - | - | 1.00 | 1.00 | 0.93 | 0.86 | 0.81 | 0.77 | 0.74 | 0.71 |

## 5.7.10 Rupture Disk Sizing (per ASME/API RP520)

*Coefficient of Discharge Method $K_d$:* Requires 8/5 rule (8 pipe diameters of inlet piping and 5 pipe diameters of outlet piping, and piping must be at least the same nominal size as the rupture disk. In addition, the rupture disk must discharge to atmosphere:

**Gas and Vapor:**

*(Sub-critical flow)(API RP520)*          *(Critical flow)(ASME)*

$$A = \frac{W_{lb/hr}}{735F_2K_d}\sqrt{\frac{TZ}{MP(P-P_e)}} \qquad A = \frac{W_{lb/hr}}{CPK_d}\sqrt{\frac{TZ}{M}}$$

$$A = \frac{W_{scfm}}{4645F_2K_d}\sqrt{\frac{TZ}{MP(P-P_e)}} \qquad A = \frac{W_{scfm}\sqrt{TZM}}{6.32K_dCP}$$

$$A = \frac{W_{scfm}}{864F_2K_d}\sqrt{\frac{TZG}{MP(P-P_e)}} \qquad A = \frac{W_{scfm}\sqrt{TZG}}{1.175K_dCP}$$

**Steam:**

(API RP520)          (ASME)

$$A = \frac{W_{lb/hr}}{51.5PK_dK_NK_{SH}} \qquad A = \frac{W_{lb/hr}}{51.5PK_dK_N}$$

**Liquid:**

(API RP520)          (ASME)
*Viscous:*           *Water:*

$$A_V = \frac{A_R}{K_V} \qquad\qquad A = \frac{W_{lb/hr}}{2407\sqrt{(P-P_e)\omega}}$$

*Non-Viscous:*

$$A_R = \frac{W_{gpm}}{38K_dK_V}\sqrt{\frac{G}{P-P_e}}$$

A = Area in square inches
    $A_R$ = Area without viscosity correction
    $A_V$ = Area with viscosity correction
W = Required relieving rate, [lb/hr (liquids and steam); scfm (gas/vapor)]
C = Gas constant (function of ratio of specific heat)
Z = Compressibility factor (use 1.0 if unknown)
G = Specific gravity
M = Molecular weight

P = Set pressure plus overpressure allowance plus atmospheric pressure (psia)

$P_e$ = Exit pressure (psia)

T = Inlet temperature (in °R)

$K_d$ = Coefficient of discharge (0.62 per ASME code)

$K_n$ = Steam correction factor

$K_{SH}$ = Superheat correction factor (for saturated use 1.0; otherwise, reference Table 5-19)

$K_v$ = Correction factor due to viscosity at flowing conditions (reference Table 5-20)

$\omega$ = Specific weight of water at inlet

k = Ratio of specific heats

$r = P_e/P$

$$F_2 = \sqrt{\left(\frac{k}{k-1}\right)(r)^{2/k}\left[\frac{1-r^{(k-1)/k}}{1-r}\right]}$$

*Rupture Disk Pressures as a % of MAWP (Figure 5-83)*

*Reference API 521*

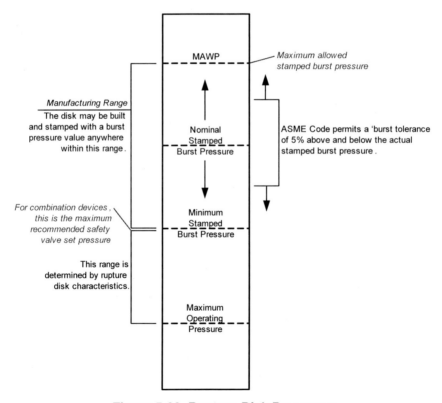

**Figure 5-83. Rupture Disk Pressures**

## 5.7.11 Selection of Pressure Relief/Safety Devices

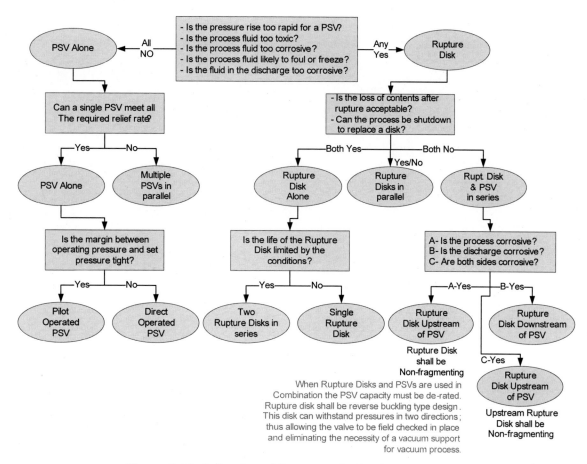

**Figure 5-84. Selection of Pressure Relief/Safety Devices**

## 5.8 RELAYS/SWITCHES

### 5.8.1 Relays

A relay is an electrically operated switch made up of an electromagnet and a set of contacts.

*Types of Relays:*

**Non-Latching:** This type has an initial position of normally closed (NC), or normally open (NO) maintained by the force of a spring or permanent magnet while no current flows. The normally closed (NC) contact will open and the normally open (NO)) contact will close while current is flowing through the relay coil. The contact is maintained by the force of the magnetic field while current flows through the coil. When the current stops, the relay reverts back to its initial NC or NO position.

**Latching**: A latching relay can have one or two coils. Latching relays have no default position and remain in their last position when the drive current stops flowing. While the relays themselves may be latching, their reset position in a module is based on the control circuitry and software. Latching relays are useful in applications where power consumption and dissipation must be limited. Once actuated, they require no current flow to maintain their position. In one-coil latching, the direction of current flow determines the position of the relay. In two-coil latching, the coil in which the current flows determines the position of the armature.

**Time Delay Relays**

*Off-Delay:* The delay feature of an off-delay relay acts when the relay is de-energized. When the coil of an off-delay relay is energized, the state of the contacts immediately changes. When the coil is de-energized, the state of the contacts will return to the shelf state AFTER a pre-determined time delay.

An off-delay relay (Figure 5-84) may have either NOTO (normally open time open) or NCTC (normally closed time close) contacts (may also be in combination).

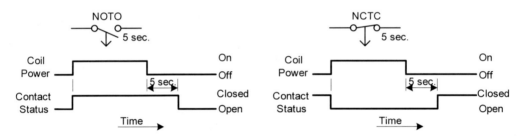

**Figure 5-84. Off-Delay Relays (NOTO and NCTC)**

*On-Delay:* The delay feature of an on-delay relay acts when the relay is energized. When the coil of an on-delay relay is energized, the state of the contacts will change AFTER a pre-determined time delay. When the coil is de-energized, the state of the contacts will immediately return to the shelf state.

An on-delay relay (Figure 5-85) may have either NOTC or NCTO contacts (may also be in combination).

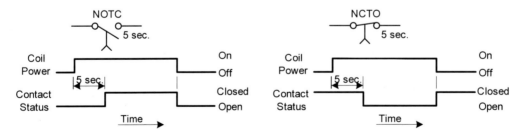

**Figure 5-85. On-Delay Relays (NOTO and NCTC)**

## 5.8.2 Switches

***Toggle Switch:*** Toggle switches are actuated by a pivoting lever placed in one of two or more positions.

**Reed Switch:** An electrical switch that is actuated by an applied magnetic field. The reed switch contains a pair (or more) of magnetically susceptible, flexible metal reeds whose end portions are separated by a small gap when the switch is open. The reeds are hermetically sealed in opposite ends of a tubular glass envelope. A magnetic field (from an electromagnet or a permanent magnet) will cause the reeds to bend, and the contacts to pull together, thus completing an electrical circuit. When the magnetic field ceases, the stiffness of the reeds causes them to separate, and open the circuit.

Mercury Switch: A mercury switch consists of a sealed glass tube containing two unconnected electrodes and a small amount of mercury. As long as the liquid metal remains on the opposite end of the tube, the electrodes remain unconnected and no current will flow. Once the tube is moved past a certain angle, the mercury pools between the two electrodes and a connection is made. The result is electricity flowing through a completed circuit. Once the liquid metal is returned to its original position, the electrical current stops immediately.

Biased (Spring Return): This type of switch contains a spring that returns the actuator to a certain position. The momentary push-button switch is a type of biased switch. The most common type is a "push-to-make" or NO) switch, which makes contact when the button is pressed and breaks when the button is released. A "push-to-break" (normally-closed, or NC) switch, on the other hand, breaks contact when the button is pressed and makes contact when it is released.

**Make before Break aka "Closed Transition Switching":** In this type of switch, the second contact is closed BEFORE the first contact is opened, causing a short overlapping period where both contacts are simultaneously closed.

**Switch Forms**

- SPST (single-pole single throw): 
- DPST (double-pole single throw): 
- DPDT (double-pole double throw): 
- Form A Contact: SPST-NO 
- Form B Contact: SPST-NC 
- Form C Contact: SPDT-NO & NC

# 6. CONTROL SYSTEMS

## 6.1 DOCUMENTATION

Documentation includes items such as P&IDs (Piping and Instrumentation Diagrams, Loop Wiring Diagrams, Logic Diagrams and Instrument Specifications). For example instrument specifications reference ISA-20 standard.

### 6.1.1 Instrumentation Identification Letters

Reference ANSI/ISA-5.1-2009 Instrumentation Symbols and Identification Standard

### Table 6-1. Instrumentation Identification Letters

Note: Numbers in parentheses refer to the preceding explanatory notes in Clause 4.2.

| | First letters (1) | | Succeeding letters (15) | | |
|---|---|---|---|---|---|
| | Column 1 | Column 2 | Column 3 | Column 4 | Column 5 |
| | Measured/ Initiating Variable | Variable Modifier | Readout/Passive Function | Output/Active Function | Function Modifier |
| A | Analysis | | Alarm | | |
| B | Burner, Combustion | | User's Choice | User's Choice | User's Choice |
| C | User's Choice | | | Control | Close |
| D | User's Choice | Difference, Differential | | | Deviation |
| E | Voltage | | Sensor, Primary | | |
| F | Flow, Flow Rate | Ratio | | | |
| G | User's Choice | | Glass, Gauge, Viewing Device | | |
| H | Hand | | | | High |
| I | Current | | Indicate | | |
| J | Power | | Scan | | |
| K | Time, Schedule | Time Rate of Change | | Control Station | |
| L | Level | | Light | | Low |
| M | User's Choice | | | | Middle, Intermediate |
| N | User's Choice | | User's Choice | User's Choice | User's Choice |
| O | User's Choice | | Orifice, Restriction | | Open |
| P | Pressure | | Point (Test | | |
| Q | Quantity | Integrate, Totalize | Integrate, Totalize | | |
| R | Radiation | | Record | | Run |

| | | | | | |
|---|---|---|---|---|---|
| **S** | Speed, Frequency | Safety | | Switch | Stop |
| **T** | Temperature | | | Transmit | |
| **U** | Multivariable | | Multifunction (21) | Multifunction | |
| **V** | Vibration, Mechanical Analysis | | | Valve, Damper, Louver | |
| **W** | Weight, Force | | Well, Probe | | |
| **X** | Unclassified | X-axis | Accessory Devices, | Unclassified | Unclassified |
| **Y** | Event, State, Presence | Y-axis | | Auxiliary Devices | |
| **Z** | Position, Dimension | Z-axis, Safety Instrumented System | | Driver, Actuator, Unclassified final control element | |

## Examples:

FIC = Flow Indicating Controller      FE = Flow Element      FT = Flow Transmitter
FIT = Flow Indicating Transmitter     FY = Flow Converter     FR = Flow Recorder
FSL = Flow Switch Low                 FAL = Flow Alarm Low    FV = Flow Valve
FQI = Flow Quantity Indicator (Totalizer)

## 6.1.2 Instrumentation Line Symbols

**For use on P&IDs.**

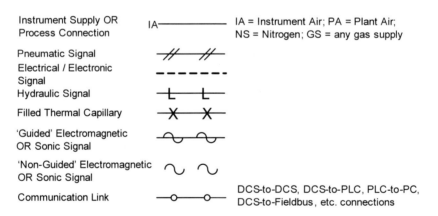

**Figure 6-1. Instrumentation Line Symbols**

## 6.1.3 Instrumentation Location Symbols

For use on P&IDs.

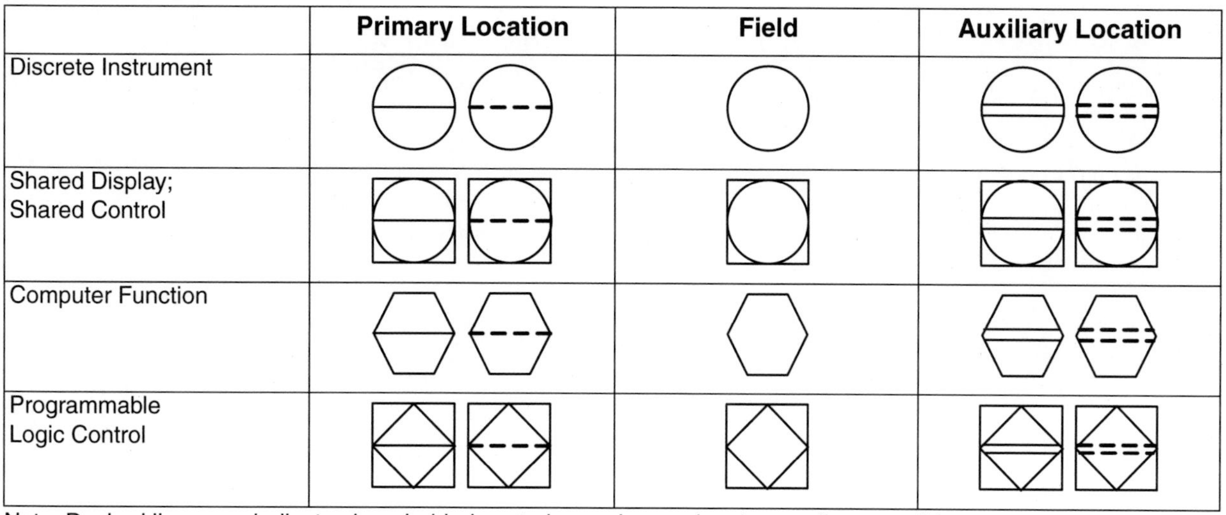

| | **Primary Location** | **Field** | **Auxiliary Location** |
|---|---|---|---|
| Discrete Instrument | | | |
| Shared Display; Shared Control | | | |
| Computer Function | | | |
| Programmable Logic Control | | | |

Note: Dashed line - - - - indicates in or behind central or main panel, not normally accessible to operator.

**Figure 6-2. Instrumentation Location Symbols**

Example:

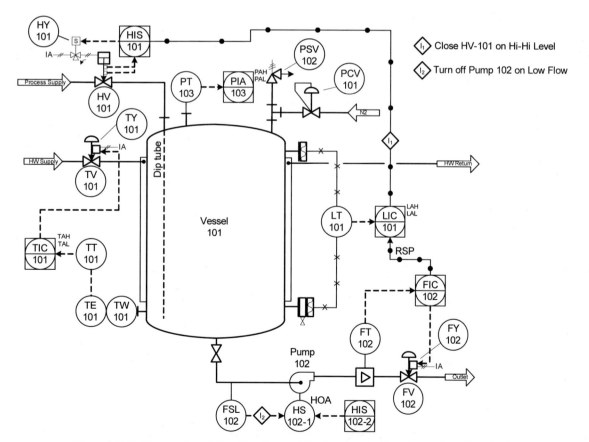

**Figure 6-3. Example of Application of Instrumentation Location Symbols**

## 6.1.4 Binary Logic Diagrams

*Reference ISA-5.2-1976 (R1992) – Binary Logic Diagrams for Process Operations Standard*

### Table 6-2. Binary Logic Diagram Symbols, Functions, and Definitions

| Function | Symbol | Definition |
|---|---|---|
| Input | ⊢ | An input to the logic sequence |
| Output | ⊣ | An Output from the logic sequence |
| AND | A, B → [A] → C | Logic output C exists if and only if all the logic inputs exist. |
| OR | A, B → [OR] → C | Logic output C exists if and only if one or more of the logic inputs exist. |
| Qualified OR | A, B → [*] → C | Logic output C exists if and only if a specified number of logic inputs exist. *indicates the following mathematic symbols shall be used as appropriate with this function: = equal to    ≠ not equal to    < less than    > greater than    ≮ not less than    ≯ not greater than    ≤ less than or equal to    ≥ greater than or equal to |
| NOT | A ⊢◯⊣ B | Logic output B exists if and only if logic input A does not exist |
| Memory (Flip-Flop) | A → [S] → C, B → [R] → D | S represents the set memory and R represents the reset memory. Logic C exists as soon as logic input A exists. C continues to exist, regardless of the subsequent state of A until the memory is reset. The memory is reset when logic input B exists. C remains terminated regardless of the subsequent state of B, until logic input A causes the memory to be set. If logic output D is used it exists when C does not exist, and D does not exist when C exists. |
| Time Element | A → [*] → B | Logic output B exists with a time relationship to logic input A. *indicates the following time parameters shall be used as appropriate with this function: DI t = Delay initiation of output for 't' time (similar to on-delay) DT t = Delay termination of output for 't' time (similar to off-delay) PO t = Pulse, output will exist immediately when input A exists for 't' time For further timing parameters, reference ISA-5.2 Standard |

## 6.1.5 Functional Diagram/Symbols

SAMA[132] (for Scientific Apparatus Makers Association) functional diagram/symbol drawings have been used in the past to represent control loops and logic. This book describes this type of diagram because some companies may still use this type of documentation. The SAMA organization no longer exists; however, SAMA PMC 22.1-1981 functional diagramming symbols have been adopted by ISA and are now included in ANSI/ISA-5.1-2009.

---

132. SAMA is an obsolete association that is referenced within this document should the reader come across this terminology within legacy documents.

## Table 6-3. General Symbols

| Function | Symbol | Function | Symbol |
|---|---|---|---|
| Measuring OR Readout | ○ | Logical AND | AND |
| Manual Signal Processing | ◇ | Logical OR | OR |
| Automatic Signal Processing | □ | Qualified Logical OR < n (LTn); = n (EQn) > n (GTn) | __n |
| Final Controlling | ▽ (trapezoid) | Logical NOT | NOT |
| Final Controlling with Positioner | (trapezoid with line) | Maintained Memory | S / R |
| Time Delay OR Pulse Duration | t (box, Optional Reset) | S; SO = Set memory R; RO = Reset memory | SO / R ; S / RO |

*Automatic Signal Processing Symbols*

## Table 6-4. Functional Diagram Automatic Signal Processing Symbols

| Function | Symbol | Function | Symbol |
|---|---|---|---|
| Controller | $\frac{\Delta}{P}$  $\frac{\Delta}{P\ I}$  $\frac{\Delta}{P\ I\ D}$  OR | Controller | $\frac{\Delta}{K}$  $\frac{\Delta}{K\ \int}$  $\frac{\Delta}{K\ \int\ \frac{d}{dt}}$ |
| Square Root Extractor | $\sqrt{\ }$ | Exponential | $X^n$ |
| Summer | $\Sigma$ | Bias | +/- |
| Multiplier | X | Divider | ÷ |
| High Selecting Low Selecting | High > < Low | Transfer | ◇ T |
| High Limiting Low Limiting | High ⋗ ⋖ Low | Auto Manual | ◇ A  ◇ M |
| Velocity Limiter | V | High/Low Signal Monitor (alarm) | H/L |
| High Signal Monitor (alarm) | H/ | Low Signal Monitor (alarm) | /L |
| Time Function | F(t) | Integrate OR Totalize | Q |

*Example Functional Diagram/Symbols*

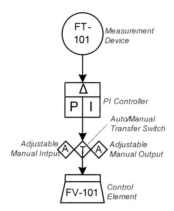

**Figure 6-4. Example Functional Diagram/Symbols (1)**

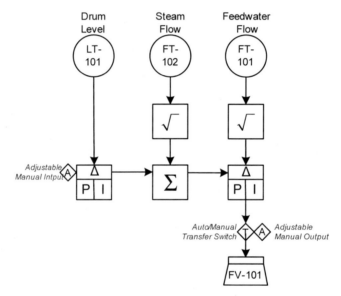

**Figure 6-5. Example Functional Diagram/Symbols (2)**

## 6.1.6 Loop Diagram

*Reference ISA-5.4-1991 Standard*
***Example Hard-wired I/O***

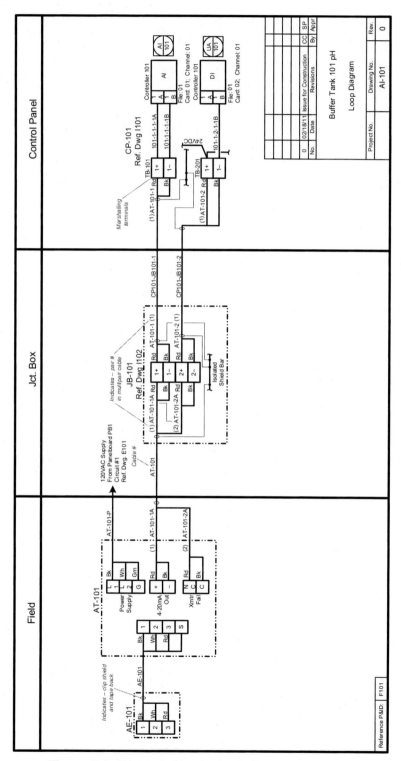

**Figure 6-6. Example Hard-wired I/O Loop Diagram**

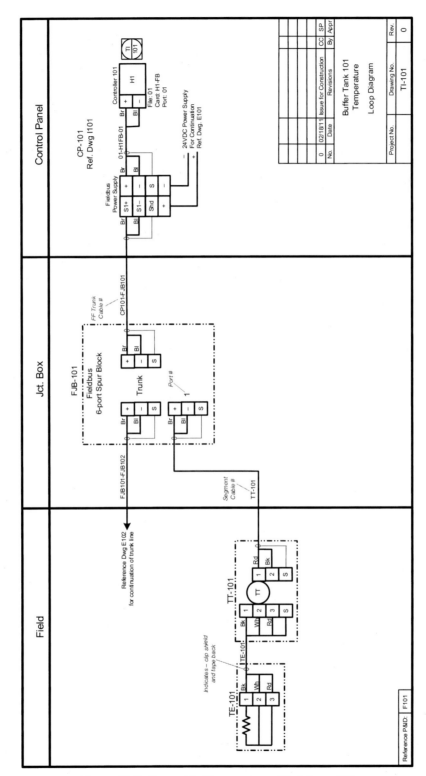

**Figure 6-7. Example** FOUNDATION **Fieldbus I/O**

## 6.2 CONTROL SYSTEM - CONTROLLER ACTIONS

The purpose of a controller is to align an operator entered setpoint with the actual process variable (simple controllers are items like a thermostat in the home and cruise control in an automobile). Reference Figure 6-8 for a block diagram of a feedback controller.

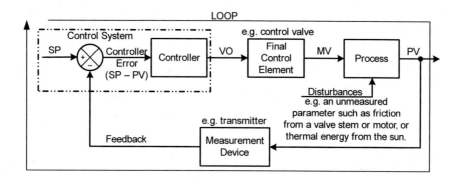

**Figure 6-8. Feedback Controller Components**

SP = Setpoint (user input defining the desired process variable to be maintained)

VO = Valve Output (aka Controller Output)(signal output to the control valve in order to make the valve physically move to the desired position)

MV = Measured Variable (value of process variable at the outlet of the final control element)

PV = Process Variable (the actual process component being measured by a transmitter such as temperature, pressure, etc.)

### 6.2.1 Terminology

*Direct Acting:* An increase in PV (process variable) causes an increase in the controller output.

*Reverse Acting:* An increase in PV (process variable) causes a decrease in the controller output.

*Valve (Process) Action:* This is not to be confused with controller actions stated above. This action defines the relationship between changes in the valve position and changes in the valve (controller) output measurement.

- An increase in valve position causes an increase in the valve output. (Direct Output; e.g., 4 mA = closed and 20 mA = open)

- An increase in valve position causes a decrease in the valve output. (Reverse Output; e.g., 4 mA = open and 20 mA = closed)

*Note: 0% on the operator display always means valve closed to the operator, and 100% always means valve full open to the operator.*

## 6.2.2 PID Control

*Modes*

**Manual:** The operator adjusts the controller setpoint/output (output tracks the setpoint).

*Setpoint Tracking:* This feature makes the SP (setpoint) in the controller track the PV (process variable). This is to enable a bumpless transfer[133] when the controller is switched to Auto from Manual.

**Auto:** The control algorithm manipulates the output to hold the process variable at its setpoint.

*PID: Proportional-Integral-Derivative*

**Proportional (P):** also referred to as GAIN.

With regard to the proportional band, the controller output is proportional to the offset error (change in process measurement so that PV no longer equals SP). With a proportional-only controller, offset error (deviation from setpoint) is always present. Reference the Integral function to see how it is used in the controller to eliminate this offset (error).

The amount added to the controller output is based on the current error. The output moves in proportion to the error between SP and PV. The higher the gain, the greater the rejection of the disturbance, and the greater the response to SP changes. *Any loop will cycle if the gain is increased far enough.*

- Proportional Gain (p-gain) is a multiplier (e.g., if the error is 10 and the gain is 0.8, then the output will move 8% in order to change the process variable value).

- Proportional Band (p-band): The size of the error that gives a 100% controller command output signal. It is a divider as a % (e.g., if the error is 10 and the p-band is 125%, then the output is $[10*(100/125)] = 8\%$, also in order to change the process variable value.

- Conversion between P-Gain & P-Band:

$$P - Band = \frac{100}{P - Gain}$$

**Integral (I):** also referred to as RESET.

The amount added to the output based on the sum of the error. It works by summing the current controller error and the integral of all previous controller errors.

The integral time is the time to repeat the change in the controller output that was due to proportional action alone. Integral action causes the controller output to change at a rate that is proportional to the error. When the output response is somewhat oscillatory, this can be

---

133. Bumpless transfer ensures that the offset error is zero when the controller is switch from manual to auto mode.

stabilized some by adding derivative action. *Any loop will cycle if you reduce the integral time far enough.*

- *Reset Time* is expressed in minutes per repeat.

- *Reset Rate* is expressed in repeats per minute. It is the number of repeats that occur during a given period of time. Doubling the reset rate will cause the output to double within a given period of time.

**Example:** Reset rate = 20 repeats per minute. What is the reset time?

$$20 \text{ repeats per minute } = \frac{60 \text{ sec (per minute)}}{20 \text{ repeats}} = 3 \text{ seconds per repeat}$$

- *Reset Windup* describes several situations in which the reset element of the controller continues to increase (or decrease) the output of the controller even when the change in output does not cause any change in the process measurement (process variable being controlled). When there is no resulting decrease in error, the controller output will continue to increase until it reaches its limit. The problem with reset windup is that when the condition causing the windup is eliminated, the output must "wind down" for a period of time before the decreasing output has any effect on the process. Most modern day controllers have an anti-reset windup feature built in to them.

**Derivative (D):** also referred to as RATE.

With derivative action, the controller output is proportional to the rate of change of the measurement or error. The controller output is calculated by the rate of change of the measurement with time. The derivative takes action to inhibit more rapid changes of the measurement than proportional action alone does. It should not be used on noisy or fast responding control loops such as flow and pressure. These typically have a steep slope in their output curve trend, which yields a large derivative. Remember from calculus, the derivative is the slope of a line tangent to a curve at a given point.

### 6.2.3 Cascade Control

With cascade control, two controllers are used, but only one process variable PV is manipulated (Figure 6-9). The primary controller (M) maintains the primary process variable PV1 at its setpoint by adjusting the setpoint RSP (remote setpoint)[134] of the secondary controller (S). The secondary controller, in turn, responds to both the output of the primary controller and the secondary process variable PV2 The inner loop is tuned first, then the outer loop is tuned. There are two distinct advantages gained with cascade control:

- Disturbances affecting the secondary variable can be corrected by the secondary controller before a pronounced influence is felt by the primary controller.

---

134. Remote setpoint indicates that the controller receives its SP from the output of another controller when placed in auto mode.

- Closing the control loop around the secondary part of the process reduces the phase lag seen by the primary controller, resulting in increased speed of response.

### Requirements for Cascade Control

The secondary loop process dynamics (how fast the PV adjusts to SP) must be at least four times as fast as primary loop process dynamics. Secondary loop must have process influence over the primary loop, reference Figure 6-9. In Figure 6-9, TIC-101 controls the process temperature by manipulating the amount of steam going through the heat exchanger. As the process temperature (PV) drops below SP, controller TIC-101 will send an output to controller's FIC-101 SP input, which in turn will raise the valve output (VO) to FV-101, allowing more steam to flow to the heat exchanger. The reverse is true for when the process temperature (PV) rises above the SP.

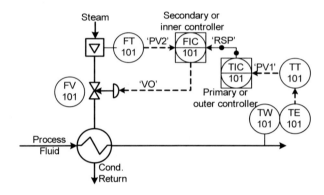

**Figure 6-9. Cascade Control**

## 6.2.4 Feedforward Control

In simple terms, feedforward control is a strategy used to compensate for disturbances in a system before they affect the controlled variable. A feedforward control system (Figure 6-10) measures a disturbance variable, predicts its effect on the process, and applies corrective action.

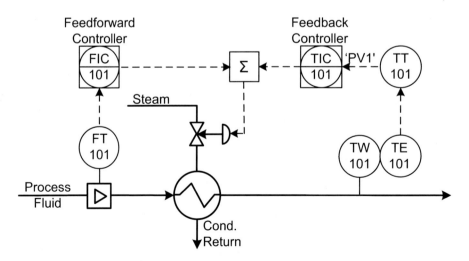

**Figure 6-10. Feedforward Control**

## 6.2.5 Ratio Control

Ratio control is used to ensure that two or more flows are kept at the same ratio, even if the flow rates are changing.

Applications of ratio control:

- Burner air/fuel ratio.

- Mixing and blending liquids.

An example of controlling an air/fuel ratio is a burner that uses natural gas. As shown in Figure 6-11 the controlled flow (combustion air) is increased or decreased to keep it at the correct ratio with the wild (uncontrolled) flow (natural gas). The combustion air flow is controlled by the natural gas flow, with a setpoint equal to the measured natural gas flow multiplied by some value (FFY-101). The result of the multiplication becomes the setpoint of the combustion air controller.

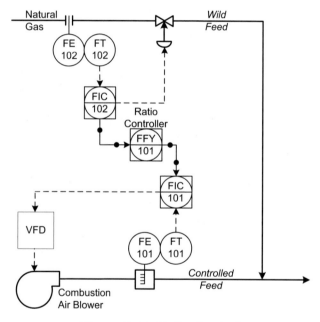

**Figure 6-11. Ratio Control**

## 6.2.6 Split Range Control

Split range control (Figure 6-12) is when the output of a controller is split to two or more control valves. Two different split-range schemes are available and are detailed in operation as follows:

- Controller output 0%: Valve A is fully open and Valve B is fully closed.

- Controller output 25%: Valve A is 75% open and Valve B is 25% open.

- Controller output 50%: Both valves are 50% open.

- Controller output 75%: Valve A is 25% open and Valve B is 75% open.

- Controller output 100%: Valve A is fully closed and Valve B is fully open.

OR

- Controller output 0%: Both valves are closed.

- Controller output 25%: Valve A is 50% open and Valve B is closed.

- Controller output 50%: Valve A is fully open and Valve B is closed.

- Controller output 75%: Valve A is fully open and Valve B is 50% open.

- Controller output 100%: Both valves are fully open.

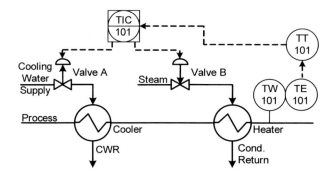

**Figure 6-12. Split Range Control**

## 6.2.7 Override Control

Override control is used to take control of an output from one loop and assign it to another, to allow the more important loop to manipulate the output (similar to high/low select). The outputs from two or more controllers are combined in a high or low selector (Figure 6-13). The output from the selector is the highest or lowest individual controller output.

## 6.2.8 Block Diagram Basics

$K_P$ = Proportional Gain

$K_I$ = Integral Gain ($K_I = K_P \div T_I$)

$T_I$ = Integral Time (time taken for proportional control action in response to step change in error signal)

$K_D$ = Derivative Gain ($K_D = K_P \div T_D$)

$T_D$ = Derivative Time (time for the proportional action to equal the derivative action for a ramp signal error)

$e(t)$ = Error signal.

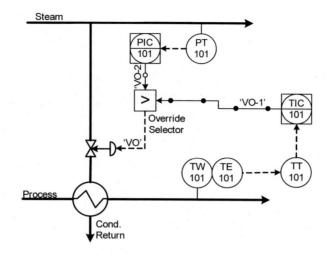

**Figure 6-13. Override Control**

## *Typical P Controller*

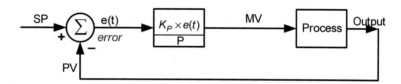

**Figure 6-14. Typical P Controller (Schematic)**

## *Typical PI Controller*

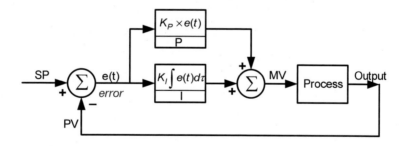

**Figure 6-15. Typical PI Controller (Schematic)**

*Typical PID Controller*

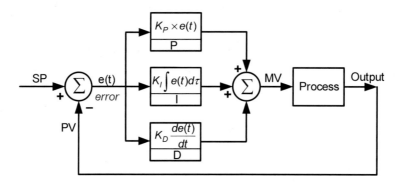

**Figure 6-16. Typical PID Controller (Schematic)**

## 6.3 CONTROLLER (LOOP) TUNING

Any tuning method applied will seek to establish a cause and effect relationship between the controller output and the process variable. See Figure 6-17.

**Proportional control ($K_p$)** will have the effect of reducing the rise time and will reduce, but never eliminate, the steady-state error ("offset").

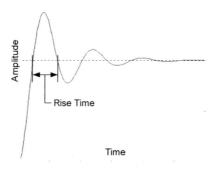

**Integral control ($K_i$)** will have the effect of eliminating the steady-state error, but it may make the transient response worse (i.e., controller will take longer to sufficiently react to the error).

**Derivative control ($K_d$)** will have the effect of increasing the stability of the system, reducing the overshoot, and improving the transient response.

### 6.3.1 Process Loop Types: Application of P, I, D

**Fast Loops: (Flow & Pressure)**

P = Little (too much will cause cycling)
I = More
D = NOT needed

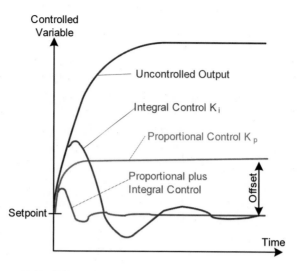

**Figure 6-17. Effects of Proportional and Integral Control**

## Slow Loops: (Temperature)

P = More
I = Some (too much will cause cycling)
D = Some

## Integrating (non self-regulating): (Level & Insulated Temperature)

P = More
I = Little (will cause cycling)
D = Must (if D is not used, loop will cycle)

## Noisy Loops: (where PV is constantly changing)

P = Little (will cause cycling)
I = Most (accumulated error)
D = Off (will cause cycling)

## 6.3.2 Tuning Map – Gain (Proportional) & Reset (Integral)

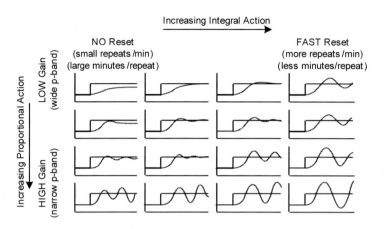

**Figure 6.18: Tuning Map – Gain and Reset**

## 6.3.3 Loop Dynamic Response

***Dead Time***[135]*:* Dead time is usually associated with the time it takes material to travel from one point to another. An example would be a well insulated flowing pipeline where the temperature is measured at two points, separated by a considerable distance. The recorded temperatures at the two measuring points would be identical, although they would be separated by the time required for the fluid to move from the upstream to the downstream point of measurement. Such a process can be described by a single parameter model that represents the dead time. Another example of dead time is analyzer to takes time to process a sample before yielding its result to back to the controller. The time from when the sample was taken to when the result is received back is a form of dead time.

***First Order Lag Filter:*** A dynamic system will come to equilibrium in five time constants. The system will reach 63.2% of equilibrium in one time constant, 63.2% of the remaining amount in one more time constant, and so on:

| Time since Step Input Change | Percentage of Steady-State Change |
|---|---|
| 1 Time Constant | 63.2% |
| 2 Time Constants | 86.5% |
| 3 Time Constants | 95.0% |
| 4 Time Constants | 98.2% |
| 5 Time Constants | 99.6% |

**Time Constant:** In general terms, the time constant t describes how fast the PV moves in response to a change in the output. The time constant t is calculated in five steps:

1. Determine $\Delta$PV, the total change that is going to occur in PV, computed as "final minus initial steady state."

---

135. Dead time is the delay from when a controller output signal is issued until when the measured process variable (PV) first begins to respond

2.  Compute the value of the PV that is 63% of the total change that is going to occur, or "initial steady state PV + 0.63(ΔPV)."

3.  Note the time when the PV passes through the 63% point of "initial steady state PV + 0.63(ΔPV)."

4.  Subtract from it the time when "the PV starts a clear response" to the step change in the output.

5.  The passage of time from step 4 minus step 3 is the process time constant τ.

### Tuning Decay Ratio Loop Response

With regards to the oscillatory loop response the decay ratio[136] is an indication of the degree of stability, the most common of which is the 1/4 decay ratio (Figure 6-19).

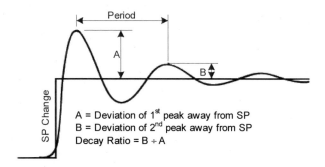

**Figure 6-19. Tuning Decay Ratio Loop Response**

### Over-Damped Loop Response

The loop response will be quite slow (Figure 6-20) because the controller output will sluggishly approach its final value so there will be no overshoot of setpoint. This result in a slow response of the controller adjusting its output to a PV change with regards to SP, which may not be acceptable to the process under control.

**Figure 6-20. Over-Damped Loop Response**

---

136. Decay ratio is the ratio of successive peaks in the PV response.

### 6.3.4 Loop Tuning Parameters

$K_p$ = proportional tuning parameter

$K_i$ = integral tuning parameter

$K_d$ = derivative tuning parameter

$K_c$ = controller gain (ratio of the change in the controller output to a change in either the controlled variable or the setpoint)

$K_u$ = ultimate gain (when the process cycles at a constant amplitude)

$P_u$ = ultimate period (aka the natural period, the period of the cycle present when the ultimate gain is in effect).

$L$ = lag time (apparent dead time)

$T_d$ = derivative time (in a PID controller it is the advance in time of the output signal for a given unit ramp signal input).

$T_i$ = integral time (in a PI controller it is the time required for the controller, while integrating the error, to cause an output change equal to the proportional change. Used in open loop tuning methods).

$\tau$ = time constant

### 6.3.5 Manual Loop Tuning

This is a closed loop tuning method. It is used when the process system MUST remain online during loop tuning (aka the trial and error method).

- With the controller in manual, adjust the controller output until the desired PV is maintained.

- Switch the controller from manual mode to auto mode.

- The $K_i$ and $K_d$ values should initially be set to zero. Increase $K_p$ until the output of the loop starts to cycle (which in turn will cause PV to cycle).

- Set the $K_p$ value to be approximately half of the value for a "1/4 decay ratio" response.

- Then increase $K_i$ until any offset is corrected within sufficient time for the process (caution: too much $K_i$ will cause instability).

- Then, if required, increase $K_d$ until the loop is sufficiently quick to reach its SP after a load disturbance (caution, too much $K_d$ will cause excessive response and overshoot).

With a fast PID loop, tuning usually overshoots slightly to reach the setpoint more quickly. However, some systems may NOT be able to accept overshoot. In this case, an "over-damped" closed-loop system is required. This will require a $K_p$ setting that is significantly less than half of the $K_p$ setting that caused the oscillation. See Table 6-5.

**Table 6-5. Effects of Increasing/Decreasing Tuning Parameters**

| Effects of Increasing/Decreasing Tuning Parameters | | | | |
|---|---|---|---|---|
| **Parameter** | **Rise Time** | **Overshoot** | **Settling Time** | **Steady State Error** |
| $K_p$ | Decrease | Increase | Small Change | Decrease |
| $K_i$ | Decrease | Increase | Increase | Eliminate |
| $K_d$ | Small Change | Decrease | Decrease | None |

### *Potential Problems with the Manual Tuning Method*

- Time consuming; many trials are needed.

- May create unstable operation or cause a hazardous situation in the process while tuning.

## 6.3.6 Closed Loop Tuning

The basic idea in the closed loop method of tuning is to get the loop to cycle without getting the process into trouble.

- Ensure that the process is running fairly smoothly. If it is not, it will be hard to differentiate between what the step change caused and what might have been happening with the process anyway.

- Ensure that the controller is in automatic.

- Use proportional only, no reset or derivative.

- Induce sustained oscillation by gradually increasing proportional parameter $K_p$.

- Note the ultimate period $P_u$ and the ultimate gain $K_u$ (this is important because the Z-N [Ziegler-Nichols[137]] correlation parameters make use of the ultimate period parameter) (Figure 6-21).

- From these parameters use the Ziegler-Nichols closed-loop correlation shown in Table 6-6 to determine the appropriate controller tuning parameters.

---

137. Ziegler-Nichols is a heuristic tuning method introduced by John G. Ziegler and Nathaniel B. Nichols in the 1940s.

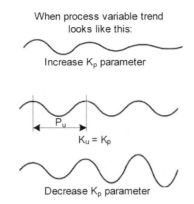

When process variable trend
looks like this:

Increase $K_p$ parameter

$P_u$

$K_u = K_p$

Decrease $K_p$ parameter

**Figure 6-21. Effects of Changing $K_p$**

**Table 6-6. Ziegler-Nichols Tuning Closed Loop Correlation**

|  | Tuning Parameters | | |
|---|---|---|---|
|  | **P** | **PI** | **PID** |
| $K_c$ | $0.5K_U$ | $0.45K_U$ | $0.6K_U$ |
| $T_I$ | NA | $0.83P_U$ | $0.5P_U$ |
| $T_D$ | NA | NA | $0.125P_U$ |

*Advantages of Closed Loop Tuning*

- Controller is operating in its normal mode, automatic

- The process variable can be kept close to the normal operating conditions, which results in less of a disruption to the process.

*Potential Problems with Closed Loop Tuning*

- May not be able to drive the process into oscillating condition

- May require several tests, thus longer testing time, than the open loop method

### 6.3.7 Open Loop Tuning

With open loop tuning, first place the controller in manual, then make a step change to the controller output. The step change must be significant enough to cause a significant change in the process variable (rule of thumb is that the signal-to-noise ratio should be greater than 5) reference Figure 6-22. Collect data and plot on trend. Then repeat, making a different step change in the opposite direction. Collect data, and plot on trend again. For further information reference Table 6-7 for the suggested Ziegler-Nichols controller parameters.

- Ensure that the process is running fairly smoothly. If it is not, it will be hard to differentiate between what the step change caused and what might have been happening with the process anyway.

- Ensure that the controller is in manual. There should be bumpless transfer from auto mode.

*Advantage of Open Loop Tuning*

- Only a single experimental test is needed

*Potential Problems with Open Loop Tuning*

- Controller is not in normal operating mode

- Results tend to be oscillatory

*Ziegler-Nichols Tuning Open Loop Correlation*

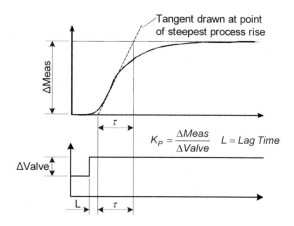

**Figure 6-22. Z-N Tuning Open Loop Correlation**

**Table 6-7. Z-N Tuning Open Loop Correlation**

|  | Tuning Parameters | | |
|---|---|---|---|
|  | **P** | **PI** | **PID** |
| Kc | $\dfrac{\tau}{K_p L}$ | $\dfrac{0.9\tau}{K_p L}$ | $\dfrac{1.2\tau}{K_p L}$ |
| $T_I$ | NA | 3.33L | 2.0L |
| $T_D$ | NA | NA | 0.5L |

## 6.3.8 Tuning Rules of Thumb

The rules of thumb in Table 6-8 are intended to give ballpark figure controller settings.

### Table 6-8. Tuning Rules of Thumb

| Loop Type | PB (%) | I (minutes) | D (minutes) |
|---|---|---|---|
| Liquid Level | < 100 | 10 | None |
| Temperature | 20 – 60 | 2-15 | I/4 |
| Flow | 150 | 0.1 | None |
| Liquid Pressure | 50 – 500 | 0.005 – 0.5 | None |
| Gas Pressure | 1 – 50 | 0.1 – 50 | 0.02 – 0.1 |
| Chromatograph | 100 - 2000 | 10 -120 | 0.1 – 20 |

PB indicates proportional band

## 6.4 FUNCTION BLOCK DIAGRAM REDUCTION ALGEBRA

A function block diagram is a block diagram that describes a function between input variables and output variables.

- A block diagram consists of unidirectional, operational blocks that represent the transfer function of the variables of interest.

- The block diagram representation of a given system often can be reduced to a simplified block diagram with fewer blocks than the original diagram.[138]

### 6.4.1 Component Block Diagram

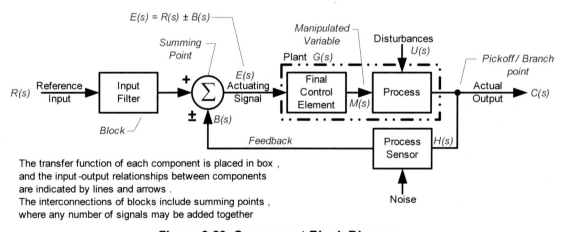

The transfer function of each component is placed in box , and the input -output relationships between components are indicated by lines and arrows .
The interconnections of blocks include summing points , where any number of signals may be added together

**Figure 6-23. Component Block Diagram**

---

138. Also known as Mason's Gain Rule, derived by Samuel Jefferson Mason. $M = \dfrac{\sum\limits_{j} M_j \Delta_j}{\Delta}$

**R(s):** An external signal applied to the feedback control system in order to command a specified action of the plant G(s).

**E(s):** The actuating signal, which is also referred to as "error." It is the sum of the reference input plus/minus the primary feedback signal.

**G(s):** The forward transfer function control element, aka the controller, which consists of the components necessary to generate the appropriate control signal to the process M(s).

**M(s):** The manipulated variable of the process.

**U(s):** An undesired input signal which may affect the value of the controlled output C(s).

**H(s):** The feedback transfer function, which is used to establish the functional relationship between the primary feedback signal B(s) and the controlled output C(s).

## 6.4.2 Basic Building Block

## 6.4.3 Elementary Block Diagrams

*Blocks in Series (Cascade)*

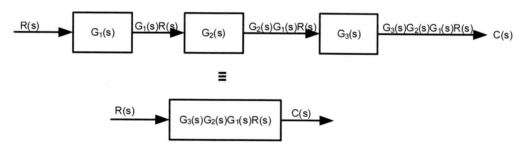

**Figure 6-24. Blocks in Series (Cascade)**

*Blocks in Parallel*

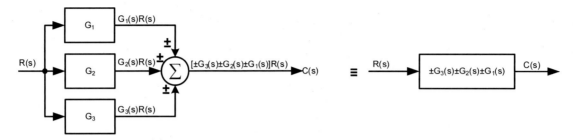

**Figure 6-25. Blocks in Parallel**

*Summing Junctions*

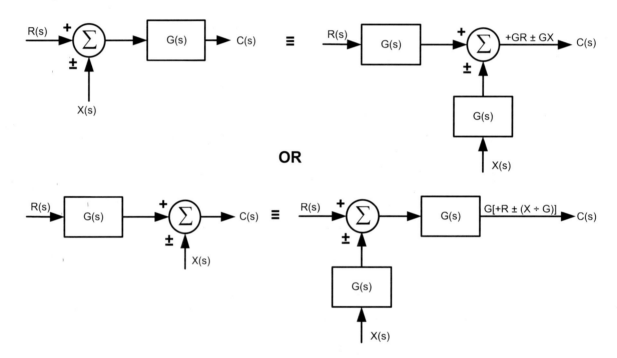

**OR**

**Figure 6-26. Summing Junctions**

*Branch Points*

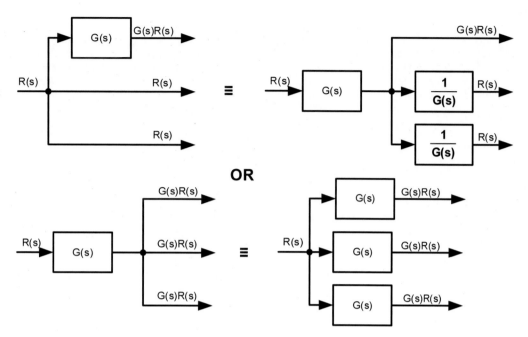

**Figure 6-27. Branch Points**

*Block Diagram Reduction Rules*

1.  Combine all series (cascade) blocks.

2.  Combine all parallel blocks.

3.  Eliminate all interior feedback loops.

4.  Shift summing points to the left.

5.  Shift takeoff points to the right.

6.  Repeat steps 1 through 5 until a canonical[139] form is obtained.

---

139. Canonical form is aka "Jordan normal form," meaning the simplest, most basic form but not the original form.

## Table 6-9. Block Diagram Reduction Manipulations

| Original Block Diagram | Manipulation | Equation | Modified Block Diagram |
|---|---|---|---|
| | Cascaded Elements | $C = (G_1 G_2)R$ | |
| | Addition or Subtraction | $C = (G_1 \pm G_2)R$ | |
| | Shifting or Pickoff Point Ahead of the Block | $C = GR_1$ | |
| | Shifting or Pickoff Point Behind the Block | $C = GR$ $R = (1/G)C$ | |
| | Shifting Summing Point Ahead of the Block | $E = GR - C$ | |
| | Shifting Summing Point Behind the Block | $E = G(R_1 - R_2)$ | |
| | Removing H from Feedback Path | | |
| | Eliminating Feedback Path | | |

## 6.5 ALARM MANAGEMENT

### 6.5.1 Good Guidelines for Alarm Management

EEMUA 191 suggests that < 1% of alarms be critical, 5% high priority, 15% medium and the remaining 80% low priority. (EEMUA = Engineering Equipment and Materials User Association) (Table 6-10).

Also reference ISA-18.2 Standard *Management of Alarm Systems for the Process Industries.*

### Table 6-10. EEMUA 191 Alarm Guidelines

| Long-Term Average Alarm Rate (in Steady Operation) | Acceptability |
|---|---|
| More than one per minute | Very likely to be unacceptable |
| One per 2 minutes | Likely to be over-demanding |
| One per 5 minutes | Manageable |
| < One per 10 minutes | Very likely to be acceptable |

| Number of Alarms Displayed in 10 Minutes Following a Major Process Upset | Acceptability |
|---|---|
| > than 100 | Definitely excessive and very likely to lead to the operator abandoning use of the system |
| 20-100 | Hard to cope with |
| < 10 | Should be manageable – but may be difficult if several of the alarms require a complex operator response |

The range of 10 -19 alarms displayed in 10 minutes following a major upset is not defined by EEMUA. It is the author's opinion that the range of 10 – 19 should also fall into the 'hard to cope with' acceptability.

### 6.5.2 Characteristics of Good System Alarms

### Table 6-11. Characteristics of Good System Alarms

| Characteristic | Reasoning |
|---|---|
| Relevant | Not spurious or of low operational value |
| Unique | Not duplicating another alarm |
| Timely | Not long before a response is needed or too late to do anything |
| Prioritized | Indicating the importance (criticality) that the operator respond/correct the problem |
| Understandable | Having a message that is clear and understandable |
| Diagnostic | Identifies the problem that has occurred |

| Advisory | Indicates the action that should be taken (corrective response) |
|---|---|
| Focusing | Draws attention to the most important issues |

## 6.5.3 Alarm Terms

- Acknowledged: Operator has confirmed recognition of the alarm

- Alarm Flood: Alarm rate is greater than one that the operator can safely handle.

- Alarm Group: Set of alarms with a common association.

- Alarm Chatter: An alarm that constantly switches between the alarm state and normal state within a short time period.

- Deadband: Change in signal away from the alarm setpoint required to clear the alarm.

- Priority: Level of alarm importance to a process.

- Stale Alarm: An alarm that stays in the alarm state for an extended period of time.

- Unacknowledged: Operator has not yet confirmed recognition of the alarm.

## 6.5.4 Alarm Review Methodology[140]

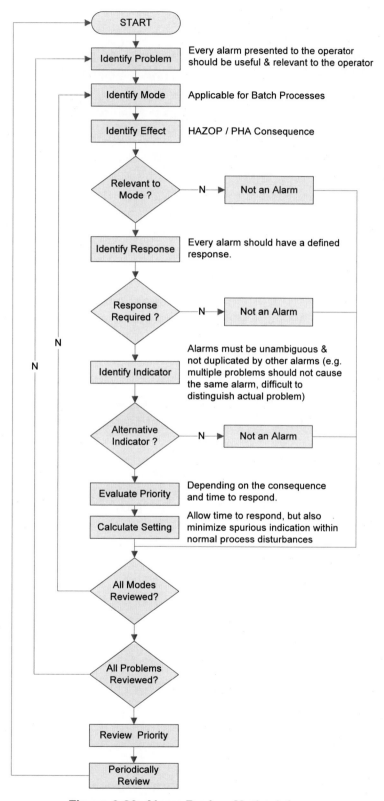

**Figure 6-28. Alarm Review Methodology**

## 6.6 TYPES OF CONTROL SYSTEM PROGRAMMING

The term programming may used interchangeably with the term configuration. The most common types of control system programming techniques are defined in IEC 61131-3:

- **Ladder Diagram (LD):** Named ladder logic due to the depiction of the logic that comes from the use of relay logic where there is a power bus on the right and left side of the diagram (similar to the rails of a ladder), with relay logic (coils and contacts) wired in between the two busses (similar to the steps of a ladder). This type of programming is popular for continuous control in PLCs but not well suited for batch control. This type of program is graphical in design.

- **Function Block Diagram (FBD):** Pre-defined function block that describes a function between input variables and output variables. These blocks may be thought of black boxes that have been pre-defined and tested by the control system vendor for functionality. However, the user does have the capability to develop their own custom function block. This type of programming is used both with PLCs and DCSs and is graphical in design.

- **Structured Text (ST):** Similar to computer programming languages with use of 'if-then-else' type statements as an example. This type of programming is textual in design.

- **Instruction List (IL):** This type of programming is textual in design and resembles assembler coding with the use of mnemonics.

- **Sequential Function Chart (SFC):** This is graphical in design and has three main components: steps, actions and transitions. Popular for use in batch processes. Reference Section 6.7.5 for further details.

Command mnemonics below are based upon the Allen Bradley Logix5000 family of controllers. Other control system vendors' mnemonics may be different but they will be functionally the same. This list is not all inclusive but does contain the majority of the instruction commands.

Note: Instruction list type programming will not be discussed within this book except to say that it is similar to assembler coding. Today's programmers tend to favor function block, ladder logic or even structured text over instruction list.

### 6.6.1 Ladder Diagram

*Basic Commands*

**XIC:** Examine If Closed (same as normally open contact)

address
─┤ ├─  1 = contact closed;  0 = contact open

---

140. Source of information: Process Automation article by David Hatch "Alarms, Prevention is Better Than Cure" in The Chemical Engineer magazine (tce today), July 2005 issue. This article is also published on the Emerson website (http://news.easydeltav.com/2005/08/).

---

**XIO:** Examine If Open (same as normally closed contact)

address
1 = contact open ; 0 = contact closed

**OTE:** Output Energize (turn on a bit when evaluation of the rung functions = true)

address

**OTL:** Output Latch (turn on a bit when evaluation of the rung functions = true, and latch the bit on) – used in conjunction with OTU function.

address
(L)

**OTU:** Output Unlatch (turn off a bit when evaluation of the rung functions = true, and latch the bit off) – used in conjunction with OTL function.

address
(U)

**OSR:** One Shot Rising (retentive input instruction that triggers an event to occur one time based on the change of state of the rung from false to true).

—[OSR]—

### Timers/Counters

**TON:** Timer On Delay (will turn an output on or off after the timer has been on for a preset time interval). Only active when rung conditions are true.

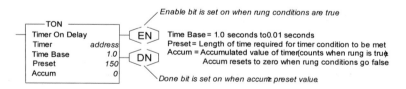

Enable bit is set on when rung conditions are true

|  | TON |  |
| --- | --- | --- |
| Timer On Delay |  | EN |
| Timer | address |  |
| Time Base | 1.0 | DN |
| Preset | 150 |  |
| Accum | 0 |  |

Time Base = 1.0 seconds to 0.01 seconds
Preset = Length of time required for timer condition to be met
Accum = Accumulated value of timer (counts when rung is true.
Accum resets to zero when rung conditions go false

Done bit is set on when accum = preset value.

**TOF:** Timer Off Delay (turn an output on or off after its rung has been off for a preset time interval). Only active (counting) when the rung conditions are false.

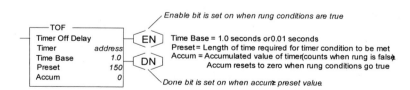

Enable bit is set on when rung conditions are true

|  | TOF |  |
| --- | --- | --- |
| Timer Off Delay |  | EN |
| Timer | address |  |
| Time Base | 1.0 | DN |
| Preset | 150 |  |
| Accum | 0 |  |

Time Base = 1.0 seconds or 0.01 seconds
Preset = Length of time required for timer condition to be met
Accum = Accumulated value of timer (counts when rung is false.
Accum resets to zero when rung conditions go true

Done bit is set on when accum = preset value.

**RTO:** Retentive Timer ON (turn an output on or off after its timer has been on for a preset time interval). Very similar to the TON and TOF command except for the fact that the Accumulated value is retained even if the rung conditions go false.

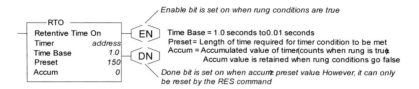

**CTU:** Count Up Instruction (counts false to true rung transitions and increments the accumulated value).

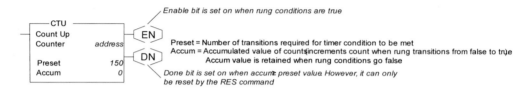

**CTD:** Count Down Instruction (counts false to true rung transitions and decrements the accumulated value).

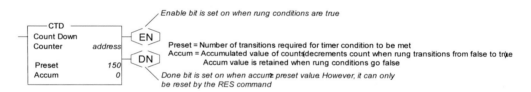

**HSC:** High Speed Counter (enabled when the rung logic is true and disabled when the rung logic is false). Used for counting where timed instances are very short and a regular CTU or CTD will not suffice. HSC is a hardware counter and runs asynchronously to the ladder program scan.

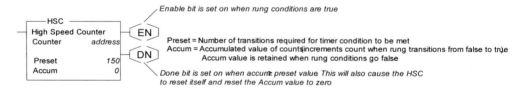

**RES:** Reset (When the RES instruction is enabled, it resets the Timer On Delay (TON), Retentive Timer (RTO), Count Up (CTU), or Count Down (CTD) instruction having the same address as the RES instruction). Do NOT use with TOF timers.

```
    address
   —(RES)—
```

## Compare Instructions

**EQU:** Equal Instruction (used to test whether two values are equal).

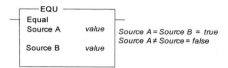

**NEQ:** Not Equal (used to test whether two values are not equal).

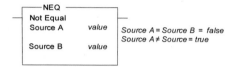

**LES:** Less Than (used to test whether one value is less than another).

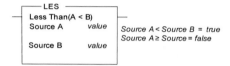

**LEQ:** Less Than or Equal (used to test whether one value is less than or equal to another).

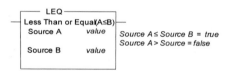

**GRT:** Greater Than (used to test whether one value is greater than another).

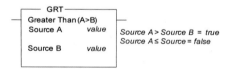

**GEQ:** Greater Than or Equal (used to test whether one value is greater than or equal to another).

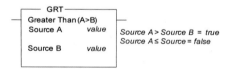

## Data Instructions

**MOV:** Move (moves a Source value to a Destination location).

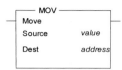

**MVM:** Masked Move (word instruction that moves data from a Source location to a Destination location).

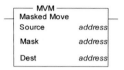

**AND:** Bitwise AND (used to perform the logic AND instruction on each bit of the value in source A with each bit of the value of source B, storing the output logic in the destination).

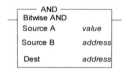

**OR:** Bitwise Inclusive OR (used to perform the OR logic on the value in source A by the bit with the value of source B and the output logic stored in the destination).

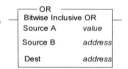

**XOR:** Bitwise Exclusive OR (used to perform the Exclusive OR logic on the value in source A by the bit with the value of source B and the output logic stored in the destination).

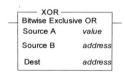

**NOT:** NOT (Inverter) (used to perform the NOT logic on the value in the source, bit by bit. The output logic value returned in the destination is the one's complement or opposite of the value in the source).

## Math Instructions

**ADD:** ADD (used to add a numerical value in source A to another numerical value in source B). The result is then placed into a destination register.

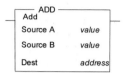

**SUB:** Subtract (used to subtract a numerical value in source B from another numerical value in source A). The result is then placed into a destination register.

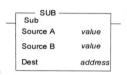

**MUL:** Multiply (used to multiply a numerical value from source A with another numerical value in source B). The result is then placed into a destination register.

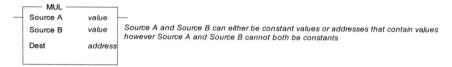

**DIV:** Divide (used to divide a numerical value from source A by another numerical value in source B). The rounded quotient is then placed into a destination register.

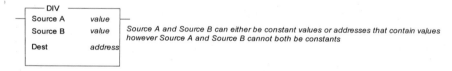

**SQR:** Square Root (When this instruction is evaluated as true, the square root of the absolute value of the source is calculated and the rounded result is placed in the destination)

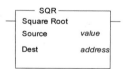

**CPT:** Compute (used to perform copy, arithmetic, logical, and conversion operations). The operation is defined in the expression and the result is written in the destination.

## Register and Sequence Instructions

**BSL:** Bit Shift Left (loads data into a bit array one bit at a time. The data is shifted through the array, then unloaded one bit at a time).

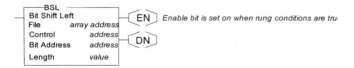

```
┌── BSL ──────────────────┐
│  Bit Shift Left         │──( EN )   Enable bit is set on when rung conditions are tru·
│  File          array address
│  Control          address
│  Bit Address      address │──( DN )
│  Length            value  │
└─────────────────────────┘
```

**BSR:** Bit Shift Right (loads data into a bit array one bit at a time. The data is shifted through the array, then unloaded one bit at a time).

```
┌── BSR ──────────────────┐
│  Bit Shift Right        │──( EN )   Enable bit is set on when rung conditions are tru·
│  File          array address
│  Control          address │──( DN )
│  Bit Address      address │
│  Length            value  │
└─────────────────────────┘
```

**SQO:** Sequencer Output (transfers 16-data to word addresses to control outputs for sequential operations).

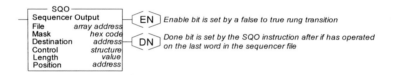

```
┌── SQO ──────────────────┐
│  Sequencer Output       │──( EN )   Enable bit is set by a false to true rung transition
│  File          array address
│  Mask             hex code
│  Destination      address │──( DN )   Done bit is set by the SQO instruction after if has operated
│  Control          structure         on the last word in the sequencer file
│  Length            value  │
│  Position         address │
└─────────────────────────┘
```

**SQC:** Sequencer Compare (used to reference data to monitor inputs).

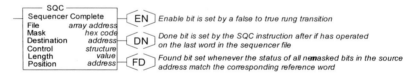

```
┌── SQC ──────────────────┐
│  Sequencer Complete     │──( EN )   Enable bit is set by a false to true rung transition
│  File          array address
│  Mask             hex code
│  Destination      address │──( DN )   Done bit is set by the SQC instruction after if has operated
│  Control          structure         on the last word in the sequencer file
│  Length            value  │
│  Position         address │──( FD )   Found bit set whenever the status of all nemasked bits in the source
└─────────────────────────┘           address match the corresponding reference word
```

**SQL:** Sequencer Load (used to store 16-bit data into a sequencer load file throughout each step of the sequencer operation).

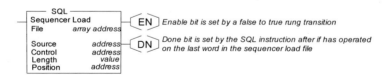

```
┌── SQL ──────────────────┐
│  Sequencer Load         │──( EN )   Enable bit is set by a false to true rung transition
│  File          array address
│                          │──( DN )   Done bit is set by the SQL instruction after if has operated
│  Source           address         on the last word in the sequencer load file
│  Control          address
│  Length            value  │
│  Position         address │
└─────────────────────────┘
```

## Program Control Instructions

**JMP:** Jump to Label (used to cause the PLC to skip over rungs). Paired with the LBL instruction.

```
    address
   —(JMP)—
```

**LBL:** Label (used by the jump JMP instruction as a target for the jump).

```
    address
   —[ LBL ]—
```

**JSR:** Jump to Subroutine (when executed, jumps to the beginning of the designated subroutine file and resumes execution at that point). Used in conjunction with SBR and RET instructions.

```
   ┌───── JSR ─────┐
  —│ Jump to Subroutine  │—
   │ SBR File Number     │
   │            address  │
   └─────────────────────┘
```

**RET:** Return from Subroutine (marks the end of the subroutine execution or the end of the subroutine file).

```
   ┌───── RET ─────┐
  —│ Return        │—
   └───────────────┘
```

**SBR:** Subroutine (the destination or target for the JSR instruction). The SBR instruction must be the first instruction on the first rung in the program files that contain the subroutine.

```
   ┌───── SBR ─────┐
  —│ Subroutine    │—
   └───────────────┘
```

**TND:** Temporary End (may be used to progressively debug a program).

```
   —(TND)—
```

**MCR:** Master Control Reset (used in pairs to create zones that clear all set outputs within that zone).

```
   —(MCR)—
```

**SUS:** Suspend (used to debug or diagnose the user program). When SUS is true, this instruction places the controller in the Suspend Idle mode.

```
   ┌───── SUS ─────┐
  —│ Suspend       │—
   │ Suspend ID   value │
   └───────────────┘
```

**OSF:** One Shot Falling (when input conditions preceding the ONS instruction on the same rung transition from true to false, the OSF instruction conditions the rung so that the output is true for one scan only).

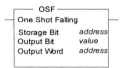

**OSR:** One Shot Rising (when input conditions preceding the ONS instruction on the same rung transition from false to true, the ONF instruction conditions the rung so that the output is true for one scan only).

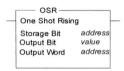

### Process Control, Message Instructions

**PID:** Proportional, Integral and Derivative Controller

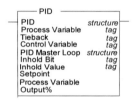

**MSG:** Send/Receive Message (when input conditions transition from false to true, the data is transferred).

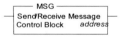

### Example Ladder Logic

Figure 6-29 is an excerpt from a small heatup subroutine from an autoclave program.

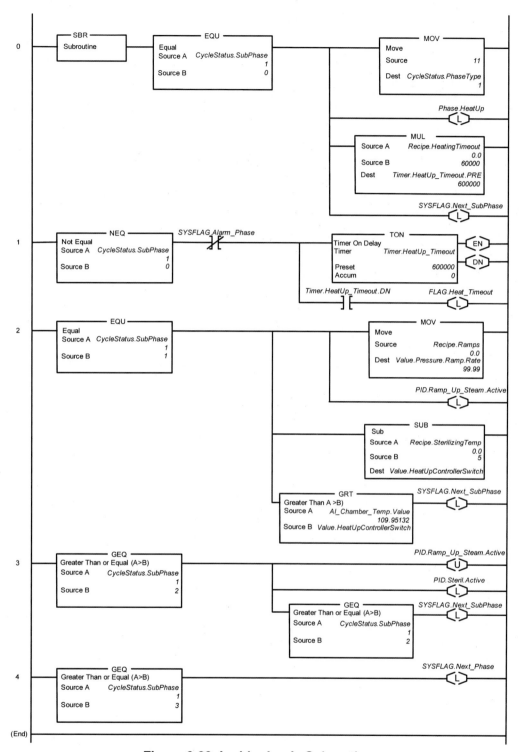

**Figure 6-29. Ladder Logic Subroutine**

## 6.6.2 Function Block Diagrams

*Reference IEC 61131-3*

Many of the functions described below function similarly to ladder logic, but they are diagrammatically shown differently.

### Input/Output Connections

**IREF:** Input Reference (input reference to supply value from an input tag or device).

**ICON:** Input Connection (input wire connections are used when the transfer of data is between function blocks that are far apart on the same documentation page or on a separate page altogether).

**OREF:** Output Reference (output reference to supply value from an output tag or device).

**OCON:** Output Connection (output wire connections are used when the transfer of data is between function blocks that are far apart on the same documentation page or on a separate page altogether).

### Basic Commands

*The Basic Commands from Ladder Logic do NOT have a Function Block equivalent, except for OSR which has an equivalent Function Block OSRI*

**OSRI:** Functionally the same as OSR.

### Timers/Counters

**TONR:** Timer ON Delay with Reset

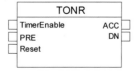

**TOF:** Timer Off Delay

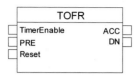

**RTOR:** Retentive Timer ON with Reset

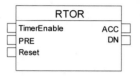

**CTUD:** Count UP DOWN Instruction

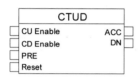

*Compare Instructions*

**EQU:** Equal Instruction

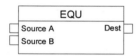

**NEQ:** Not Equal

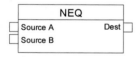

**LES:** Less Than

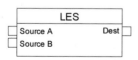

**LEQ:** Less Than or Equal

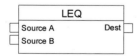

**GRT:** Greater Than

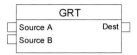

**GEQ:** Greater Than or Equal

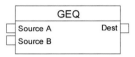

*Data Instructions*

**MVMT:** Masked Move with Target

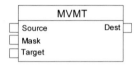

**AND:** Bitwise AND

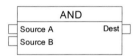

**OR:** Bitwise Inclusive OR

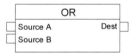

**XOR:** Bitwise Exclusive OR

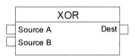

**NOT:** NOT (Inverter)

## Math Instructions

### ADD: Add

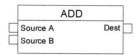

### SUB: Subtract

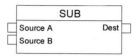

### MUL: Multiply

### DIV: Divide

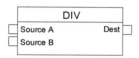

### SQR: Square Root

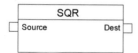

## Program Control Instructions

### JSR: Jump to Subroutine

### RET: Return from Subroutine

### SBR: Subroutine

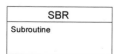

*Process Control*

**PI:** Proportional plus Integral Controller

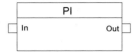

**PIDE:** Enhanced PID Controller

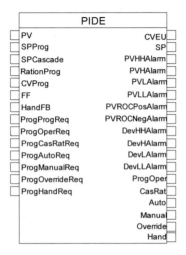

## 6.6.3 Structured Text

Many of the functions described below function similar to ladder logic, but they are diagrammatically shown differently. Tabs and carriage returns have no effect on the logic; they just aid readability.

*Basic Commands*

**XIC:** Examine If Closed

```
IF data_bit THEN
<statement.;
ENDIF;
```

**XIO:** Examine If Open

```
IF NOT data_bit THEN
<statement.;
ENDIF;
```

**OTE:** Output Energize

    data_bit [:=}
    BOOL_expression

**OTL:** Output Latch

    IF BOOL_expression THEN
    data_bit:= 1;
    ENDIF;

**OTU:** Output Unlatch

    IF BOOL_expression THEN
    data_bit:= 0;
    ENDIF;

**OSRI:** One Shot Rising with Input

    OSRI (OSRI_tag);

*Timers/Counters*

**TONR:** Timer ON Delay with Reset

    TONR (TONR_tag);:

**TOFR:** Timer OFF Delay with Reset

    TOFR (TOFR_tag);:

**RTOR:** Retentive Timer ON with Reset

    RTOR (RTOR_tag);:

**CTUD:** Count UP DOWN Instruction

    CTUD (CTUD_tag);

*Compare Instructions*

**EQU:** Equal Instruction

    IF sourceA = sourceB THEN
    <statement>.;

**NEQ:** Not Equal

IF sourceA <> sourceB THEN
<statement>.;

**LES:** Less Than

IF sourceA < sourceB THEN
*<statement>.;*

**LEQ:** Less Than or Equal

IF sourceA <= sourceB THEN
*<statement>.;*

**GRT:** Greater Than

IF sourceA > sourceB THEN
*<statement>.;*

**GEQ:** Greater Than or Equal

IF sourceA >= sourceB THEN
*<statement>.;*


*Data Instructions*

**MOV: M**ove

*dest := source*

**MVM:** Masked Move

Dest:= (Dest AND NOT
(Mask))
OR (Source AND Mask);

**AND:** Bitwise AND

dest:= sourceA AND sourceB

**OR:** Bitwise Inclusive OR

*dest:= sourceA OR sourceB*

**XOR:** Bitwise Exclusive OR

*dest:= sourceA XOR sourceB*

**NOT:** NOT (Inverter)

    dest:= NOT source

*Math Instructions*

**ADD:** Add

    dest:= sourceA + sourceB

**SUB:** Subtract

    dest:= source A – sourceB

**MUL:** Multiply

    *dest:= sourceA * sourceB*

**DIV:** Divide

    *dest:= sourceA/sourceB*

**SQR:** Square Root

    *dest:= SQRT(source)*

**CPT:** Compute

    *destination:= numeric_expression;*

*Program Control Instructions*

**JSR:** Jump to Subroutine

    JSR (RoutineName
    InputCount,
    InputPar, ReturnPar);

**RET:** Return from Subroutine

    RET (ReturnPar);

**SBR:** Subroutine

    SBR (InputPar);

**TND:** Temporary End

TND ();


*Process Control, Message Instructions*

**PI:** Proportional plus Integral Controller

PI (PI_tag);

**PID:** Proportional, Integral and Derivative Controller

PID (PID,
ProcessVariable, Tieback,
ControlVariable,
PIDMasterLoop,
InholdBit,
InholdValue);

**PIDE:** Enhanced PID Controller

PIDE (PIDE_tag);


## 6.7 BATCH CONTROL

Batch control is a process that utilizes recipes to make products in vessels rather than a continuous series processes that flow from one to the other.

*Reference ISA-88 Standard*

ISA-88 batch standard defines control activities, and control functions that provide a means to process finite quantities of input materials to an ordered set of processing activities over a finite period of time using one or more pieces of equipment. ISA-88 also defines hierarchical recipe management and process segmentation frameworks, which separates products from the processes that make them. The standard enables reuse and flexibility of equipment and software, and provides a structure for coordinating and integrating recipe-related information across traditional ERP, MES and control domains.

*Key Concepts of ISA-88*

- Object Oriented Design

    - Reusable standard modules (i.e., Libraries)
    - Makes validation easier

- Equipment/Process Separation

- Equipment Control provides process independent services
- Process Control uses equipment services to make product using defined rules and parameters (i.e., Recipes)
- The equipment and process must be independent of each for plant flexibility

• Equipment Entity Concept

- Automation is part of the equipment, not a stand-alone, separate function

## 6.7.1 Automation Pyramid

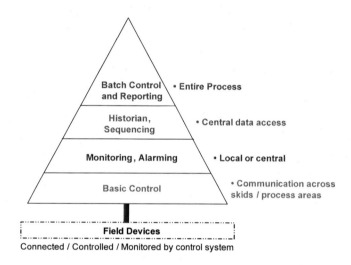

**Figure 6-30. Automation Pyramid**

## 6.7.2 Physical Model

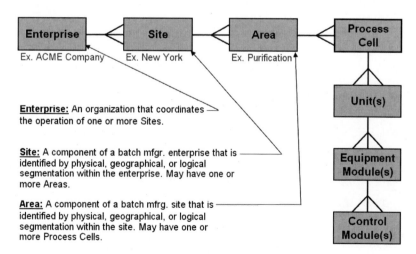

**Figure 6-31. ISA-88 Physical Model**

*Terminology*

**Process Cell:** Contains all of the units, equipment modules, and control modules, logically grouped, that are required to make one or more batches. A good example of a process cell is a process train. A process train is set of equipment grouped together to make one final product (Figure 6-32).

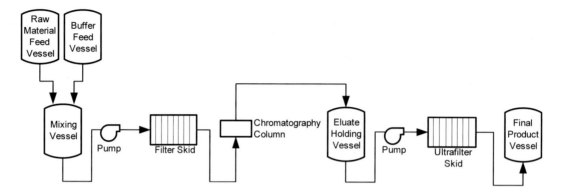

**Figure 6-32. Example Process Train**

A batch may not necessarily utilize all of the equipment located within the train, however no train may utilize equipment outside the boundary of its process cell. The order of equipment utilized by a batch is called the "path." More than one batch and more than one product are permitted to use the train simultaneously.

Reference Figure 6-33 For an example for use of the following terms (unit, EM, CM):

**Unit:** Contains equipment modules and control modules. A unit is a collection of equipment (vessels, pumps and instruments, etc.) that work together to accomplish a well defined set of tasks. In addition to normal processing, a unit must be able to respond to abnormal situations. Batching can NOT occur without units as batching occurs in the units (units run the recipes).

Examples of units include:

- Mixing Tanks

- Reactors

**Equipment Module (EM):** The equipment module may be made up of control modules and subordinate equipment modules. Equipment modules group devices for performing one or more specific processing activities.

Examples of equipment modules include:

- Weigh Tanks

- Transfer Panels

- Pump Discharge/Recirculate

**Control Module (CM):** Collection of sensors, actuators, other control modules and associated processing equipment that is operated as a single entity with regard to control functions. This is the direct connection level to the process.

Examples of control modules include:

- PID Controllers

- Analog Input/Output

- Discrete Input/Output

- Discrete Valves

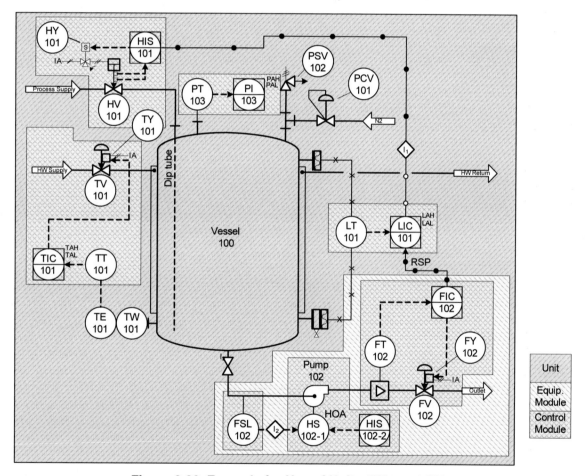

**Figure 6-33. Example for Use of Units, EMs and CMs**

## 6.7.3 Procedural Model

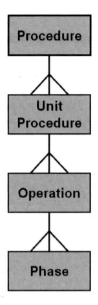

*In its simplest form, the recipe structure is one to one, i.e., 1PR → 1UP → 1OP → 1 Phase. However, more complex logic, such as a 'CIP Skid' recipe that can clean multiple different vessels, has a one to many structure (PR → xUP → xOP → xPhase).*

**Figure 6-34. ISA-88 Procedural Model**

*Terminology*

**Procedure:** Defines the strategy for carrying out a major processing operation (e.g., a batch). A procedure consists of an ordered set of unit procedures.

**Unit Procedure:** A contiguous production sequence of operations.

**Operation:** An ordered set of phases carried to completion within a single unit. Operations usually involve taking the material being processed through some type of physical, chemical, or biological change.

**Phase:** The smallest element of procedural control that can accomplish process-oriented tasks. Phases perform unique and generally independent, basic process-oriented functions. Simply put, phases are the workhorses of recipes. All other elements (procedures, unit procedures, and operations) simply group, organize, and direct phases. The phase is a series of steps (Sequential Function Chart or SFC) that cause one or more equipment- or process-oriented actions, for example, filling a tank or agitating the contents. The phase logic defines the states of the phase (running, holding, restarting, aborting, and stopping) and the logic associated with each state. Reference Table 6-12 for phase commands.

### Table 6-12. Phase Commands

| Command | Description |
|---------|-------------|
| Abort | Invokes aborting logic. "Abort" is an emergency stop. |
| Hold | Used for a phase in the running or restarting state. "Hold" invokes the holding logic to temporarily stop the phase and proceed to held state, from which the "Restart" command can be used.<br>Alarms/Failure monitoring may trigger Hold conditions. |
| Reset | Transitions the phase from an aborted, complete, or stopped state to the idle state. |
| Restart | Used for a phase in the held state to invoke "Restarting" logic. |
| Start | From the idle state, the Start command invokes the running logic. |
| Stop | Invokes the stopping logic. |

## 6.7.4 Recipes

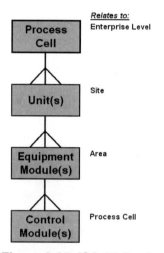

**Figure 6-35. ISA-88 Recipes**

**General Recipe:** An enterprise level recipe that serves as the basis for lower-level recipes. It identifies raw materials, their relative quantities, and required processing, but without specific regard to a particular site or the equipment available at that site.

**Site Recipe:** Specific to a particular site. It is the combination of site-specific information and a general recipe. It is usually derived from a general recipe to meet the conditions found at a particular manufacturing location and provides the level of detail necessary for site-level, long-term production scheduling. However, it may also be created directly without the existence of a general recipe

**Master Recipe:** Level of recipe that is targeted to a process cell or a subset of the process cell equipment. The master recipe level is a required recipe level, because without it no control recipes can be created and, therefore, no batch can be produced.

**Control Recipe:** Starts as a copy of the master recipe and is then modified as necessary with scheduling and operational information to be specific to a single batch. It also provides the level of detail necessary to initiate and monitor equipment procedural entities in a process cell.

*Recipe Components*

**Header:** Administrative information (recipe ID, author name, version number, revision history, approvals, etc.) and a process summary.

**Equipment Requirements:** Information about specific equipment necessary to make a batch or a specific part of the batch. (Since general and site recipes don't usually call out specific equipment, the requirements in these recipes are typically described in general terms.)

**Procedure:** See terminology in the previous section.

**Formula:** Describes recipe process inputs (e.g., ingredients, quantities), process parameters (e.g., processing temperature, ingredient transfer rate, mixing speed), and process outputs (the product and its quantity resulting from a single batch). Formulas can be used to distinguish between different products or different product "grades" defined by the same procedure.

**Other Information:** Anything that doesn't fit in the other categories (this can be miscellaneous operator instructions, safety comments, etc.).

## 6.7.5 Sequential Function Chart (SFC)

*Reference IEC-61131-3 Standard.*

SFCs (Figure 6-36) are a series of steps and transitions. Steps are represented by boxes and transitions by vertical lines with crosses attached. Each step contains a set of actions that affect the process. At any given time, one or more of the steps and transitions can be active. Each time the SFC scans, the active steps and transitions are evaluated. When a transition evaluates as TRUE (for example, the transition condition is met), the steps prior to the transition are made inactive and the step(s) following the transition become active.

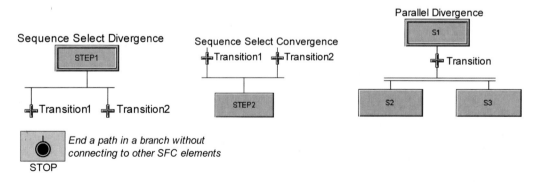

**Figure 6-36. Sequential Function Chart**

**Example** of a "one to many" recipe structure for a procedure containing multiple unit procedures in both parallel and selection-based (Figure 6-37):

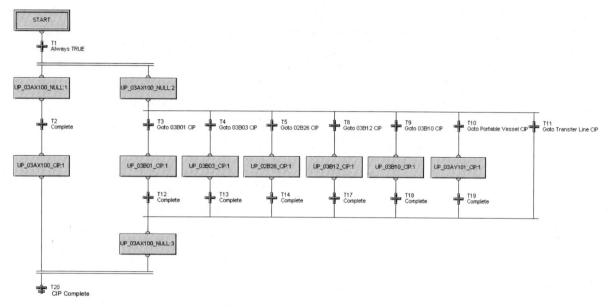

Figure 6-37. "One to Many" Recipe Structure

## 6.8 ADVANCED CONTROL TECHNIQUES

This is a very high level description of three types of advanced control techniques.

### 6.8.1 Fuzzy Logic

Fuzzy logic attempts to mimic human control logic. Fuzzy logic applies numerical weighting parameters so that the control system can determine what is considered a significant error and a significant rate of change of error. Fuzzy logic applies "If-Then" type rules rather than just "True or False" type rules.

To create Fuzzy Logic:

- Determine the appropriate process inputs.

- Determine the cause and effect action of the system with "fuzzy rules."

- Develop controller logic to act on the inputs and determine the appropriate controller output, considering each process input separately. The control logic then becomes "If-Then" statements within in the controller logic. In the controller logic, a weighted average is used to combine the various actions called for by the individual process inputs into one controller output acting on the process system to be controlled.

## 6.8.2 Model Predictive Control

Model predictive control is a mathematical model that is used to predict the result of a sequence of controller variable (SP, P, I and D) manipulations. It uses the mathematical model and multiple process input and output measurements (MIMO) to predict system behavior. These predictions can then be used to optimize the process control over a period of time. Model predictive control is very specific to the process it is modeling. Unlike ratio or cascade control setups, where it is simple to implement and change set points in various situations, model predictive control will model one specific process and optimize it. The block diagram (Figure 6-38) shows an example of model predictive feed-forward control.

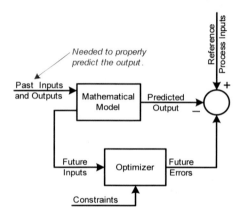

**Figure 6-38. Model Predictive Feedforward Control**

## 6.8.3 Artificial Neural Networks

An artificial neural network is developed with a systematic step by step procedure which optimizes a criterion known as the "learning rule." The input/output training data is fundamental for neural networks because it conveys the information that is necessary to discover the optimal operating point. The network system is a structure that receives an input, processes the data, and provides an output.

An artificial neural network consists of three main components:

- Neurons

    - Input Neurons
    - Hidden Neurons
    - Output Neurons

- Weights

    - Negative weight reflects an inhibitory connection
    - Positive weight reflects an excitatory connection

- Learning Rules

Inputs are provided to the input neurons, such as equipment parameters, and outputs are provided by the output neurons. These outputs may be a measurement of the performance of the process. The network is trained by establishing the weighted connections between input neurons and output neurons via the hidden neurons. Weights are continuously modified until the neural network is able to predict the outputs from a given set of inputs within an acceptable user defined error level.

Once the neural network is sufficiently trained, a general model is created for the relationship between inputs and outputs. If an element of the neural network should fail, the network can continue without any problem by its parallel nature. The block diagram Figure 6-39) shows an example of an artificial neural feed-forward network.

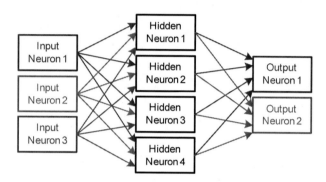

**Figure 6-39. Artificial Neural Feed-forward Network**

## 6.9 EXAMPLE PROCESS CONTROLS

### 6.9.1 Boiler Control

*Three-Element Control:* A common application of cascade control combined with feed forward control is level control in a boiler (Figure 6-40). The term "three-element control" refers to the number of process variables that are measured to effectively control the boiler feedwater valve. These three measured PVs are:

- Steam Drum Level

- Boiler Feedwater Flow to Steam Drum

- Steam Flow Exiting Steam Drum

When steam vapor is generated in the risers of the boiler, some of the water volume in the tubes is displaced by steam bubbles. These steam bubbles have a smaller density than the boiler feed water. As more and more of these steam bubbles are generated, the level indicator does not give a true indication of the inventory of boiler feed water in the system.

As steam demand is increased, this effect becomes increasingly more pronounced, and the level in the steam drum rises rather than decreases, as one would expect. This phenomenon is known as "Shrink and Swell." Under these circumstances, a simple level controller would cut back rather than increase the amount of boiler feed water to the steam drum. Consequently, a more sophisticated control scheme is required. This control scheme is the three-element control for the boiler feedwater.

With this type of control, steam flow rate is summed with the output of the steam drum level controller. This signal then serves as the RSP for the boiler feedwater flow controller. The level controller is tuned proportional only, and the boiler feed water controller is tuned proportional-integral. With this type of control, when steam demand increases, boiler feedwater flow increases as well.

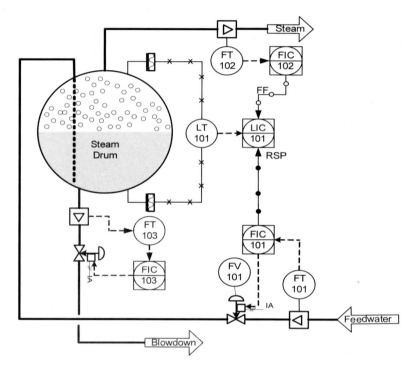

**Figure 6-40. Three-Element Control**

*Four Element Control:* Same as three element control, except that blowdown is also taken into consideration. The reason for blowdown is that dissolved solids and particles enter a boiler through the make-up water, but remain behind when steam is generated. During continuous boiler operation the concentration of solids builds up, and a concentration level is finally reached where operation of the boiler becomes impossible. Blowdown frequency is dependent upon the quality of water being used as boiler feedwater. Hard water contains dissolved minerals and would require more frequent blowdown than soft water. If the solids are not purged from the boiler they can lead to the following:

- Scale formation in the boiler

- Carryover into the steam

- Corrosion of the boiler

- Embrittlement of the boiler

## 6.9.2 Distillation Column Control

The most important output of a distillation column is the distillate flow. In distillation column control (Figure 6-41), all controllers work together to ensure good quality distillate as the output from the column. The liquid mixture that is to be processed by the column is known as the "feed" and is usually introduced somewhere near the middle of the column to a tray known as the feed tray. The feed tray divides the column into a top (enriching or rectification) section and a bottom (stripping) section. The feed flows down the column where it is collected at the bottom in the reboiler.

Heat is supplied to the reboiler to generate vapor. The source of heat input can be any suitable fluid, although in most chemical plants it is steam. The vapor raised in the reboiler is re-introduced into the unit at the bottom of the column. The liquid removed from the reboiler is known as the "bottoms product" or simply "bottoms" (low boilers).

The vapor moves up the column, and as it exits the top of the unit, it is cooled by a condenser. The condensed liquid is stored in a holding vessel known as the reflux drum. Some of this liquid is recycled back to the top of the column, and this is called the "reflux." The condensed liquid that is removed from the system is known as the "distillate" or "top product."

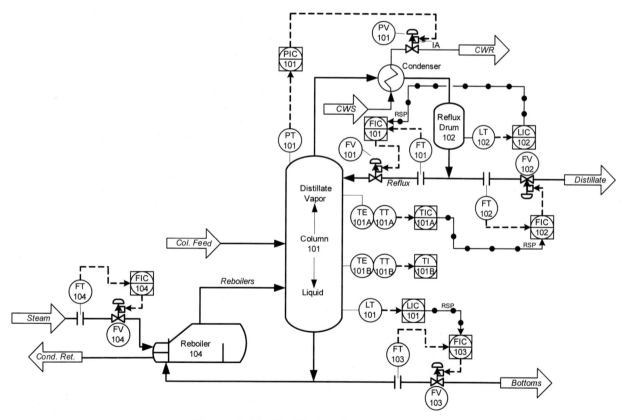

**Figure 6-41. Distillation Column Control**

### 6.9.3 Burner Combustion Control

Burner combustion is controlled to ensure a proper air to gas ratio mix to ensure complete combustion. A ratio control strategy (Figure 6-42) can play a fundamental role in the safe, compliant and profitable operation of fired heaters, boilers, furnaces and similar fuel burning processes. This is because the air-to-fuel ratio in the combustion zone of these processes directly impacts fuel combustion efficiency and environmental emissions. If the air-to-fuel ratio is extremely lean (excessive gas) then an explosive environment may be created.

A requirement for ratio control implementation is that both the fuel feed rate and the combustion air feed rate are measured and are available as process variable (PV) signals.

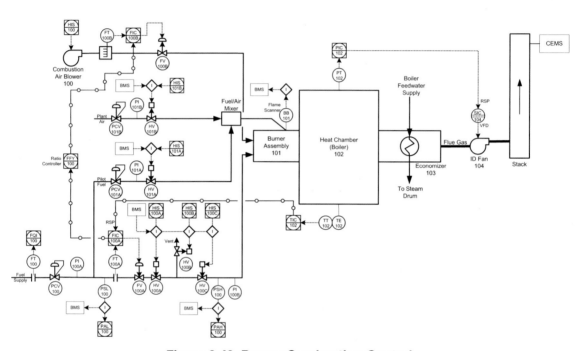

**Figure 6-42. Burner Combustion Control**

An air/fuel ratio that is too small leads to incomplete combustion of the fuel. As the availability of oxygen decreases, noxious exhaust gases, including carbon monoxide, will form first. As the air/fuel ratio decreases further, partially burned and unburned fuel can appear in the exhaust stack, often revealing itself as smoke and soot. Carbon monoxide and partially burned and unburned fuel are all poisons whose release is regulated by the government. Incomplete combustion also means that there is wasted expensive fuel. Fuel that does not burn to provide useful heat energy, including carbon monoxide that could yield energy as it converts to carbon dioxide, literally flows up the exhaust stack as lost profit.

As the air/fuel ratio increases above that needed for complete combustion, the extra nitrogen and unneeded oxygen absorb heat energy, decreasing the temperature of the flame and gases in the combustion zone. As the operating temperature drops, the system is less capable of extracting useful heat energy for the intended application. When the air/fuel ratio is too high, a surplus of hot air is produced. This hot air simply carries its heat energy up and out the exhaust stack as lost profit.

# 7. ISA-95

*The international standard for the integration of enterprise and control systems.*

Premise of ISA-95:

- Provide open information exchange across manufacturing production and business planning systems.

- Provide integrated real-time manufacturing applications.

## *Definitions*

**ERP:** Enterprise Resource Planning. This is at the company management level. ERP specifically describes software that is used to connect many aspects of a business. It consists of central functions such as accounting/finance, sales/marketing, purchasing, and human resources. ERP may also contain planning functions for the evaluation of medium and long term material requirements.

**MES:** Manufacturing Execution System This is the production level management. It may consist of maintenance management systems, laboratory information management systems, electronic document management systems, learning management systems, CAD, etc.

## 7.1 ISA-95 HIERARCHY MODEL

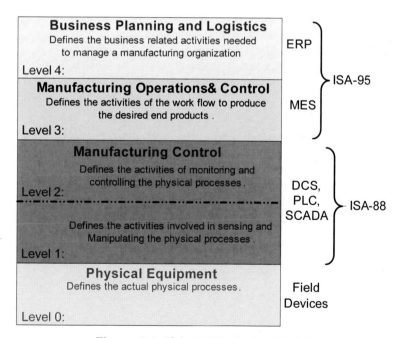

**Figure 7-1. ISA-95 Hierarchy Model**

## 7.2 ISA-88 PHYSICAL MODEL AS IT PERTAINS TO ISA-95

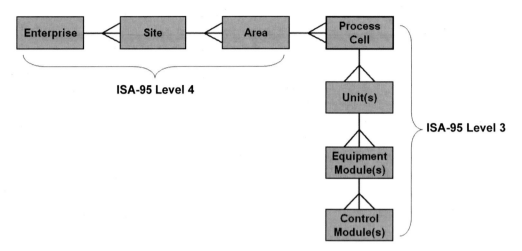

Figure 7-2. ISA-88 Physical Model As It Pertains to ISA-95

## 7.3 LEVELS 4–3 INFORMATION EXCHANGE

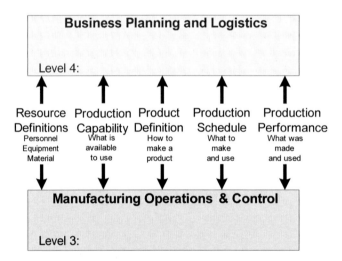

Figure 7-3. Levels 4–3 Information Exchange

# 8. Hazardous Areas and Safety Instrumented Systems

Hazardous area is an area where the potential for an flammable/explosive atmosphere may exist, therefore any equipment (e.g., instrumentation, electrical components) must not have the capability of producing sufficient energy to ignite the flammable/explosive atmosphere. This section will discuss techniques to prevent the equipment from providing such energy needed to ignite the flammable/explosive area. The best technique of all that is not discussed is to NOT locate any equipment in the hazardous area, but this is not always possible.

Safety instrumented systems are designed for use to mitigate the level of risk for a process with respect to maintaining safe control a process that may get out of control and cause damage to personnel, equipment, the environment, and possibly the surrounding community.

## 8.1 Hazardous Areas

- *Reference NEC Articles 500 through 505.*

- *OSHA 1910.307 Hazardous (classified) locations also defines requirements for hazardous areas (covers the requirements for electric equipment and wiring in locations that are classified depending on the properties of the flammable vapors, liquids or gases, or combustible dusts or fibers that may be present therein and the likelihood that a flammable or combustible concentration or quantity is present)*

### 8.1.1 NEC Articles 500–504

*Definitions*

**Class I** Locations are those in which flammable gases, flammable liquid-produced vapors or combustible liquid-produced vapors are or may be present in the air in quantities sufficient to produce explosive or ignitable mixtures.

**Class II** Locations are those that are hazardous because of the presence of combustible dust.

**Class III** Locations are those that are hazardous because of the presence of easily ignitable fibers or materials producing combustible flyings.

**Explosion-Proof:** Apparatus enclosed in a case that is capable of withstanding an explosion of a specified gas or vapor that may occur within it and of preventing the ignition of a specified gas or vapor surrounding the enclosure by sparks, flashes or explosion of gas or vapor within, and that operates at such an internal temperature that a surrounding flammable atmosphere will not be ignited thereby.

*The difference between "explosion-proof" and "flameproof":* "Explosion-Proof" enclosures are constructed to withstand (i.e., stay intact) 4 times the explosive pressures of the gases. "Flameproof" enclosures are only required to withstand 1.5 times the explosive pressure.

**Hermetically Sealed:** Equipment sealed against the entrance of an external atmosphere, where the seal is made by fusion process such as soldering, brazing and welding or the fusion of glass to metal.

**Nonincendive Equipment:** Equipment having electrical/electronic circuitry that is incapable, under NORMAL operating conditions, of causing ignition of a specified flammable gas-air, vapor-air or dust-air mixture due to arcing or thermal means.

**Intrinsically Safe:** Equipment and wiring which is incapable of releasing sufficient electrical or thermal energy under normal OR abnormal conditions to cause ignition of a specific hazardous atmospheric mixture in its most easily ignited concentration. A common protective device used in intrinsically safe circuits is a Zener diode barrier.

Conductors for intrinsically safe circuits shall NOT be placed in any raceway, cable tray or cable with conductors of any non-intrinsically safe circuit. *Exception:* where conductors of intrinsically safe circuits are separated from non-intrinsically safe circuit conductors by a distance of at least 2 inches and secured by a grounded metal partition or an approved insulating partition.

**Flash Point:** The lowest temperature at which a liquid gives off sufficient vapor at its surface to form a flammable or explosive mixture in the atmosphere.

**Auto-Ignition Temperature:** The minimum temperature for self-sustained combustion of a substance without the need of an external ignition source such as a spark or flame.

**Lower Explosive Limit (LEL):** The minimum concentration of a flammable gas or vapor that will support a flame when exposed to a source of ignition. Below this level the mixture would be considered too lean to sustain combustion.

**Upper Explosive Limit (UEL):** The maximum concentration of a flammable gas or mixture that will combust. Above this level the mixture would be considered too rich to sustain combustion.

**Flammable Liquid:** One defined as one having a flash point below 100°F.

- Class IA Flammable Liquid: Flash point below 73°F, and a boiling point below 100°F

- Class IB Flammable Liquid: Flash point below 73°F, and a boiling point ≥ 100°F

- Class IC Flammable Liquid: Flash point ≥ 73°F, and a boiling point below 100°F

**Combustible Liquid:** One defined as one having a flash point ≥ 100°F.

- Class IA Combustible Liquid: Flash point ≥ 100°F, and below 140°F

- Class IB Combustible Liquid: Flash point ≥ 140°F and below 200°F

- Class IC Combustible Liquid: Flash point ≥ 200°F

### Class I Locations

Atmospheric hazards are divided into four groups: A, B, C, D, but also into two divisions.

## Groups

*Group A:* Acetylene

*Group B:* Flammable gas, flammable liquid-produced vapors or combustible liquid-produced vapors mixed with air in sufficient quantities that may burn or explode, having a Maximum Experimental Safe Gap (MESG) value ≤ 0.45 mm or a minimum igniting current ratio (MIC ratio) of ≤ 0.40.

*Group C:* Flammable gas, flammable liquid-produced vapors or combustible liquid-produced vapors mixed with air in sufficient quantities that may burn or explode, having a Maximum Experimental Safe Gap (MESG) value ≥ 0.45 mm and ≤ 75 mm or a minimum igniting current ratio (MIC ratio) of ≥ 0.40 and ≤ 0.80.

*Group D:* Flammable gas, flammable liquid-produced vapors or combustible liquid-produced vapors mixed with air in sufficient quantities that may burn or explode, having a Maximum Experimental Safe Gap (MESG) value > 0.75 mm or a minimum igniting current ratio (MIC ratio) of > 0.80

## Divisions

*Division 1:* Covers locations where flammable gas, flammable liquid produced vapors or combustible liquid produced vapors are or may exist under normal operating conditions, under frequent repair or maintenance operations or where breakdown or faulty operation of process equipment might also cause simultaneous failure of electrical equipment.

*Division 2:* Covers locations where flammable gas flammable liquid-produced vapors or combustible liquid-produced vapors are handled, processed, or used, but in which the liquids, vapors or gases will normally be confined within a closed container or a closed system from which they can escape under accidental rupture or breakdown, or in case of abnormal operation, OR locations in which flammable gas, flammable liquid-produced vapors or combustible liquid-produced vapors are prevented by positive mechanical ventilation (purge) and which may become hazardous due to the failure of the positive mechanical ventilation equipment. Areas adjacent to Division 1 locations, into which gases might occasionally flow, would also be Division 2, unless such flow is prevented by adequate positive pressure ventilation from a source of clean air, and effective safeguards against ventilation failure are provided.

### Class II Locations

Atmospheric hazards cover three groups of combustible dusts:

*Group E:* Metallic dusts including aluminum, magnesium and their commercial alloys or other combustible dusts whose particle size, abrasiveness and conductivity present similar hazards in the use of electrical equipment.

*Group F:* Carbonaceous dusts that have more than 8% total entrapped volatiles. (Coal, carbon black, charcoal and coke dust are examples of carbonaceous dusts).

*Group G:* Combustible dusts not included in Group E or F, including flour, grain, wood, plastic and chemicals.

Whether an area is Division 1 or Division 2 depends on the quantity of dust present, except that for Group E there is only Division 1.

## Class III Locations

Atmospheric hazards cover locations where combustible fibers/flyings are present but not likely to be in suspension in air in quantities sufficient to produce ignitable mixtures. Division 1 is where they are manufactured; and Division 2 is where they are stored.

## Maximum Surface Temperature

Apparatus for use in a hazardous area must not have a surface temperature > the auto-ignition temperature (see definition above). Apparatus is therefore marked with a maximum surface temperature or "T" rating (Table 8-1).

### Table 8-1. Classification of Maximum Surface Temperature

| Maximum Temperature | | N.A. Temperature Class (T Code) | IEC/CENELEC Temp. Codes |
|---|---|---|---|
| °C | °F | | |
| 450 | 842 | T1 | T1 |
| 300 | 572 | T2 | T2 |
| 280 | 536 | T2A | |
| 260 | 500 | T2B | |
| 230 | 446 | T2C | |
| 215 | 419 | T2D | |
| 200 | 392 | T3 | T3 |
| 180 | 356 | T3A | |
| 165 | 329 | T3B | |
| 160 | 320 | T3C | |
| 135 | 275 | T4 | T4 |
| 120 | 248 | T4A | |
| 100 | 212 | T5 | T5 |
| 85 | 185 | T6 | T6 |

## Appropriate Protection Techniques

### Table 8-2. Appropriate Protection Techniques

| Protection Technique | Hazardous Area | | | |
|---|---|---|---|---|
| | Class I | | Class II | |
| | Div 1 | Div 2 | Div 1 | Div 2 |
| Dust-ignitionproof equipment | X | X | ✓ | ✓ |
| Explosionproof equipment | ✓ | ✓ | ✓ | ✓ |
| Hermetically sealed equipment | X | ✓ | ✓ | ✓ |
| Intrinsically safe systems | ✓ | ✓ | ✓ | ✓ |
| Nonincendive circuits and components | X | ✓ | ✓ | ✓ |
| Oil immersed equipment | X | ✓ | ✓ | ✓ |
| Purged and pressurized systems | ✓ | ✓ | ✓ | ✓ |

## Sealing Between Electrical Systems and Flammable or Combustible Process Fluids

Reference ANSI/ISA-12.27.01-2011 - *Requirements for Process Sealing Between Electrical Systems and Flammable or Combustible Process Fluids* for further details.

### Secondary Process Seals (pressurized systems)

Secondary process seals are used to prevent combustible/flammable fluids from travelling to a non-classified electrical system via the conduit system, should the primary seal at the field device fail. Many manufacturers include an integral rupture 'indication sensor' that provides can provide immediate notification of a potentially explosive seal rupture. The sealing devices are typically, filled with Chico®[141] compound once the cables/conductors are in place.

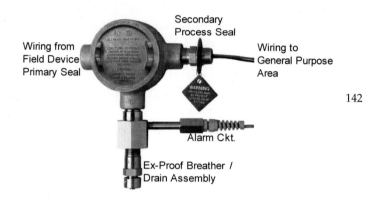

142

### Conduit Seals (atmospheric pressure systems)

Conduit seals are installed in conduit runs to prevent the passage of gases, vapors or flames from one section of an electrical installation to another section via the conduit. Sealing fittings are designed to be filled with a chemical compound (Chico®) after the wires have been pulled.

Sealing fittings are required at each entrance to an enclosure housing an arcing or sparking device when used in Class I, Division 1 and 2 hazardous locations. The fittings are to located as close as practical and, in no case, no more than 18 inches from such enclosures.

Sealing fittings are also required in conduit systems upon transition from Class I, Division 1 or Division 2 hazardous locations to general purpose locations.

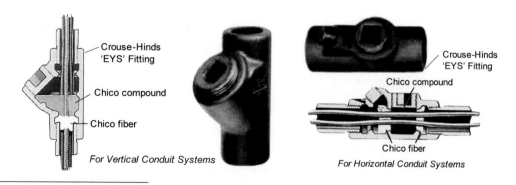

---

141. Chico compound expands slightly as it hardens to a dense, strong mass that is suitable for Class I, Division 1, Groups C & D and Class II, Division 1, Groups E, F & G hazardous applications.
142. Crouse-Hinds (Cooper Industries) secondary process seal.

---

## Purged & Pressurized Systems

NFPA 496 standards require that the enclosure air volume be exchanged a minimum of four times every hour. *Reference NFPA 496.*

### Type X Purge

Type X pressurization reduces the hazard classification from Division 1 to non-classified. Type X purging requires that if the enclosure pressure is lost, the power supply is automatically disconnected on loss of purge pressure and a re-purge is required before the supply is restored.

*Class I:* In Class I applications the Type X purge system operates by forcing air or an inert gas through the enclosure until all of the hazardous gas is removed (purging). After purging, constant pressure is maintained within the enclosure and equipment inside the enclosure is energized. If enclosure pressure drops below its minimum, enclosure power is cut and a purging sequence begins. Requires a minimum of 0.1 inch WC pressurization before power can be applied.

*Class II:* In Class II applications the enclosure must be free of all combustible material. It is pressurized before powering the equipment inside. There is no purging of the Class II, Type X system. Requires a minimum of 0.5 inch WC pressurization before power can be applied.

*Vents* are required for most Type X purge enclosures (Figure 8-1[143]).

---

143. Type X purge from Pepperl-Fuchs.

---

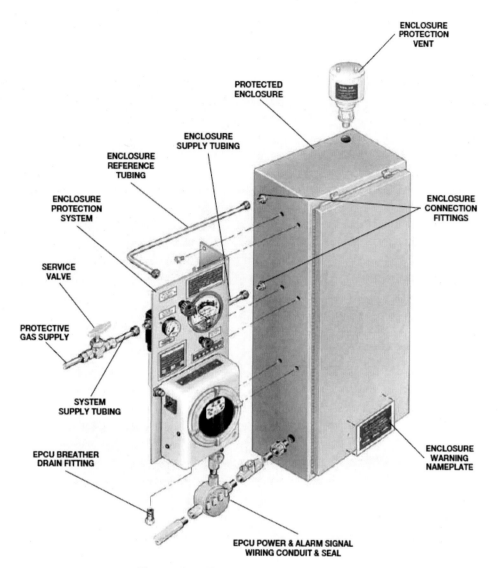

ENCLOSURE
PROTECTION
VENT

PROTECTED
ENCLOSURE

ENCLOSURE
SUPPLY TUBING

ENCLOSURE
REFERENCE
TUBING

ENCLOSURE
PROTECTION
SYSTEM

ENCLOSURE
CONNECTION
FITTINGS

SERVICE
VALVE

PROTECTIVE
GAS SUPPLY

SYSTEM
SUPPLY TUBING

EPCU BREATHER
DRAIN FITTING

ENCLOSURE
WARNING
NAMEPLATE

EPCU POWER & ALARM SIGNAL
WIRING CONDUIT & SEAL

**Figure 8-1. Type X Purge Enclosure**

**Type Y Purge**

Type Y pressurization reduces the hazard classification from Division 1 to Division 2. Type Y purging does not require supply disconnection on loss of pressure but the equipment in the purged enclosure must be suitable for Division 2.

*Class I:* Note: All equipment inside the enclosure (Figure 8-2[144]) must be Division 2 rated. For Class I applications this system operates by forcing air or an inert gas through the enclosure for a specified time until all of the hazardous gas has been removed (purging). After purging, a constant pressure with the enclosure is maintained and the equipment with the enclosure can be energized. Loss of safe pressure requires immediate attention if power is not discontinued to the enclosure. Requires a minimum of 0.1 inch WC pressurization before power can be applied.

*Class II:* In Class II applications no purging is performed but the enclosure must be cleaned of all combustible material, then pressurized before energizing the equipment with the enclosure. Loss of safe pressure requires immediate attention if power is not discontinued to the enclosure. Requires a minimum of 0.5 inch WC pressurization before power can be applied.

**Type Z Purge**

Type Z pressurization reduces the hazard classification from Division 2 to non-classified. Type Z purging, because of the lower level of risk in Division 2, requires only an indication of loss of purge pressure.

Z-Purge has the same components as those of a Y-Purge, except instead of reducing the hazard classification from Division 1 to Division 2, it reduces it from Division 2 to non-hazardous.

*Wiring Methods*

**Class I, Division 1**

The following wiring methods are permitted in Class I, Division 1 locations:

- Threaded rigid metal conduit or threaded intermediate metal conduit (IMC).

    - Exception: Plastic/fiberglass conduit that is encased in concrete with a minimum envelope cover of 2 inches and with the last 24 inches of conduit to the emergence point being either rigid metal or IMC.

- Type MI (mineral insulated) cable with termination fittings listed for the location (MI = made from copper conductors inside a copper sheath, insulated by inorganic magnesium oxide powder). Must be installed with sufficient support to avoid tensile stress on the termination fittings.

- In industrial establishments with restricted public access, where the conditions of maintenance and supervision ensure that only qualified persons service the installation: Type MC-HL cable listed for use in Class I, Zone 1 or Division 1 locations with a gas/vaportight continuous corrugated metallic sheath, an overall jacket of suitable polymeric material, a separate equipment grounding conductor(s) and provided with termination fittings listed for the application. (MC-HL is a metal clad cable such as IAC, interlocked armor cable and CLX, metal clad armored cable).

144. Type Y purge from Pepperl-Fuchs.

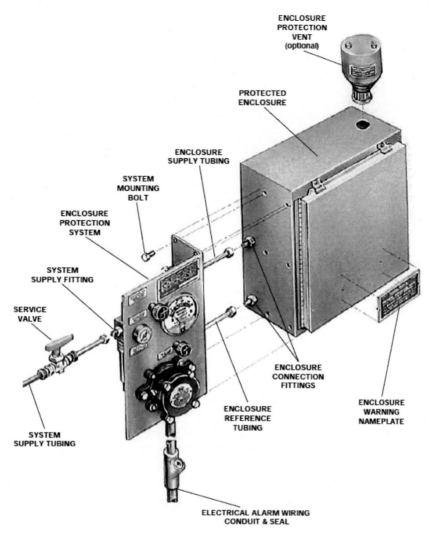

**Figure 8-2. Type Y Purge Enclosure**

- In industrial establishments with restricted public access, where the conditions of maintenance and supervision ensure that only qualified persons service the installation: Type ITC-HL cable listed for use in Class I Zone 1 or Division 1 locations with a gas/vaportight continuous corrugated metallic sheath, an overall jacket of suitable polymeric material, a separate equipment grounding conductor(s) and provided with termination fittings listed for the application. (ITC-HL is a metal clad instrument tray cable such as IAC and CLX).

**Class I, Division 2**

The following wiring methods are permitted in Class I, Division 2 locations:

Threaded rigid metal conduit or threaded intermediate metal conduit (IMC).

- Enclosed gasketed busways or wireways.

- Type PLTC (Power Limited Tray Cable). Must installed with sufficient support to avoid tensile stress on the termination fittings.

- Type ITC Cable (Instrumentation Tray Cable). Must installed with sufficient support to avoid tensile stress on the termination fittings.

- Type MI, MC, MV or TC cable with termination fittings, or in cable tray systems. Must be installed with sufficient support to avoid tensile stress on the termination fittings. (MI = Mineral Insulated; MC = Metal Clad; MV = Medium Voltage; TC = Tray Cable)

## 8.1.2 NEC Article 505 (Class I, Zone 0, Zone 1 and Zone 2 Locations)

*Definitions*

**AEx:** North American version of the IEC Ex Marking Scheme

**EEx:** Additional E in EEx implies the use of a CENELEC standard and should be used on a product that is ATEX approved.

**ATEX:** Atmosphère Explosible (French)

**Zone 0:** A place in which an explosive atmosphere consisting of a mixture with air of flammable substances in the form of gas, vapor or mist is presently continuously or for long periods of time.

**Zone1:** A place in which an explosive atmosphere consisting of a mixture with air of flammable substances in the form of gas, vapor or mist is likely to occur in normal operation.

**Zone 2:** A place in which an explosive atmosphere consisting of a mixture with air of flammable substances in the form of gas, vapor or mist is not likely to occur in normal operation but, if it does occur, will persist for a short period only.

**Zone 20:** A place in which an explosive atmosphere in the form of a cloud of combustible dust in air is present continuously, or for long periods or frequently.

**Zone 21:** A place in which an explosive atmosphere in the form of a cloud of combustible dust in air is likely to occur occasionally in normal operation.

**Zone 22:** A place in which an explosive atmosphere in the form of a cloud of combustible dust in air is not likely to occur in normal operation but, if it does occur, will persist for a short period only.

**Encapsulation "m":** Type of protection where electrical parts that could ignite an explosive atmosphere by either sparking or heating are enclosed in a compound in such a way that this explosive atmosphere cannot be ignited.

**Flameproof "d":** Type of protection where the enclosure can withstand an internal explosion of flammable mixture that has penetrated into the interior, without suffering damage and without causing ignition, through any joints or structural openings in the enclosure, of an external gas atmosphere consisting of one or more of the gases or vapors for which it is designed.

**Increased Safety "e":** Type of protection applied to electrical equipment that does not produce arcs or sparks in normal service and under specified abnormal conditions, in which additional measures are applied so as to give increased security against the possibility of excessive temperatures and of the occurrence of arcs and sparks. This is done by reducing and controlling working temperatures, ensuring that the electrical connections are reliable, increasing insulation effectiveness, and reducing the probability of contamination by dirt and moisture ingress.

**Intrinsic Safety "i":** Type of protection where any spark or thermal effect is incapable of causing ignition of mixture of flammable or combustible material in air under prescribed conditions.

**Oil Immersion "o":** Type of protection where electrical equipment is immersed in a protective liquid in such a way that an explosive atmosphere that may be above the liquid or outside the enclosure cannot be ignited. This is an old technique, primarily used for switchgear. The spark is formed under oil and venting is controlled.

**Powder Filling "q":** Type of protection where electrical parts capable of igniting an explosive atmosphere are fixed in position and completely surrounded by filling material (glass or quartz powder) to prevent ignition of an external explosive atmosphere. It is primarily of use where the incendive action is the abnormal release of electrical energy by the rupture of fuses or failure of components such as capacitors. Usually it is used for components inside Ex "e" or Ex "n" apparatus and for heavy duty traction batteries.

**Pressurization "p":** Type of protection for electrical equipment that uses the technique of guarding against the ingress of the external atmosphere, which may be explosive, into an enclosure by maintaining a protective gas therein at a pressure above that of the external atmosphere.

**Type of Protection "n":** Type of protection where electrical equipment, in normal operation, is not capable of igniting a surrounding explosive atmosphere and a fault capable of causing ignition is not likely to occur.

*Type of Protection Designations*

### Table 8-3. Type of Protection Designations

| Designation | Technique | Zone(s) | Protection |
|---|---|---|---|
| Ex d | Flameproof enclosure | 1, 2 | Intended to prevent an ignition from escaping outside the equipment |
| Ex q | Powder filled | 1, 2 | |
| Ex nC | Sparking equipment in which contacts are suitably protected other than by restricted breathing | 2 | |
| Ex e | Increased safety | 1, 2 | Intended to prevent a potential ignition from occurring |
| Ex nA | Non-sparking equipment | 2 | |
| Ex ia | Intrinsic safety (2 fault) | 0, 1, 2 | Intended to limit the ignition energy of the equipment |
| Ex Ib | Intrinsic safety (1 fault) | 1, 2 | |
| Ex ic | Intrinsic safety (energy limited apparatus) | 2 | |

| Ex m | Encapsulation | 1, 2 | Intended to prevent the explosive atmosphere from contacting the ignition source |
| Ex ma | Encapsulation | 0, 1, 2 | |
| Ex mb | Encapsulation | 1, 2 | |
| Ex nR | Restricted breathing enclosure | 2 | |
| Ex o | Oil immersion | 1, 2 | |
| Ex px | Pressurization | 1, 2 | |
| Ex py | Pressurization | 1, 2 | |
| Ex pz | Pressurization | 2 | |

## Expanded Markings

- ### US (NEC 500)

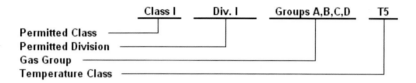

- ### US (NEC 505)

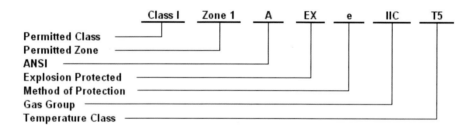

- ### IEC Ex Marking

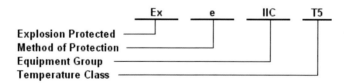

- ### ATEX & CE Marking

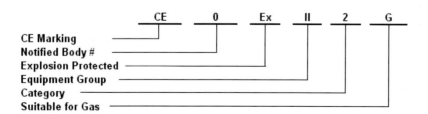

- **Additional Marking**

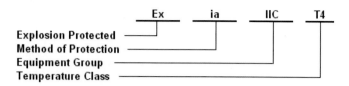

## 8.2 SAFETY INSTRUMENTED SYSTEMS (SIS)

*Reference ISA-84 Standard*

A Safety Instrumented System (SIS) plays a vital role in providing a protection layer around industrial process systems. The SIS's purpose is to take the process to a safe state when predetermined setpoints have been exceeded or when safe operating limits have been transgressed.

### 8.2.1 Safety Integrity Level (SIL)

Defined as a relative level of risk reduction provided by a safety function, or as a means to specify a target level of risk reduction. In simple terms, SIL is a measurement of the safety risk of a given process. The assignment of a certain SIL is a decision reached through process evaluations such as PHA (Process Hazard Analysis); HAZOP (Hazard and Operability Study) and FMEA (Failure Mode and Effects Analysis).

Four SILs are defined, with SIL4 being the most dependable (i.e., available to function when called upon to function) and SIL1 being the least. A SIL is a way to indicate the tolerable failure rate of a particular safety function. Each level indicates an order of magnitude of risk reduction. See Table 8-4.

### Table 8-4. Safety Integrity Levels

| SIL | ISA-84 | IEC 61508 | Availability | PFD Avg. | Risk Reduction | Qualitative Consequence |
|-----|--------|-----------|--------------|----------|----------------|-------------------------|
| 4 | NO | Yes | > 99.99% | $10^{-5}$ to < $10^{-4}$ | 100,000 to 10,000 | Potential community fatalities |
| 3 | Yes | Yes | 99.9% | $10^{-4}$ to < $10^{-3}$ | 10,000 to 1,000 | Potential multiple on site fatalities |
| 2 | Yes | Yes | 99 to 99.9% | $10^{-3}$ to < $10^{-2}$ | 1,000 to 100 | Potential major on site injuries |
| 1 | Yes | Yes | 90 to 99% | $10^{-2}$ to < $10^{-1}$ | 100 to 10 | Potential minor on site injuries |

PFD = Probability of Failure on Demand; Risk Reduction (RRF) = 1 ÷ PFD

## 8.2.2 Relationship between a Safety Instrumented Function and Other Functions

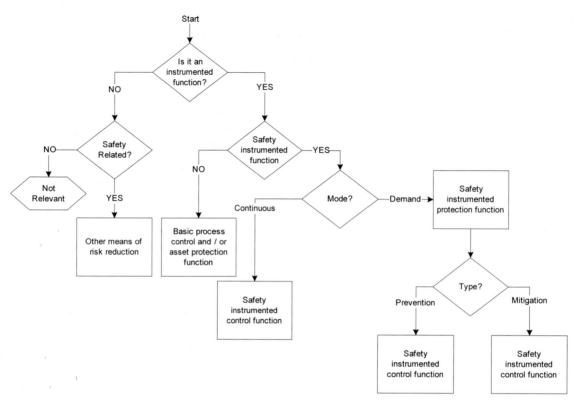

**Figure 8-3. Relationship between a Safety Instrumented Function and Other Functions**

## 8.2.3 Definitions

**Operating Measures:** During normal process operations, the process parameters fluctuate slightly around their normal operating setpoint, while staying within their desirable range. Equipment which allows the operation to stay within this desirable range is classified as "Operating Measures." Thus, it ensures that safety equipment that can override the operating measure is rarely activated. *Reference Figure 8-4 – Point 1*

**Safety Measures:** If a process parameter exceeds the desirable range and enters the permissible failure range, additional protection measures are activated. These additional measures are termed "Safety Measures" since they are determined by the PHA to be important to the safety of the process. The primary purpose of safety measures is to monitor the process for undesirable conditions and alert the operator and/or take automatic action to bring the process back to the desirable range. The equipment serves to maintain normal operation, initiates interlocks, if necessary, and thus avoids the activation of the protection measure. *Reference Figure 8-4 – Point 2*

**Protection Measures:** If the process continues its excursion away from the desirable range, it may approach the non-permissible failure range. If a process parameter enters the non-permissible failure range, it could possibly cause damage and the parameter becomes a process safety parameter. Any measure which is designed to activate just prior to the process entering the non-permissible failure range is considered a "Protection Measure." The protection measure should act upon the direct cause of the potential damage. *Reference Figure 8-4 – Point 3*

**Mitigating Measures:** If the process parameter continues its excursion and actually enters the non-permissible failure range, then "Mitigating Measures" may be required. The mitigating measures consist of measures designed to limit the extent of the damage after the incident has occurred. *Reference Figure 8-4 - Top*

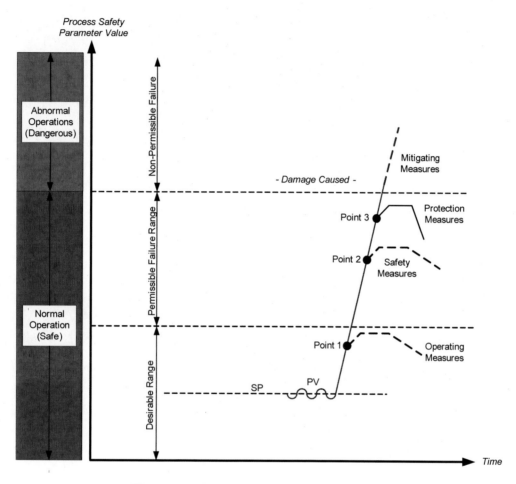

**Figure 8-4. SIS Measures Defined and Related**

## 8.2.4 Layers of Protection

*As defined by ISA-84 Part 3*

A protection layer (Figure 8-5) consists of a grouping of equipment and/or administrative controls that function in concert with other protection layers to control or mitigate process risk.

Per ISA-84, the protection layer shall reduce the identified risk by at least a factor of 10.

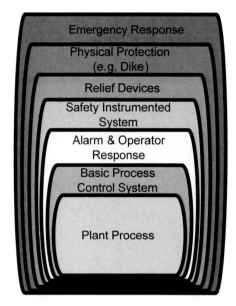

**Figure 8-5. Protection Layers**

*Layer of Protection Analysis (LOPA)*

LOPA (Figure 8-6) is a useful tool often used to assess the risk of major accidents. LOPA uses a multi-functionality team similar to what is employed in a HAZOP.

## 8.3 DETERMINING PFD (PROBABILITY OF FAILURE ON DEMAND)

Failure Rate = # of Failures/Unit of Time (expressed in hours)

A constant Failure Rate is assumed for the normal life of the device.

**Mean Time To Failure (MTTF):** Mean time until a device's first failure (assumes that the device can NOT be repaired nor resume any of its normal operation).

MTTF = 1/Failure Rate

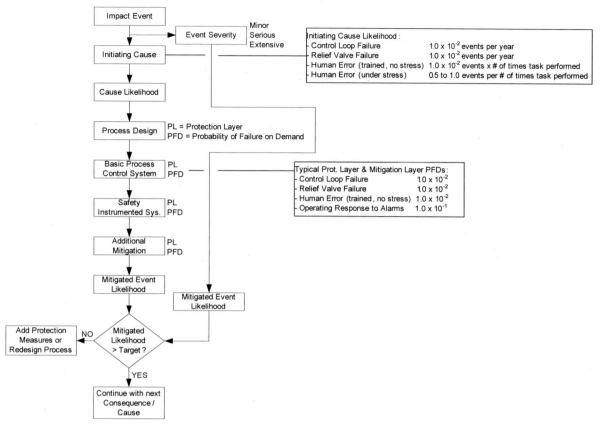

**Figure 8-6. Layer of Protection Analysis**

## 8.3.1 Basic Reliability Formulas

### Table 8-5. Basic Reliability Formulas

| Configuration | MTTF$_{sp}$ | PFD |
|---|---|---|
| 1oo1 (1 out of 1) | $\dfrac{1}{\lambda_s}$ | $\lambda_{du} \times \dfrac{TI}{2}$ |
| 1oo2 (1 out of 2) | $\dfrac{1}{2\lambda_s}$ | $\left[ \left(\lambda_{du}\right)^2 \times \dfrac{TI^2}{3} \right]$ |
| 2oo2 (2 out of 2) | $\left[ \dfrac{1}{2\left(\lambda_s\right)^2 \times MTTR} \right]$ | $\left(\lambda_{du}\right)^2 \times TI$ |
| 2oo3 (2 out of 3) | $\left[ \dfrac{1}{6\left(\lambda_s\right)^2 \times MTTR} \right]$ | $\left(\lambda_{du}\right)^2 \times TI^2$ |

XooX = Voting logic architecture of the safety system

$\lambda$ = Failure Rate

MTTR = Mean Time to Repair

TI = Test Interval

$\lambda_s$ = Safe Failure = $MTTF_s/(MTTF_s + MTTR)$

$\lambda_{du}$ = Dangerous Undetected Failure = $MTTF_d/(MTTF_d + TI/2 + MTTR)$

## 8.3.2 Architectures

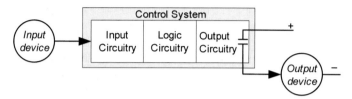

**Figure 8-7. 1oo1 (One out of One)**

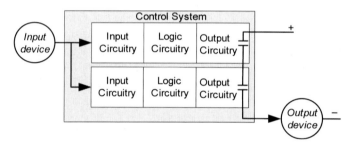

**Figure 8-8. 1oo2 (One out of Two)**

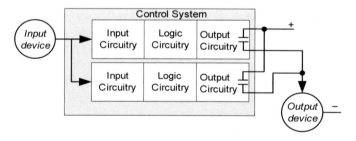

**Figure 8-9. 2oo2 (Two out of Two)**

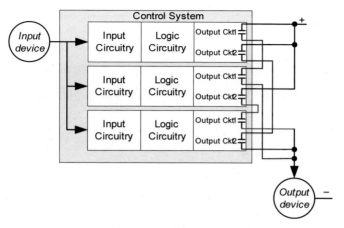

**Figure 8-10. 2oo3 (Two out of Three)**

# 9. CODES, STANDARDS AND REGULATIONS

## 9.1 STANDARDS LISTINGS

### 9.1.1 ISA

(International Society of Automation, a professional organization for automation practitioners)

*Note: RP indicates Recommended Practice and TR indicates Technical Report.*

- **ISA-2 – Manometer Tables**

  Presents abbreviations and fundamental conversion factors commonly used in manometry, recommended definitions of pressure in terms of a column of mercury or water, and, for a large number of liquids, tables of various pressures indicated by, or equivalent to, heights of columns at various temperatures.

- **ISA-5.1 – Instrument Symbols and Identification**

  The purpose of this standard is to establish a uniform means of designating instruments and instrumentation systems used for measurement and control. Reference Section 6.1 within this book.

- **ISA-5.2 – Binary Logic Diagrams for Process Operations**

  The purpose of this standard is to provide a method of logic diagramming of binary interlock and sequencing systems for the startup, operation, alarm, and shutdown of equipment and processes in the chemical, petroleum, power generation, air conditioning, metal refining, and numerous other industries. Reference Section 6.1.4 within this book.

- **ISA-5.3 – Graphic Symbols for Distributed Control/Shared Display Instrumentation, Logic and Computer Systems.**

  The purpose of this standard is to establish documentation for that class of instrumentation consisting of computers, programmable controllers, minicomputers and microprocessor based systems that have shared control, shared display or other interface features.

- **ISA-5.4 – Instrument Loop Diagrams**

  The purpose of this standard is to provide guidelines for the preparation and use of instrument loop diagrams in the design, construction, startup, operation, maintenance, and modification of instrumentation systems. Reference Section 6.1.6 within this book.

- **ISA-5.5 – Graphic Symbols for Process Displays**

  The purpose of this standard is to establish a system of graphic symbols for process displays that are used by plant operators, engineers, etc. for process monitoring and control. The system is intended to facilitate rapid comprehension by the users of the information that is conveyed through displays, and to establish uniformity of practice throughout the process industries.

- **ISA-7 – Quality Standard for Instrument Air**

  This standard establishes a standard for instrument quality air.

- **ISA-12 – Electrical Equipment for Hazardous Locations (Reference Section 8.1 within this book)**

  This ISA standard category (Hazardous Locations) contains the following: ISA-12.00.02; ISA-12.01.01; ISA-12.10; ISA-12.12.01; ISA-12.13.01; ISA-12.13.04; ISA-12.20.01; ISA-12.27.01; ISA-RP12.2.02; ISA-RP.12.03; ISA-RP12.4; ISA-RP12.06.01; ISA-RP12.13.02; ISA-TR12.06.01; ISA-TR12.2.02; ISA-TR12.13.01; ISA-TR12.13.02; ISA-TR12.13.03; ISA-TR12.21.01; ISA-TR12.24.01.

- **ISA-18.1 – Annunciator Sequences and Specifications**

  This standard is primarily for use with electrical annunciators that call attention to abnormal process conditions by the use of individual illuminated visual displays and audible devices

- **ISA-18.2 – Management of Alarm Systems for the Process Industries**

  This standard addresses the development, design, installation, and management of alarm systems in the process industries.

- **ISA-20 – Specification Forms for Process Measurement and Control Instruments, Primary Elements, and Control Valves.** This category also contains ISA-TR20.00.01.

- **ISA-RP31.1 – Specification, Installation, and Calibration of Turbine Flowmeters**

- **ISA-37 – Measurement Transducers**

  This ISA standard category contains the following: ISA-37.1; ISA-37.2; ISA-37.5; ISA-37.6; ISA-37.8; ISA-37-10; ISA-37.12; ISA-37.16.01.

- **ISA-RP42.00.01 – Nomenclature for Instrument Tube Fittings**

  This Recommended Practice defines nomenclature for the tube fittings most commonly used in instrumentation.

- **ISA-50 – Signal Compatibility of Electrical Instruments**

This ISA standard category contains the following: ISA-50.00.01; ISA-TR50.02 Parts 3 & 4; ISA-TR50.02 Part 9.

- **ISA-51.1 – Process Instrumentation Terminology**

- **ISA-TR52.00.01 – Recommended Environments for Standards Laboratories**

- **ISA-RP60 – Control Center Facilities**

  This recommended practice category contains the following: ISA-RP60.1; ISA-RP60.2; ISA-RP60.3; ISA-RP60.4; ISA-RP60.6; ISA-RP60.8; ISA-RP60.9; ISA-RP60.11.

- **ISA-67 – Nuclear Power Plant Standards**

  This ISA standard category contains the following: ISA-67.01.01; ISA-67.02.01; ISA-67.03; ISA-67-.04.01; ISA-67.04.02; ISA-67.06.01; ISA-67./14.01; ISA-TR67.04.08; ISA-TR67.04.09.

- **ISA-71 – Environmental Conditions for Process Measurement and Control**

  This ISA standard category contains the following: ISA-71.01; ISA-71.02; ISA-71.03; ISA 71.04

- **ISA-RP74.01 – Application and Installation of Continuous-Belt Weighbridge Scales**

- **ISA-75 – Control Valve Standards**

  This ISA standard category includes the following: ISA-75.01.01; ISA-75.02.01; ISA-75.05.01; ISA-75.07; ISA-75.08.01; ISA-75-08.02; ISA-75-08.03; ISA-75-08.04; ISA-75.08.05; ISA-75.08.06; ISA-75.08.07; ISA-75.08.08; ISA-75.10.01; ISA-75.10.02; ISA-75.11.01; ISA-75.13.01; ISA-75.17; ISA-75.17; ISA-75.19.01; ISA-75.26.01; ISA-RP75.23; ISA-TR75.04.01; ISA-TR75.25.02.

- **ISA-RP76.0.01 – Analyzer System Inspection and Acceptance**

- **ISA-77 – Fossil Fuel Power Plant Standards**

  This ISA standard category includes the following: ISA-77.13.01; ISA-77.14.01; ISA-77.20; ISA-77.41.01; ISA-77.42.01; ISA-77.43.01; ISA-77.44.01; ISA-77.70.02; ISA-77.82.01; ISA-RP77.60.02; ISA-RP77.60.05; ISA-TR77.60.04; ISA-TR77.70.01.

- **ISA-82.03 – Safety Standard for Electrical and Electronic Test, Measuring, Controlling, and Related Equipment**

- **ISA-84 – Functional Safety: Safety Instrumented Systems for the Process Industry Sector**

  This ISA standard category includes the following: ISA-84.00.01 Part 1; ISA-84.00.01 Part 2; ISA-84.00.01 Part 3; ISA-TR84.00.02 Part 1; ISA-TR84.00.02 Part 2; ISA-TR84.00.02 Part 3; ISA-TR84.00.02 Part 4; ISA-TR84.00.02 Part 5; ISA-TR84.00.03; ISA-TR84.00.04 Part 1;

ISA-TR84.00.04 Part 2; ISA-TR84.00.05; ISA-TR84.00.06; ISA-R84.00.07. See Section 8.2 of this book.

- **ISA-88 – Batch Control Systems**

  This ISA standard category includes the following: ISA-88.00.01; ISA-88.00.02; ISA-88.00.03; ISA-88.00.04; ISA-TR88.00.02; ISA-TR88.03; ISA-TR88.95.01. See Section 6.7 of this book.

- **ISA-91.00.01 – Identification of Emergency Shutdown Systems and Controls that Are Critical to Maintaining Safety in Process Industries**

- **ISA-92 – Performance Requirements for Toxic Gas Detectors**

  This ISA standard category includes the following: ISA-92.00.01; ISA-92.0.01; ISA-92.02.01; ISA-92.03.01; ISA-92-.04.01; ISA-92.06.01; ISA-RP92.0.02; ISA-RP92.02.02; ISA-RP92.03.02; ISA-RP92.04.02; ISA-RP92.06.02; ISA-TR92.06.03.

- **ISA-93.00.01 – Standard Method for the Evaluation of External Leakage of Manual and Automated On-Off Valves**

- **ISA-95 – Enterprise/Control Integration**

  This ISA standard category includes the following: ISA-95.00.01; ISA-95.00.02; ISA-95.00.03; ISA-95.00.05. See Section 7 of this book.

- **ISA-96.02.01 – Guidelines for the Specification of Electric Valve Actuators**

- **ISA-98.00.01 – Qualifications and Certification of Control System Technicians**

- **ISA-99 – Security for Industrial Automation and Control Systems**

  This ISA standard category contains the following: ISA-99.00.01; ISA-99.02.01; ISA-TR99.00.01.

- **ISA-100.11a – Wireless systems for Industrial Automation: Process Control and Related Applications**

## 9.1.2 ASME

(American Society of Mechanical Engineers)

*Boiler & Pressure Vessel Code (Section VIII Division 1)*

*Note: Many relief valve manufacturers have a condensed version of the ASME Section VIII, Division 1 Code available on their website.*

- **UG-125 (General)**

- **UG-126 (Pressure Relief Valves)**

- UG-127 (Non-reclosing Pressure Relief Devices)

- UG-128 (Liquid Relief Valves)

- UG-129 (Marking)

- UG-130 (Use of Code Symbol Stamp)

- UG-131 (Certification of Capacity of Pressure Relief Valves)

- UG-132 (Certification of Capacity of Safety and Relief Valves in Combination with Non-reclosing Pressure Relief Devices)

- UG-133 (Determination of Pressure Relieving Requirements)

- UG-134 (Pressure Setting of Pressure Relief Devices)

- UG-135 (Installation)

- UG-136 (Minimum Requirements for Pressure Relief Valves)

### 9.1.3 API

(American Petroleum Institute)

- Recommended Practice 520 Part 1 (Relief Valve Sizing and Selection)

- Recommended Practice 520 Part 2 (Relief Valve Installation)

- Recommended Practice 521 (Guide for Pressure Relief and Depressuring Systems)

- Recommended Practice 526 (Flanged Steel Relief Valves)

- Recommended Practice 527 (Seat Tightness of Pressure Relief Valves)

### 9.1.4 NFPA

(National Fire Protection Agency is an agency whose mission is to reduce the worldwide burden of fire and other hazards on the quality of life by providing and advocating consensus codes and standards, research, training and education: The standards/codes listed below are applicable in the control systems arena.)

- NFPA 70 – National Electrical Code

- NFPA 70E – Standard for Electrical Safety in the Workplace

- NFPA 77 – Static Electricity

- NFPA 79 – Industrial Machinery

- **NFPA 496 – Standard for Purged and Pressurized Enclosures for Electrical Equipment**

- **NFPA 497 – Recommended Practice for the Classification of Flammable Liquids, Gases, or Vapors and of Hazardous (Classified) Locations for Electrical Installations in Chemical Process Areas**

- **NFPA 780 – Lightning Protection**

## 9.1.5 IEC

(International Electrotechnical Commission, a not-for-profit, non-governmental international standards organization that prepares and publishes International Standards for all electrical, electronic and related technologies)

- **IEC 61508 – Functional safety of electrical/electronic/programmable electronic safety-related systems**

- **IEC 61511 – Functional safety - safety instrumented systems for the process industry sector**

## 9.1.6 CSA

(Canadian Standards Association, a not-for-profit association composed of representatives from government, industry, and consumer groups)

- **C22.1 – Canadian National Electrical Code**

**The following three sections 9.1.7 "UL", 9.1.8 "FM", and 9.1.9 "CE" providing listing/approval information that certifies the components listed/approved meet minimum safety standards as set forth by their organizations.**

## 9.1.7 UL

(Underwriters Laboratories, a U.S. privately owned and operated, independent, third party product safety testing and certification organization)

Requiring the use of UL listed components in your control system design ensures that a sample of the manufacturer's components have been tested and certified to meet UL's safety requirements.

## 9.1.8 FM

Factory Mutual FM Approvals certifies products and services with a focus on:

- Objectively testing property loss prevention products and services and certifying those that meet rigorous loss prevention standards.

- Encouraging the development and use of FM Approved products and services that improve and advance property loss prevention practices.

As with the UL statement above, Requiring the use of FM approved components in your control system design ensures that a sample of the manufacturer's components have been tested and certified to meet FM's safety requirements.

## 9.1.9 CE

The "Conformité Européenne" **CE marking** (also known as **CE mark**) is a mandatory conformity mark on many products placed on the single market in the European Economic Area (EEA). The CE marking certifies that a product has met EU consumer safety, health or environmental requirements.

There is no 3rd party testing. This approach to conformity enables manufacturers to use what is called "SELF DECLARATION," where the manufacturer himself declares conformity by signing the "Declaration of Conformity (DOC)" and then affixes the CE Mark on his product by following a seven step procedure.

## 9.1.10 ANSI

(American National Standards Institute) ANSI is a private non-profit organization that oversees the development of voluntary consensus standards for products, services, processes, systems, and personnel in the United States. The organization also coordinates U.S. standards with international standards so that American products can be used worldwide.

## 9.1.11 IEEE

(Institute of Electrical & Electronic Engineers) IEEE is an international non-profit, professional organization for the advancement of technology related to electricity.

## 9.1.12 AICHE (American Institute of Chemical Engineers)

## 9.1.13 OSHA

(United States Occupational Safety and Health Administration) Its mission is to prevent work-related injuries, illnesses, and deaths by issuing and enforcing rules (called standards) for workplace safety and health. The agency is headed by a Deputy Assistant Secretary of Labor. Standards are covered under 29CFR Part 1910.

Part 1910 Occupational Safety and Health Standards (list of applicable subparts):

- **Subpart E – Means of Egress**

  This subpart focuses on design and construction of exit routes; maintenance, safeguards and operational features for exit routes; emergency action plans; and fire prevention plans.

- **Subpart H – Hazardous Materials**

  This subpart focuses on compressed gases; flammable and combustible liquids; and storage and handling of hazardous material.

- **Subpart I – Personal Protective Equipment**

  This subpart focuses on eye/face protection; respiratory protection; head protection; foot protection; hand protection; and electrical protective equipment.

- **Subpart L – Fire Protection**

  This subpart focuses on portable fire extinguishers; standpipe and hose systems; automatic sprinkler systems; fixed extinguishing systems (dry, foam, and gas); and fire detection systems.

- **Subpart R – Special Industries**

  1910.269 Electric Power Generation, Transmission, and Distribution. This section covers the operation and maintenance of electric power generation, control, transformation, transmission, and distribution lines and equipment

- **Subpart S – Electrical**

  - DESIGN SAFETY STANDARDS FOR ELECTRICAL SYSTEMS

    - 1910.302 Electric utilization systems. Defines the provisions of §1910.302 through §1910.308

    - *1910.303 General requirements.* Defines working clearances; identification of disconnecting means; guarded live parts; and entrance and workspace access.

    - *1910.304 Wiring design and protection.* Defines receptacle and cord connectors; temporary wiring installations; and overcurrent protection.

    - *1910.305 Wiring methods, components, and equipment for general use.* Defines raceways, cable trays, etc.; cabinets, boxes and fittings; appliances; motors; capacitors; and storage batteries.

    - *1910.306 Specific purpose equipment and installations.* Defines electric signs and outline lighting; disconnecting means for crane and monorail hoists; elevators, escalators, etc.; electric welders; and lighting fixtures.

    - *1910.307 Hazardous (classified) locations.* (Reference Section 8.1 within this book)

    - *1910.308 Special systems.* Defines above ground wiring methods; mobile and portable equipment; emergency illumination; separation from conductors of other circuits; and grounding.

  - SAFETY-RELATED WORK PRACTICES

    - 1910.332 Training. Defines the training requirements contained in this section apply to employees who face a risk of electric shock that is not reduced to a safe level by the electrical installation requirements of §1910.303 through 1910.308.

    - *1910.333 Selection and use of work practices.* Defines working on or near exposed live parts.

    - 1910.334 Use of equipment.

    - *1910.335 Safeguards for personnel protection.* Defines use of personnel protective equipment (PPE).

# 9.2 NEC

## 9.2.1 Allowable Conduit Fill

### Table 9-1. NEC Allowable Conduit Fill

| RGS Size | Nominal ID | 2-Cond. 31% Fill Sq. In. | >2 Cond. 40% Fill Sq. In. | 1-Cond 53% Fill Sq. In. |
|---|---|---|---|---|
| 3/4"C | 0.836" | 0.17 | 0.22 | 0.29 |
| 1"C | 1.063" | 0.275 | 0.355 | 0.47 |
| 1.5"C | 1.624" | 0.765 | 0.829 | 1.098 |
| 2"C | 2.083" | 1.056 | 1.363 | 1.806 |
| 3"C | 3.090" | 2.325 | 3 | 3.975 |
| 4"C | 4.050" | 3.993 | 5.153 | 6.828 |

To determine appropriate conduit size, add the area of all the cables to be installed in said conduit and then determine from Table 9-1 what size conduit is required.

## 9.2.2 Wiring Methods

**Conduit Bends: There shall be no more than the equivalent of four quarter bends (360° total) between pull points.**

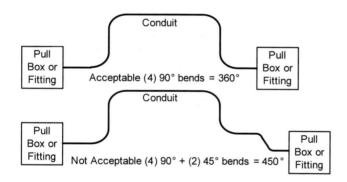

*Conduit Types*

Example of popular conduit body fitting types:[145]

C-Fitting

Purpose is to provide a straight cable pulling point in a conduit run

LB-Fitting  LL-Fitting  LR-Fitting

Purpose of L fitting is to provide a 90° pulling point in a conduit run at a box or device.
A way to remember where the cover is located on each type, is to grip the fitting by the short end and point the long end away from you like a gun. Wherever the cover opening is, that is the type of fitting (B – back; L – left; R – Right)

T-Fitting

Purpose is to provide a splitting point in a conduit run for two or more cables, or to provide a splice point as in a lighting circuit.

---

145. Reference Cooper Crouse-Hinds for example conduit body fitting technical information.

## Article 342 (Intermediate Metal Conduit – Type IMC)

Minimum size permitted = 1/2 inch; maximum size permitted = 4 inches

*Installation – Supports:*

- Conduit shall be supported at intervals not exceeding 10 feet or as shown in Table 9-2.

**Table 9-2. IMC Conduit Support Intervals**

| Conduit Size | Max Distance between Supports |
|---|---|
| ½ - ¾" | 10ft |
| 1" | 12ft |
| 1¼" - 1½" | 14ft |
| 2 - 2½" | 16ft |
| ≥ 3" | 20ft |

- Vertical straight risers shall be supported at intervals not exceeding 20 feet

- Conduit shall be securely fastened within 3 feet of each outlet box, jct box, device box, cabinet, conduit body or other conduit termination

*Advantages compared with RMC:*
Lighter weight due to decreased wall thickness

*Disadvantages compared with RMC:*
Less protection to physical damage due to decreased wall thickness

## Article 346 (Rigid Metal Conduit – Type RMC)

Minimum size permitted = 1/2 inch; maximum size permitted = 6 inches

*Installation – Supports:*

- Conduit shall be supported at intervals not exceeding 10 feet or shown in Table 9-3.

**Table 9-3. RMC Conduit Support Intervals**

| Conduit Size | Max distance between Supports |
|---|---|
| ½ - ¾" | 10ft |
| 1" | 12ft |
| 1¼" - 1½" | 14ft |
| 2 - 2½" | 16ft |
| ≥ 4" | 20ft |

- Vertical straight risers shall be supported at intervals not exceeding 20 feet

- Conduit shall be securely fastened within 3 feet of each outlet box, jct box, device box, cabinet, conduit body or other conduit termination

## Article 348 (Flexible Metal Conduit – Type FMC) (Figure 9-1)

Minimum size permitted = 3/8 inch; maximum size permitted = 4 inches. *May be referred to as "Greenfield" conduit.*

*Installation*

- Shall not be installed where subject to physical damage or high temperature applications

*Supports*

- Horizontal runs shall be securely fastened every 4 1/2 feet

- Securely fastened within 12 inches of each box, cabinet, conduit body or other conduit termination.

**Figure 9-1. Type FMC Conduit**

## Article 350 (Liquidtight Flexible Metal Conduit – Type LFMC) (Figure 9-2)

Minimum size permitted = 1/2 inch; maximum size permitted = 4 inches

*Installation:*

- Shall not be installed where subject to physical damage or high temperature applications

*Supports:*

- Horizontal runs shall be securely fastened every 4 1/2 feet

- Securely fastened within 12 inches of each box, cabinet, conduit body or other conduit termination.

**Figure 9-2. Type LFMC Conduit**

## Article 352 (Rigid Polyvinyl Chloride Conduit – Type PVC)

Minimum size permitted = 1/2 inch; maximum size permitted = 6 inches

*Installation:*

- Schedule 40 is suitable for underground use by direct burial or encasement in concrete; it is also suitable for above ground applications (indoors & outdoors) exposed to sunlight and weather, but NOT subject to physical damage.

- Schedule 80 has reduced cross-sectional area available for wire fill and is suitable for use wherever schedule 40 conduit may be used; it may also be used in areas subject to physical damage and for installation on poles.

- Unless marked for higher temperature PVC conduit is intended for use with ≤ 75°C cable within buildings. Where encased in concrete in trenches outside of buildings it is suitable for use with ≤ 90°C cable.

- Is NOT suitable for use in hazardous classified areas or in areas where ambient temperatures are in excess of 50°C.

*Supports:*

- Conduit shall be supported at intervals not exceeding 3 feet or shown in Table 9-4.

### Table 9-4. Type PVC Conduit Support Intervals

| Conduit Size | Max distance between Supports |
|:---:|:---:|
| ½ - 1" | 3ft |
| 1¼" - 2" | 5ft |
| 2½" - 3" | 6ft |
| 3½" - 5" | 7ft |
| 6" | 8ft |

- Securely fastened within 3 feet of each outlet box, jct box, device box, cabinet, conduit body or other conduit termination

## Article 353 (High Density Polyethylene Conduit – Type HDPE)

Minimum size permitted = 1/2 inch; maximum size permitted = 6 inches

The following bullets are not code requirements, just items to consider for the environment:

- Pretty much the same as PVC conduit but more environmentally friendly

- Being non-chlorinated, requiring fewer additives, and having a much higher recycling rate, it is considered a more benign plastic than PVC. PVC is more resistant to combustion, but smolders at a lower temperature than HDPE and releases toxic hydrochloric gases before combustion

## Article 356 (Liquidtight Flexible Nonmetallic Conduit – Type LFNC) (Figure 9-3)

Minimum size permitted = 1/2 inch; maximum size permitted = 4 inches. *May be referred to as "Sealtight" conduit*

*Installation:*

- Shall not be installed where subject to physical damage or in high temperature applications

*Supports:*

- Horizontal runs shall be securely fastened every 6 feet

- Securely fastened within 3 feet of each box, cabinet, conduit body or other conduit termination.

**Figure 9-3. Type LFNC Conduit**

## Article 358 (Electrical Metallic Tubing – Type EMT)

Minimum size permitted = 1/2 inch; maximum size permitted = 6 inches

*Installation:*

- Shall not be installed where subject to physical damage, in hazardous classified area.

- Shall not be used for the support of lighting fixtures or other equipment

*Supports:*

- Conduit shall be supported at intervals not exceeding 10 feet

- Securely fastened within 3 feet of each outlet box, jct box, device box, cabinet, conduit body or other conduit termination

## Article 360 (Flexible Metallic Tubing – Type FMT) (Figure 9-4)

Minimum size permitted = 1/2 inch; maximum size permitted = 3/4 inch

Maximum length permitted = 6 feet *This type is typically used as "fixture whip"*

*Installation:*

- Shall not be installed in hoistways, storage battery areas, areas subject to physical damage, hazardous classified areas

**Figure 9-4. Type FMT Conduit**

**Article 362 (Electrical Non-Metallic Tubing – Type ENT)** (Figure 9-5)

Minimum size permitted = 1/2 inch; maximum size permitted = 2 inches. *Very similar to "Innerduct"*

*Installation:*

- Shall not be installed in exposed locations, underground, areas subject to physical damage, hazardous classified areas, theaters or areas where ambient temperatures are in excess of 50°C.

- Shall not be used for the support of lighting fixtures or other equipment

**Figure 9-5. Type ENT Conduit**

**Article 376 (Metal Wireways)** (Figure 9-6)

- Very similar to auxiliary gutters[146] except that the limit for 30 current carrying conductors does not apply. However, the maximum fill of **20%** of the cross-sectional area does apply.

- Derating factors from article 310.15(B)4 apply if there are more than 30 current carrying conductors within the wireway.

---

146. Auxiliary gutter is a sheet metal enclosure used to supplement wiring spaces at meter centers, distribution centers, switchboards, and similar points of wiring systems.

**Figure 9-6. Metal Wireway**

## Article 378 (Non-Metallic Wireways)

- Very similar to metal wireways including the maximum fill of 20% of the cross-sectional, however the derating factors from article 310.15(B)4 apply to all the applicable current carrying conductors within the wireway.

## Article 384 (Strut-Type Channel Raceway) (Figure 9-7)

*Typically used to support and supply power to lighting fixtures*

- Number of conductors shall not exceed the % fill allowable shown in Table 9-5.

### Table 9-5. Article 384 Maximum Percent Allowable Fill

| Channel Size | Area (in²) | 40% Area* (in²) | 25% Area** (in²) |
|---|---|---|---|
| 1 5/8" x 1 3/16" | 0.887 | 0.355 | 0.222 |
| 1 5/8" x 1" | 1.151 | 0.460 | 0.288 |
| 1 5/8" x 1 3/8" | 1.677 | 0.671 | 0.419 |
| 1 5/8" x 1 5/8" *** | 2.028 | 0.811 | 0.507 |
| 1 5/8" x 2 7/16" | 3.169 | 1.267 | 0.792 |
| 1 5/8" x 3¼" | 4.308 | 1.723 | 1.077 |
| 1½" x ¾" | 0.849 | 0.340 | 0.212 |
| 1½" x 1½" | 1.828 | 0.731 | 0.457 |
| 1½" x 1 7/8" | 2.301 | 0.920 | 0.575 |
| 1½" x 3" | 3.854 | 1.542 | 0.964 |

\*Raceways with external joiners shall use a 40% fill calculation
\*\*Raceways with internal joiners shall use a 25% fill calculation
\*\*\*Indicates standard Unistrut size.

*Fill Calculation:* $n = \dfrac{ca}{wa}$

Where n = number of wires
   ca = channel area in square inches
   wa =wire area

**Figure 9-7. Strut Type Channel Raceway**

**Articles 386 & 388 (Surface Mounted Raceways)** *aka Wiremold* (Figure 9-8)

- Consists of a surface-mounted assembly of metal backing (providing mechanical support) and capping (providing a protective covering); used for electric wiring for branch circuits or feeder conductors

- The number, type and sizes of conductors permitted to be installed in a listed surface metal raceway are typically marked on the raceway or on the packaging. This information is usually available in detail from the manufacturer's catalog.

**Figure 9-8. Metallic (left) and Non-metallic (right) Surface Mounted Raceways**

**Article 392 (Cable Trays)**

*Cable Installation:*

- Multi-Conductor Cables: Shall be permitted to be installed in cable trays provided the cable is listed for use in cable trays (e.g., type TC); if installed outdoors then the cable must also be listed as sunlight resistant.

- Single Conductor Cables: Size #1/0 AWG or larger shall be permitted to be installed in cable trays provided the cable is listed for use in cable trays (e.g., type TC); if installed outdoors then the cable must also be listed as sunlight resistant.

  - Equipment Grounding Conductor Cables: Single-Conductor Equipment Grounding Conductor Cables Size #4 AWG or larger shall be permitted to be installed in cable trays (insulated covered or bare).

  - Where 1/C #1/0 thru 4/0 are installed in ladder type tray maximum allowable rung spacing shall be 9 inches.

- Number of Multi-Conductor Cables (≤ 2 kV) Permitted in Cable Tray:

  - Where all of the cables are ≥ 4/0: The sum of all diameters of all cables shall not exceed the cable tray width, and the cables shall be installed in a single layer (i.e., not stacked).

  - Where all of the cables are < 4/0: The sum of the cross-sectional areas of all cables shall not exceed the maximum allowable cable fill as shown in Table 9-6 (columns 1, 3, 5 or 7) for appropriate cable tray width.

  - Where 4/0 or larger cables are mixed with < 4/0: The sum of the cross-sectional areas of all cables smaller than 4/0 shall not exceed maximum allowable fill area as shown in Table 9-6 (columns 2, 4, 6 or 8) for appropriate cable tray width. The 4/0 and larger cables shall be placed on top of them.

### Table 9-6. Article 392 Maximum Allowable Fill

| Inside Width of Cable Tray (inches) | Maximum Allowable Fill Area for Multi-Conductor Cables | | | | | | | |
| | Ladder or Ventilated | | Solid Bottom | | Ventilated Channel | | Solid Channel | |
| | Column 1 (in²) | Column 2 (in²) | Column 3 (in²) | Column 4 (in²) | Column 5 One Cable (in²) | Column 6 > 1 Cable (in²) | Column 7 One Cable (in²) | Column 8 > 1 Cable (in²) |
|---|---|---|---|---|---|---|---|---|
| 2 | - | - | - | - | - | - | 1.3 | 0.8 |
| 3 | - | - | - | - | 2.3 | 1.3 | 2.0 | 1.1 |
| 4 | - | - | - | - | 4.5 | 2.5 | 3.7 | 2.1 |
| 6 | 7.0 | 7 – (1.2 x Sd) | 5.5 | 5.5 – (1.2 x Sd) | 7.0 | 3.8 | 5.5 | 3.2 |
| 9 | 10.5 | 10.5 – (1.2 x Sd) | 8.0 | 87 – (1.2 x Sd) | - | - | - | - |
| 12 | 14.0 | 14 – (1.2 x Sd) | 11.0 | 11 – (1.2 x Sd) | - | - | - | - |
| 18 | 21.0 | 21 – (1.2 x Sd) | 16.5 | 16.5 – (1.2 x Sd) | - | - | - | - |
| 24 | 28.0 | 28 – (1.2 x Sd) | 22.0 | 22 – (1.2 x Sd) | - | - | - | - |
| 30 | 35.0 | 35 – (1.2 x Sd) | 27.5 | 27.5 – (1.2 x Sd) | - | - | - | - |
| 36 | 42.0 | 42 – (1.2 x Sd) | 33.0 | 33 – (1.2 x Sd) | - | - | - | - |

Sd indicates "Sum of Diameters of the cables located in the cable tray"

*Grounding:*

Metallic cable trays shall be permitted to be used as equipment grounding conductors where continuous maintenance and supervision ensure that qualified persons are servicing the installed cable tray system and that the cable tray complies with the minimum metal requirements shown in Table 9-7; in addition, cable tray sections, fittings and connected raceways are bonded using bolted mechanical connectors or sufficiently sized bonding jumpers.

## Table 9-7. Minimum Metal Requirements

| Max Overcurrent Protective Device Rating of any Cable Circuit in the Tray System (amps) | Minimum Cross-Sectional Area of Metal | |
|---|---|---|
| | Steel* Cable Trays (in$^2$) | Aluminum** Cable Trays (in$^2$) |
| 60 | 0.2 | 0.2 |
| 100 | 0.4 | 0.2 |
| 200 | 0.7 | 0.2 |
| 400 | 1.0 | 0.4 |
| 600 | 1.5* | 0.4 |
| 1000 | - | 0.6 |
| 1200 | - | 1.0 |
| 1600 | - | 1.5 |
| 2000 | - | 2.0** |

*Steel cable trays shall not be used as equipment grounding conductors for circuits with ground-fault protection above 600amps
**Aluminum cable trays shall not be used as equipment grounding conductors for circuits with ground-fault protection above 2000amps

### (NEC Article 392) Cable Tray Types

**Ladder Cable Tray** provides:

- Solid side rail protection and system strength with smooth radius fittings and a wide selection of materials and finishes.

- Maximum strength for long span applications

- Standard widths of 6,12,18, 24, 30, and 36 inches

- Standard depths of 3, 4, 5, and 6 inches

- Standard lengths of 10, 12, 20 and 24 feet

- Rung spacing of 6, 9, 12, and 18 inches

Ladder cable tray (Figure 9-9) is generally used in applications with intermediate to long support spans, 12 feet to 30 feet.

**Figure 9-9. Ladder Cable Tray**

**Solid Bottom Cable Tray** provides:

- Nonventilated continuous support for delicate cables with added cable protection available in metallic and fiberglass.

- Solid bottom metallic with solid metal covers for nonplenum rated cable in environmental air areas

- Standard widths of 6, 12, 18, 24, 30, and 36 inches

- Standard depths of 3, 4, 5, and 6 inches

- Standard lengths of 10, 12, 20 and 24 feet

Solid Bottom cable tray (Figure 9-10) is generally used for minimal heat generating electrical or telecommunication applications with short to intermediate support spans of 5 feet to 12 feet.

**Figure 9-10. Solid Bottom Cable Tray**

**Trough Cable Tray** provides:

- Moderate ventilation with added cable support frequency and with the bottom configuration providing cable support every 4 inches. Available in metal and nonmetallic materials.

- Standard widths of 6, 12, 18, 24, 30, 36 inches

- Standard depths of 3, 4, 5, and 6 inches

- Standard lengths of 10, 12, 20 and 24 feet

- Fixed rung spacing of 4 inches on center

Trough cable tray (Figure 9-11) is generally used for moderate heat generating applications with short to intermediate support spans of 5 feet to 12 feet.

**Channel Cable Tray** provides:

- An economical support for cable drops and branch cable runs from the backbone cable tray system.

---

**Figure 9-11. Trough Cable Tray**

- Standard widths of 3, 4, and 6 inches in metal systems and up to 8 inches in nonmetallic systems.

- Standard depths of 1 1/4–1 3/4 inches in metal systems and 1, 1 1/8, 1 5/16 and 2 3/16 inches in nonmetallic systems

- Standard lengths of 10, 12, 20 and 24 feet

Channel cable tray (Figure 9-12) is used for installations with limited numbers of tray cable when conduit is undesirable. Support frequency of every 5 to 10 feet.

**Figure 9-12. Channel Cable Tray**

**Wire Mesh Cable Tray** provides:

- A job site, field adaptable support system primarily for low voltage, telecommunication and fiber optic cables. These systems are typically steel wire mesh, zinc plated.

- Standard widths of 2, 4, 6, 8, 12, 16, 18, 20, and 24 inches

- Standard depths of 1, 2, and 4 inches

- Standard length of about 10 feet (118 inches)

Wire Mesh cable tray (Figure 9-13) is generally used for telecommunication and fiber optic applications and is installed on short support spans, 4 to 8 feet.

**Single Rail Cable Tray**

These aluminum systems are the fastest systems to install and provide the maximum freedom fort cable to enter and exit the system. Single Rail cable tray provides:

- Single hung or wall mounted systems in single or multiple tiers.

---

**Figure 9-13. Wire Mesh Cable Tray**

- Standard widths of 6, 9, 12, 18, and 24 inches.

- Standard depths of 3, 4, and 6 inches.

- Standard lengths of 10 and 12 feet.

Single Rail cable tray (Figure 9-14) is generally used for low voltage and power cable installations where maximum cable freedom, side fill, and speed to install are factors.

**Figure 9-14. Single Rail Cable Tray**

# 9.3 NEMA/IEC-IP ENCLOSURE CLASSIFICATIONS

## 9.3.1 NEMA Designations (Non-Hazardous)

### Table 9-8. NEMA Designations (Non-Hazardous)

| NEMA | IP Equiv | NEMA Definition |
|------|----------|-----------------|
| 1 | IP10 | Enclosures constructed for indoor use to provide a degree of protection to personnel against incidental contact with the enclosed equipment and to provide a degree of protection against falling dirt. |
| 2 | IP11 | Enclosures constructed for indoor used to provide a degree of protection to personnel against incidental contact with the enclosed equipment, to provide a degree of protection against falling dirt, and to provide a degree of protection against dripping and light splashing of liquids. |
| 3 | IP54 | Enclosures constructed for either indoor or outdoor used to provide a degree of protection to personnel against incidental contact with the enclosed equipment; to proved a degree of protection against falling dirt, rain, sleet, snow, and windblown dust; and that will undamaged by external formation of ice on the enclosure. |
| 3R | IP14 | Enclosures constructed for either indoor or outdoor used to provide a degree of protection to personnel against incidental contact with the enclosed equipment; to provide a degree of protection against falling dirt, rain, sleet, and snow; and that will be undamaged by external formation of ice on the enclosure. |
| 3S | IP54 | Enclosures constructed for either indoor or outdoor use to provide a degree of protection to personnel against incidental contact with the enclosed equipment; to provide a degree of protection against falling dirt, rain, sleet, snow, and windblown dust; and in which the external mechanism(s) remain operable when ice laden |
| 4 | IP56 | Enclosures constructed for either indoor or outdoor use to provide a degree of protection to personnel against incidental contact with the enclosed equipment; to provide a degree of protection against falling dirt, rain, sleet, snow, windblown dust, splashing water, and hose-directed water, and corrosion; and that will be undamaged by the external formation of ice on the enclosure. |
| 4X | IP56 | Enclosures constructed for either indoor or outdoor use to provide a degree of protection to personnel against incidental contact with the enclosed equipment; to provide a degree of protection against falling dirt, rain, sleet, snow, windblown dust, splashing water, hose-directed water, and corrosion; and that will be undamaged by thee external formation of ice on the enclosure. |
| 5 | IP52 | Enclosures constructed for indoor use to provide a degree of protection to personnel against incidental contact with the enclosed equipment; to provide a degree of protection against falling dirt; against settling airborne dust, lint, fibers, and flyings; and to provide a degree of protection against dripping and light splashing of liquids. |
| 6/6P | IP67 | Enclosures constructed for either indoor or outdoor use to provide a degree of protection to personnel against incidental contact with the enclosed equipment; to provide a degree of protection against falling dirt; against hose-directed water and the entry of water during occasional temporary submersion at a limited depth; and that will be undamaged by the external formation of ice on the enclosure. |
| 12/12K | IP52 | Enclosures constructed (without knockouts) for indoor use to provide a degree of protection to personnel against incidental contact with the enclosed equipment; to provide a degree of protection against falling dirt; against circulating dust, lint, fibers, and flying; and against dripping and light splashing of liquids |
| 13 | IP54 | Enclosures constructed for indoor use to provide a degree of protection to personnel against incidental contact with the enclosed equipment; to provide a degree of protection against falling dirt; against circulating dust, lint, fibers, and flyings; and against the spraying, splashing, and seepage of water, oil, and noncorrosive coolants. |

## 9.3.2 NEMA Designations (Hazardous)

- **Type 7 and 10** enclosures are designed to contain an internal explosion without causing an external hazard.

- **Type 8** enclosures are designed to prevent combustion through the use of oil immersed equipment.

- **Type 9** enclosures are designed to prevent the ignition of combustible dust.

### Table 9-9. NEMA Designations (Hazardous)

| NEMA | NEMA Definition |
|------|-----------------|
| 7 | Enclosures constructed for indoor use in hazardous (classified) locations classified as Class I, Division 1, Groups A, B, C or D. |
| 8 | Enclosures constructed for either indoor or outdoor use in hazardous (classified) locations classified as Class I, Division 1, Groups A, B, C or D. |
| 9 | Enclosures constructed for indoor use in hazardous (classified) locations classified as Class II, Division 1, Groups E, F or G. |
| 10 | Enclosures constructed to meet the requirements of the Mine Safety and Health Administration. |

## 9.3.3 IEC-IP (Ingress Protection)

### Table 9-10. IEC-IP (Ingress Protection Standard 60529)

| FIRST DIGIT | | |
|-------------|--------------------------|-----------------|
| **Level** | **Object Size Protection** | **Effective Against** |
| 0 | NA | No protection against contact and ingress of objects |
| 1 | > 50 mm | Any large surface of the body, such as the back of a hand, but no protection against deliberate contact with a body part |
| 2 | > 12.5 mm | Fingers or similar objects |
| 3 | > 2.5 mm | Tools, thick wires, etc. |
| 4 | > 1 mm | Most wires, screws, etc. |
| 6 | Dust Protected | Ingress of dust is not entirely prevented, but it must not enter in sufficient quantity to interfere with the satisfactory operation of the equipment; complete protection against contact |
| 6 | Dust Tight | No ingress of dust; complete protection against contact |
| **FIRST DIGIT** | | |
| **Level** | **Protect Against** | **Details** |
| 0 | not protected | NA |
| 1 | dripping water | Dripping water (vertically falling drops) shall have no harmful effect. |
| 2 | dripping water when tilted up to 15° | Vertically dripping water shall have no harmful effect when the enclosure is tilted at an angle up to 15° from its normal position |
| 3 | spraying water | Water falling as a spray at any angle up to 60° from the vertical shall have no harmful effect |
| 4 | splashing water | Water splashing against the enclosure from any direction shall have no harmful effect |

| 5 | water jets | Water projected by a nozzle against enclosure from any direction shall have no harmful effects |
| 6 | powerful water jets | Water projected in powerful jets against the enclosure from any direction shall have no harmful effects |
| 7 | immersion up to 1m | Ingress of water in harmful quantity shall not be possible when the enclosure is immersed in water under defined conditions of pressure and time (up to 1m of submersion). |
| 8 | immersion beyond 1m | The equipment is suitable for continuous immersion in water under conditions which shall be specified by the manufacturer.<br>NOTE: Normally, this will mean that the equipment is hermetically sealed. However, with certain types of equipment, it can mean that water can enter but only in such a manner that it produces no harmful effects |

## 9.4 NFPA 70E ELECTRICAL SAFETY IN THE WORKPLACE:

The purpose of this standard is to provide a practical safe working area for employees relative to the hazards arising from the use of electricity.

NFPA 70E is subdivided into three chapters:

- Chapter 1 – Safety Related Work Practices

- Chapter 2 – Safety Related Maintenance Practices

- Chapter 3 – Safety Requirements for Special Equipment

We will be discussing Chapter 1: Safety Related Work Practices.

Table 9-11 shows the effects of electric shock on the human body:

### Table 9-11. How Electrical Current Affects the Human Body

| Current Level | Probable Effects on the Human Body |
|---|---|
| 0.001 A | Perception level. Slight tingling sensation. Still dangerous under <u>certain conditions</u>. |
| 0.005 A | Slight shock felt; not painful but disturbing. Average individual can let go. However, strong <u>involuntary reactions</u> to shocks in this range may lead to injuries. |
| 0.006 to 0.016 A | Painful shock, begin to lose muscular control. Commonly referred to as the freezing current or "let-go" threshold. |
| 0.017 to 0.099 A | Extreme pain, respiratory arrest, severe <u>muscular contractions</u>. Individual cannot let go. <u>Death is possible</u>. |
| 0.1 to 2.0 A | Ventricular fibrillation (uneven, uncoordinated pumping of the heart.) Muscular contraction and nerve damage begins to occur. <u>Death is likely</u>. |
| > 2.0 A | Cardiac arrest, internal organ damage, and severe burns. <u>Death is most probable</u>. |

Table is from www.OSHA.gov

Notes:

1. Because women's frames and body parts, such as hands and fingers, are often smaller than men's, women tend to suffer damage at lower amounts of current density or exposure.

2. "Let-Go" threshold is the point where a person cannot open their hand and thus cannot release their grip.

3. 'GFCI' activation is ~5 mA

## 9.4.1 Shock Hazard Analysis

Shock Hazard Analysis: shall determine the voltage to which personnel will be exposed, boundary requirements and the personal protective equipment necessary in order to minimize the potential for electrical shock to personnel.

- *Arc Flash Boundary:* The distance from exposed live parts from which a person could receive a second degree burn (1.2 cal/cm$^2$) should a flash occur. It is assumed that a second degree burn is curable and will not result in death.

- *Shock Protection Boundaries* (Figure 9-15) are applicable to the situation in which approaching personnel are exposed to energized electrical conductors or circuit parts. (See Table 9-12 for distances associated with various system voltages.)

  - *Limited Approach Boundary:* An approach limit at a distance from an energized electrical conductor or circuit part within which a shock hazard exists. No untrained personnel may approach any closer to the energized item than this boundary unless escorted by a qualified individual.

  - *Restricted Approach Boundary:* An approach limit at a distance from an exposed energized electrical conductor or circuit part within which there is an increased risk of shock, due to electrical arc-over combined with inadvertent movement, for personnel working in close proximity to the electrical conductor or circuit part. To cross the Restricted Approach Boundary into the Restricted Space, the qualified person, who has completed required training, must wear appropriate personal protective equipment (PPE). Also, he must have a written approved plan for the work that he will perform and plan the work to keep all parts of the body out of the Prohibited Space

  - *Prohibited Approach Boundary:* An approach limit at a distance from an exposed energized electrical conductor or circuit part within which work is considered the same as making contact with the electrical conductor or circuit part. Only qualified personnel wearing appropriate personal protective equipment (PPE), having specified training to work on energized conductors or components, and a documented plan justifying the need to perform this work may cross the boundary and enter the Prohibited Space

  - *Flash Protection Boundary:* An approach limit at a distance from a prospective arc source within which a person could receive a second degree burn if an electric arc flash were to occur. NFPA 70E establishes the default flash protection boundary at 4 feet for low voltage (< 600 V) systems where the total fault exposure is less than 5000 ampere-seconds (fault current in amperes multiplied by the upstream device clearing time in seconds).

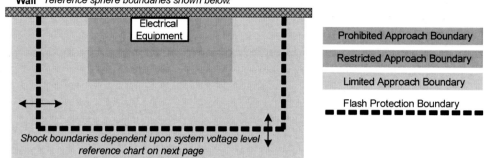

Note: If wall is not present then the boundaries required are 360°, reference sphere boundaries shown below.

Wall

Electrical Equipment

Prohibited Approach Boundary

Restricted Approach Boundary

Limited Approach Boundary

Flash Protection Boundary

Shock boundaries dependent upon system voltage level reference chart on next page

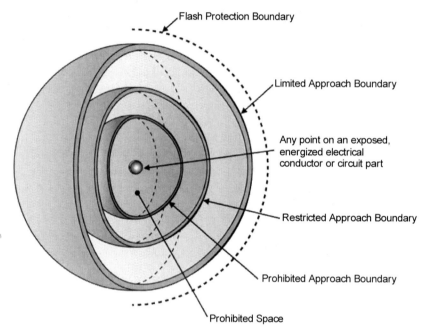

**Figure 9-15. Shock Protection Boundaries Diagram**

**Table 9-12. Shock Protection Boundaries**

| Nominal System Voltage Range Phase – Phase | Limited Approach Boundary | | Restricted Approach Boundary | Prohibited Approach Boundary |
| --- | --- | --- | --- | --- |
| | Exposed Moveable Conductor | Exposed Fixed Circuit Part | | |
| < 50 | Not Specified | Not Specified | Not Specified | Not Specified |
| 50 to 300 | 10'-0" | 3'-6" | Avoid Contact | Avoid Contact |
| 301 to 750 | 10'-0" | 3'-6" | 1'-0" | 0'-1" |
| 751 to 15kV | 10'-0" | 5'-0" | 2'-2" | 0'-7" |
| 15.1kV to 36kV | 10'-0" | 6'-0" | 2'-7" | 0'-10" |
| 36.1kV to 46kV | 10'-0" | 8'-0" | 2'-9" | 1'-5" |
| 46.1kV to 72.5kV | 10'-0" | 8'-0" | 3'-3" | 2'-2" |
| 72.6kV to 121kV | 10'-8" | 8'-0" | 3'-4" | 2'-9" |
| 138kV to 145kV | 11'-0" | 10'-0" | 3'-10" | 3'-4" |
| 161kV to 169kV | 11'-8" | 11'-8" | 4'-3" | 3'-9" |
| 230kV to 242kV | 13'-0" | 13'-0" | 5'-8" | 5'-2" |
| 345kV to 362kV | 15'-4" | 15'-4" | 9'-2" | 8'-8" |
| 500kV to 550kV | 19'-0" | 19'-0" | 11'-10" | 11'-4" |
| 765kV to 800kV | 23'-9" | 23'-9" | 15'-11" | 15'-5" |

## 9.4.2 Arc Flash Hazard Analysis

**Arc Flash Hazard Analysis:** shall determine the arc flash protection boundary and the personal protective equipment that people within the arc flash protection boundary shall use.

- The arc flash hazard analysis shall take into consideration the design of the overcurrent protective device and its opening time, including its condition of maintenance.

- The arc flash hazard analysis shall NOT be required where <u>all</u> of the following conditions exist:

  - The circuit is rated 240 V or less
  - The circuit is supplied by one transformer
  - The transformer supplying the circuit is rated less than 125 KVA.

*The arc flash boundary is defined as the distance at which the worker is exposed to 1.2 cal/cm$^2$ for 0.1 second. One calorie is the energy required to raise one gram of water one °C at one atmosphere pressure. 2$^{nd}$ degree burns (Figure 9-16[147]) occur at 1.2 cal/cm$^2$, which is comparable to holding your hand several inches above a disposable lighter.*

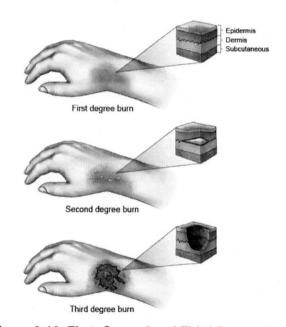

**Figure 9-16. First, Second and Third Degree Burns**

**Arc Flash** is an electrical explosion due to a fault condition or short circuit when either a phase to ground or phase to phase conductor is connected and current flows through the air.

---

147. Image obtained from www. quailridgestudios.com.

---

The cause of the arc normally burns away during the initial flash and the arc fault is then sustained by the establishment of a highly-conductive plasma. The plasma will conduct as much energy as is available and the only energy conductance limiting factor is the impedance of the arc. This massive energy discharge burns the bus bars with temperatures that may reach 35,000°F, which in turn vaporizes the copper and thus causing an explosive volumetric increase. The arc blast is conservatively estimated as an expansion of 40,000 to 1. This fiery explosion devastates everything in its path, creating deadly shrapnel as it dissipates.

The arc fault current is usually much less than the available bolted fault current[148] and below the rating of circuit breakers. Unless these devices have been selected to handle an arc fault condition, they will not trip and the full force of an arc flash will occur. The electrical equation for energy is volts x current x time. The transition from arc fault to arc flash takes a finite time, increasing in intensity as the pressure wave develops. The challenge is to sense the arc fault current and shut off the voltage in a timely manner before it develops into a serious arc flash condition.

## 9.4.3 Protective Clothing Characteristics

**Table 9-13. Protective Clothing Characteristics as Recommended by NFPA 70E**

| Hazard/Risk Category | Clothing Description | Reqd. Min. Arc Rating of PPE |
|---|---|---|
| 0 | Nonmelting materials (i.e., untreated cotton, wool rayon or silk or blends of these materials with a fabric weight of at least 4.5 oz/yard | N/A |
| 1 | Arc rated FR shirt and FR pants or FR coverall | 4 cal/cm$^2$ |
| 2 | Arc rated FR shirt and FR pants or FR coverall | 8 cal/cm$^2$ |
| 3 | Arc rated FR shirt and FR pants or FR coverall, and arc flash suit selected so that the system arc rating meets the required minimum. | 25 cal/cm$^2$ |
| 4 | Arc rated FR shirt and FR pants or FR coverall, and arc flash suit selected so that the system arc rating meets the required minimum. | 40 cal/cm$^2$ |

---

148. Bolted fault current is a term used to identify the amount of current available if two busses of different phases or polarities where physically bolted together.

---

## 9.4.4 Personal Electrical Shock Protection Equipment

### Table 9-14. Rubber Glove Labeling Chart

| Class | Label Color | Conventional Work Position for Worker | Voltage Rating (max. use) |
|---|---|---|---|
| 00 | Beige | Ground, Structure or Basket | 500 (ac) 750 (dc) |
| 0 | Red | Ground, Structure or Basket | 1000 (ac) 1500 (dc) |
| 1 | White | Structure or Basket | 7500 (ac) 11250 (dc) |
| 2 | Yellow | Electrically Isolated Basket or Platform | 17000 (ac) 25500 (dc) |
| 3 | Green | Electrically Isolated Basket or Platform | 26500 (ac) 39750 (dc) |
| 4 | Orange | Electrically Isolated Basket or Platform | 36000 (ac) 54000 (dc) |

NOTE: Leather outer glove protectors must be worn over the insulating rubber gloves to provide the needed mechanical protection against cuts, abrasions and punctures!!![149]

**Figure 9-17. Category 3 & 4 Flash Suit**

## 9.4.5 Qualified Personnel

Because crossing the Prohibited Approach Boundary and entering Prohibited Space is considered the same as making contact with exposed energized conductors or circuit parts the qualified personnel must do the following:

- Have specified training to work on energized conductors or circuit parts

- Have a documented plan justifying the need to work that close

---

149. Image obtained from www.aplussafety.net.

- Perform risk analysis

- Have the plan and risk analysis approved by authorized management

- Use PPE that is appropriate for working on exposed energized conductors or circuit parts, and is rated for the voltage and energy level involved.

### 9.4.6 Label Requirements

Switchboards, panelboards, industrial control panels, meter socket enclosures and motor control centers in other than dwelling occupancies, which are likely to require examination, adjustment, servicing or maintenance while energized, shall be field marked to warn qualified persons of potential electric arc flash hazards (Figure 9-18). The marking shall be located so as to be clearly visible to qualified persons before examination, adjustment, servicing or maintenance of the equipment. Label information must include:

- Flash Hazard Category

- Working Distance

- Boundaries: The limited, restricted and prohibited approach boundaries stated on an electrical safety warning label.

- PPE Required

- Incident Energy

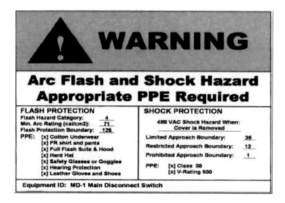

**Figure 9-18. Field Marking**

## 9.5 LIGHTNING PROTECTION

Some common power system lightning protection terminology:

- **Withstand Voltage:** The level of voltage that electrical equipment can handle without failure or disruptive discharge.

- **Transient Insulation Level (TIL):** An insulation level specified in terms of the crest value (aka peak value) of the *withstand voltage*, for a specified waveform.

- **Lightning Impulse Insulation Level:** An insulation level specified for the crest value of a lightning impulse withstand voltage.

- **Basic Lightning Impulse Insulation Level (BIL):** An insulation level specified in terms of the crest value of the standard lightning impulse.

- **Standard Lightning Impulse:** A full impulse with a front time of 1.2 μs ± 30% and a half-time value of 50 μs ± 20% (normally termed a 1.2/50 impulse). The front time ($T_1$) is the time required to reach the crest value, see Figure 9-19.

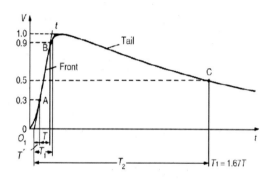

**Figure 9-19. Lightning Impulse: Voltage vs. Time**

## 9.5.1 Lightning Protection (NFPA 780; UL96 & 96A; LPI 17S; IEEE Std 487)

Systems of ~230 KV or less cannot be adequately insulated against direct lightning strikes because of the high voltage and current of lightning (several million volts & ~30 KA). To protect these systems a combination of shielding and grounding must be used (Figure 9-20).

Indirect lightning strikes result in voltage surges by means of conductive coupling through the soil, or by means of inductive and capacitive coupling to the lines. To protect such systems, surge arresters are typically used.

## 9.5.2 Facility Lightning Protection per NFPA 780

- Ordinary Structures ≤ 75 feet above grade shall be protected with Class I materials (Table 9-16).

- Ordinary Structures > 75 feet above grade shall be protected with Class II materials (Table 9-17).

- If part of a structure exceeds 75 feet above grade (e.g., a steeple) and the remaining portion is ≤ 75 feet above grade, only the portion of the structure exceeding 75 feet above grade shall be protected with Class II materials.

**Figure 9-20. Typical Lightning Protection Measures.**
**Approximate Range of Lightning Voltage: 10 MV to 1000 MV**
**Approximate Range of Lightning Current: 1000 A to 200,000 A**

- System components shall be either aluminum or copper, however, no aluminum shall be installed within two feet of grade.

### Table 9-15. Minimum Class I Materials Requirements

| Type of Conductor | Parameter | Copper | Aluminum |
|---|---|---|---|
| Air terminal, solid | Diameter | 3/8" | ½" |
| Air terminal, tubular | Diameter<br>Wall Thickness | 3/8"<br>0.033" | 3/8"<br>0.064" |
| Main conductor, cable | Size of Each Strand<br>Cross Sectional Area | #17AWG<br>57,400 CM | #14AWG<br>98,600 CM |
| Main conductor, solid strip | Thickness<br>Cross Sectional Area | 0.051"<br>57,400 CM | 0.064"<br>98,600 CM |
| Bonding conductor, cable | Size of Each Strand<br>Cross Sectional Area | #17AWG<br>26,240 CM | #14AWG<br>41,100 CM |
| Bonding conductor, solid strip | Thickness<br>Width | 0.051"<br>½" | 0.064"<br>½" |

### Table 9-16. Minimum Class II Materials Requirements

| Type of Conductor | Parameter | Copper | Aluminum |
|---|---|---|---|
| Air terminal, solid | Diameter | ½" | 3/8" |
| Main conductor, cable | Size of Each Strand<br>Cross Sectional Area | #15AWG<br>115,000 CM | #13AWG<br>192,000 CM |
| Main conductor, solid strip | Thickness<br>Cross Sectional Area | 0.064"<br>115,000 CM | 0.1026"<br>192,000 CM |
| Bonding conductor, cable | Size of Each Strand<br>Cross Sectional Area | #17AWG<br>26,240 CM | #14AWG<br>41,100 CM |
| Bonding conductor, solid strip | Thickness<br>Width | 0.051"<br>½" | 0.064"<br>½" |

*Strike Termination Devices*

- The tip of the air terminal shall be NLT 10 inches above the object or area it is to support.

---

- Air terminals exceeding 24 inches in height above the area or object they are to protect shall be supported at a point NLT 1/2 their height.

## Conductors

- Conductors shall maintain a horizontal or downward coursing and free from "U" or "V" type pockets.

- Conductor bends – No bend of a conductor shall form an angle < 90°, nor shall it have a radius of < 8 inches

- Down Conductors:

    - Down conductors shall be as widely separated as possible.
    - At least two down conductors shall be provided on any kind of structure.
    - Structures exceeding 250 feet in perimeter shall have a down conductor for every 100 feet of perimeter.
    - Total number of down conductors on flat or gently sloping roofs shall be such that the average distance between all conductors does not exceed 100 feet.
    - Irregularly shaped structures shall have additional down conductors as necessary to provide a two-way path from each strike device.
    - Down conductor shall be protected for a minimum of 6 feet above grade. If metallic guards are used, they shall be bonded at each end.

## Grounding Electrodes

- Grounding electrodes shall be copper-clad steel, solid copper or stainless steel (ground rods, radials, plate electrode).

- Ground rods shall be NLT (not less than) 1/2 inch in diameter and 8 feet in length and shall be free of any paint or other non-conductive coatings.

- Ground rods shall extend vertically NLT 10 feet into the earth.

# 10. SAMPLE PROBLEMS

1. The vessel below is filled with fuel oil at 60°F with a specific gravity of 0.893. The pressure in psig on the transmitter diaphragm connected to the bottom nozzle is most nearly equal to:

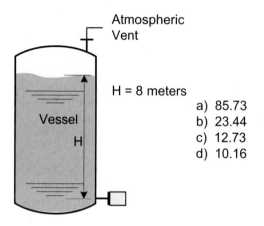

Atmospheric Vent

H = 8 meters

Vessel

H

a) 85.73
b) 23.44
c) 12.73
d) 10.16

2. At 506°C, a type K thermocouple with a 0°C reference junction will produce a millivolt output that is most nearly equal to:

   a. 10.7
   b. 27.7
   c. 20.9
   d. 14.3

3. A Wheatstone bridge is connected as shown below to measure the unknown resistance, $R_X$. $R_{AD}$ is manipulated until $V_{BD} = 0$ V. The indicator on $R_{AD} = 30 \ \Omega$ at this point. What is most nearly the value of $R_X$ at this point?

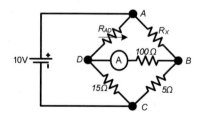

10V    $R_{AD}$    $R_X$
D    (A)    100 Ω    B
15Ω    5Ω
C

   a. 10 Ω
   b. 15 Ω
   c. 30 Ω
   d. 40 Ω

4. According to ASME Code, Section VIII, Paragraph UG-127 what capacity derating percentage must be applied for a combination relief valve rupture disk assembly that has not been flow test certified by the manufacturer as an assembly?

    a. 5%
    b. 0%
    c. 10%
    d. 20%

5. According to ASME Code, Section VIII, when are pressure relief valves required to have lifting levers on air, hot water and steam service?

    a. > 140°F
    b. > 125°F
    c. > 200°F
    d. > 212°F

6. According to NFPA 496 what shall be the constantly maintained positive pressure above the surrounding atmosphere for pressurized system enclosures?

    a. 0.1" WC
    b. 1.0" WC
    c. 10" WC
    d. 1 psig

7. According to NFPA 496 what type of pressurizing system requires that a cutoff switch be incorporated to de-energize power automatically upon failure of the protective gas supply to maintain positive pressure? (Assume no equipment within the enclosure would develop temperatures higher than marked temperature class (T-Code) upon failure of the protective gas supply.)

    a. Type X pressurizing system
    b. Type Y pressurizing system
    c. Type Z pressurizing system
    d. All of the above

8. According to NFPA 497 which classification below best describes ethanol in which sufficient concentrations of flammable vapors exist under normal operating conditions?

    a. Class I, Division 2, Group D
    b. Class I, Division 1, Group D
    c. Group I, Division 1, Group C
    d. Class I, Division 1, Group C

9. According to NFPA 499 to which combustible dust group does wheat reside?

    a. Group D
    b. Group E
    c. Group F
    d. Group G

10. According to NFPA 70E which term below is best described by the statement "The minimum distance from the energized item where unqualified personnel may safely stand. No untrained personnel may approach any closer to the energized item than this boundary unless escorted by a qualified individual"?

    a. Limited Approach Boundary
    b. Restricted Approach Boundary
    c. Prohibited Approach Boundary
    d. Flash Protection Boundary

11. According to OSHA 1910, which subpart describes the electrical safe work requirements?

    a. Subpart E
    b. Subpart H
    c. Subpart S
    d. Subpart Z

12. The logic circuit shown below is equivalent to what type of basic logic gate?

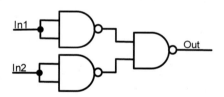

    a. AND
    b. XOR
    c. NAND
    d. OR

13. The flow rate in GPM for a turbine meter with a K factor of 3.133 and a pulse rate of 2 pulses/sec is most nearly equal to:

    a. 0.7 GPM
    b. 94 GPM
    c. 38 GPM
    d. 25 GPM

14. The flow rate in GPM for a vortex meter with the following parameters is most nearly equal to what value?

Pulses per second = 3
Bluff body width = 1.2"
Avg flow velocity = 5 ft/sec
Internal diameter of the meter = 3"

    a. 22 GPM
    b. 110 GPM
    c. 250 GPM
    d. 3100 GPM

15. The flow of a process fluid in a 3-inch pipe is measured with an orifice plate and differential pressure transmitter. At a flow rate of 80 gpm the differential pressure is 30 inches WC. At a differential pressure of 40 inches WC what would the flow rate be most nearly equal to?

    a. 92 GPM
    b. 69 GPM
    c. 107 GPM
    d. 60 GPM

16. Given the following data for process liquid flow, what would the orifice bore diameter be most nearly equal to?

Flow rate:   0 to 100 GPM
Pipe size:   3-inch schedule 40 carbon steel
Process fluid at 20°F and 50 psia
Differential Pressure: 100 inches WC
$\beta = 0.65$

    a. 1.0"
    b. 1.25"
    c. 2.5"
    d. 2.0"

17. A magnetic flowmeter would NOT be a good choice to measure the flow rate for which of the following process media?

    a. Waste water
    b. Hydrochloric Acid
    c. Sodium Hydroxide
    d. Glacial Acetic Acid

18. Water is flowing at a rate 150 gpm with a temperature of 125°F through a 4-inch schedule 40 carbon steel pipe. What is the maximum fluid velocity most nearly equal to?

    a. 3.8 ft/sec
    b. 2.8 ft/sec
    c. 0.3 ft/sec
    d. 4.5 ft/sec

Reference the diagram below for questions 19 and 20:

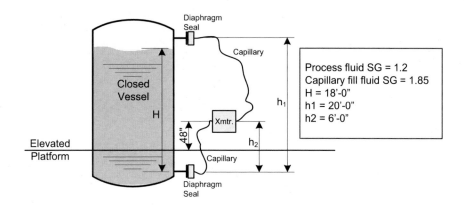

19. What should the span of the transmitter be set to based upon the information given in the diagram above?

    a. 216″
    b. 259″
    c. 288″
    d. 444″

20. What should the transmitter zero be set to based upon the information given in the diagram above?

    a. −260″
    b. −444″
    c. −133.2″
    d. −177″

21. The tank shown below is level measured utilizing a differential pressure transmitter and bubbler tube assembly. If the tank is filled to 50% (15'-0") with a process fluid of SG = 0.9, what would the pressure gauge reading (in psig) on the bubbler assembly be most nearly equal to?

    a. 6.5 psig
    b. 13.5 psig
    c. 11.7 psig
    d. 5.8 psig

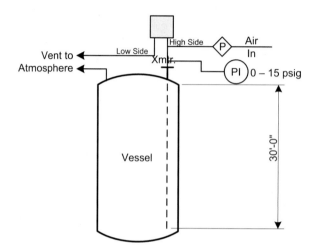

22. According to API 521 what is the maximum relieving pressure, as a % of MAWP, for fire sizing a single relief valve?

    a. 100%
    b. 110%
    c. 120%
    d. 121%

23. A Class IV valve shutoff classification corresponds to which of the following?

    a. 0.001% of rated valve capacity
    b. 0.01% of rated valve capacity
    c. 0.05% of rated valve capacity
    d. 0.1% of rated valve capacity

24. Which valve body style would be best suited to control the flow of nickel slurry catalyst into a reactor?

    a. Butterfly valve
    b. Ball valve
    c. Angle valve
    d. Split body globe valve

25. What would the maximum process liquid flow rate (in gpm) be most nearly equal to for a control valve with a $C_V$ of 15, an inlet pressure of 50 psig and an outlet pressure of 40 psig, and the process fluid SG = 1.15?

    a. 130 gpm
    b. 44 gpm
    c. 90 gpm
    d. 60 gpm

26. Which type of large motor design should never be allowed to operate with no load connected to the motor?

    a. Squirrel cage motor
    b. Shunt motor
    c. Compound motor
    d. Series motor

27. What is the synchronous speed of a 4-pole AC 60 Hz induction motor?

    a. 900 rpm
    b. 1200 rpm
    c. 1800 rpm
    d. 3600 rpm

28. Which NEMA motor design exhibits the highest slip %?

    a. Design A
    b. Design B
    c. Design C
    d. Design D

29. What Variable Frequency Drive design type would be best utilized in a centrifugal pump motor application?

    a. Constant torque
    b. Variable torque
    c. Constant HP
    d. Regenerative

30. What is the DC voltage drop most nearly equal to for a circuit which utilizes #16 AWG wiring (uncoated-stranded copper from NEC Chapter 9, Table 8) and where the field device, which draws 200 mA, is located 750 feet from the 24 VDC power supply?

    a. 1.47 V
    b. 1.50 V
    c. 1.52 V
    d. 1.59 V

31. A Safety Instrumented System having which SIL level is least likely to experience a failure of a safety function?

    a. SIL 1
    b. SIL 2
    c. SIL 3
    d. SIL 4

32. What NEMA designated enclosure type is designed to prevent the ignition of combustible dust?

    a. NEMA 4
    b. NEMA 7
    c. NEMA 9
    d. NEMA 12

33. According to the NEC, what is the maximum % of cross-sectional area fill allowed for a wireway?

    a. 20%
    b. 40%
    c. 50%
    d. 53%

34. A Cat3 cable can support up to what maximum data rate?

    a. 10 Mbps
    b. 16 Mbps
    c. 100 Mbps
    d. 1000 Mbps

35. Which OSI model layer provides switching and routing technologies?

    a. Physical Layer
    b. Transport Layer
    c. Network Layer
    d. Application Layer

36. Which disk array setup provides for 100% data redundancy?

    a. RAID 1
    b. RAID 3
    c. RAID 5
    d. RAID 6

37. What is the maximum allowable FOUNDATION Fieldbus segment length utilizing a #18 AWG shielded twisted-pair cable?

    a. 200 meters
    b. 400 meters
    c. 1200 meters
    d. 1900 meters

38. What is the best wavelength analytical solution to find the concentration of a gas that absorbs light at a wavelength of 5000 nanometers?

    a. Near Infrared (NIR)
    b. Mid Infrared (MIR)
    c. Far Infrared (FIR)
    d. Ultraviolet (UV)

39. Of the following level measurement techniques, which would be the best solution for measuring phosgene in a vessel under vacuum conditions?

    a. Ultrasonic
    b. Differential Pressure
    c. Conductivity
    d. Nuclear

40. Of the following types of control, which will have the effect of eliminating the steady-state error?

    a. Proportional Control
    b. Integral Control
    c. Derivative Control
    d. Cascade Control

# 11. SAMPLE PROBLEMS - SOLUTIONS

1.  The vessel below is filled with fuel oil at 60°F with a specific gravity of 0.893. The pressure in psig on the transmitter diaphragm connected to the bottom nozzle is most nearly equal to: *(answer d)*

$$(8m)\left(\frac{12"}{1ft}\right)\left(\frac{3.28083ft}{1m}\right)\left(\frac{1psig}{27.7"WC}\right)(0.893) = 10.154psig$$

2.  At 506°C, a type K thermocouple with a 0°C reference junction will produce a millivolt output is most nearly equal to: *(answer c)*

    From the K °C Table:

| °C | 0 | 1 | 2 | 3 | 4 | 5 | 6 | 7 | 8 | 9 |
|---|---|---|---|---|---|---|---|---|---|---|
| | Thermocouple voltage in absolute millivolts | | | | | | | | | |
| 500 | 20.640 | 20.683 | 20.725 | 20.768 | 20.811 | 20.853 | 20.896 | 20.938 | 20.981 | 20.024 |

3.  A Wheatstone bridge is connected as shown below to measure the unknown resistance, $R_X$. $R_{AD}$ is manipulated until $V_{BD} = 0$ V. The indicator on $R_{AD} = 30\ \Omega$ at this point. What is most nearly the value of $R_X$ at this point? *(answer a – 10 Ω)*

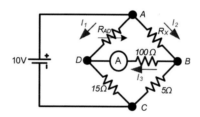

$Since\ V_{BD} = 0, \quad then\ I_3 = 0$

$V_{AB} = V_{AD} \quad \therefore \quad 30\Omega(I_1) = R_X(I_2)$

$I_1 = \dfrac{10V}{R_{AD} + 15\Omega} = \dfrac{10V}{45\Omega} = 0.222A \quad \therefore \quad V_{DC} = V_{BC} = (I_1)15\Omega = (0.222A)(15\Omega) = 3.33V$

$\therefore I_2 = \dfrac{V_{BC}}{5\Omega} = \dfrac{3.33V}{5\Omega} = 0.666A$

$R_X = \dfrac{V_{AB}}{I_2} = \dfrac{V_{AD}}{I_2} = \dfrac{(30\Omega)(0.222A)}{0.666A} = 10\Omega$

4.  According to ASME Code, Section VIII, Paragraph UG-127 what capacity derating percentage must be applied for a combination relief valve/rupture disk assembly that has not been flow test certified by the manufacturer as an assembly? *(answer c – 10%)*

5.  According to ASME Code, Section VIII, when are pressure relief valves required to have lifting levers on air, hot water and steam service? *(answer a - > 140°F)*

6. According to NFPA 496 what shall be the constantly maintained positive pressure above the surrounding atmosphere for pressurized system enclosures? *(answer - 0.1 inch WC)*

7. According to NFPA 496 what type of pressurizing system requires that a cutoff switch be incorporated to de-energize power automatically upon failure of protective gas supply to maintain positive pressure? (Assume no equipment within the enclosure would develop temperatures higher than marked temperature class (T-Code) upon failure of the protective gas supply.) *(answer a – X-pressurizing system, since the qualifier of NO equipment developing temperatures higher than T-code was included, this eliminated Y-pressurizing system)*

8. According to NFPA 497 which classification below best describes ethanol in which sufficient concentrations of flammable vapors exist under normal operating conditions? *(answer b – Class I, Division I, Group D)*

9. According to NFPA 499 to which combustible dust group does wheat reside? *(answer – Group G, note that group D is not even part of the combustible dust groups)*

10. According to NFPA 70E which term below is best described by the statement "The minimum distance from the energized item where unqualified personnel may safely stand. No untrained personnel may approach any closer to the energized item than this boundary unless escorted by a qualified individual"? *(answer a - Limited Approach Boundary)*

11. According to OSHA 1910 which subpart describes the electrical safe work requirements? *(answer c – Subpart S)*

12. The logic circuit shown below is equivalent to what type of basic logic gate? *(answer d – OR)*

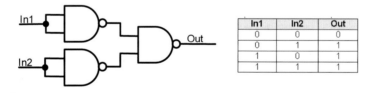

| In1 | In2 | Out |
|-----|-----|-----|
| 0 | 0 | 0 |
| 0 | 1 | 1 |
| 1 | 0 | 1 |
| 1 | 1 | 1 |

13. The flow rate in GPM for a turbine meter with a K factor of 3.133 and a pulse rate of 2 pulses/sec is most nearly equal to? *(answer c – 38.3GPM)*

$$Q_{GPM} = \frac{60 f_{Hz}}{K_{F,gal}} = \frac{60(2)}{3.133} = 38.3 gpm$$

14. The flow rate in GPM for a vortex meter with the following parameters is most nearly equal to? *(answer b – 110 GPM)*

Pulses per second = 3
Bluff body width = 1.2"

---

Avg flow velocity = 5 ft/sec
Internal diameter of the meter = 3"

$$Q = AV_{avg} = \left[(\pi)\left(\frac{3"}{2}\right)^2\right]\left(\frac{60"}{sec}\right) = \frac{424.4 in^3}{sec}$$

*to convert to GPM:*

$$\left(\frac{60 sec}{1 min}\right)\left(\frac{1 gal}{231 in^3}\right)\left(\frac{424.4 in^3}{sec}\right) = 110.23 GPM$$

15. The flow of a process fluid in a 3-inch pipe is measured with an orifice plate and differential pressure transmitter. At a flow rate of 80 gpm the differential pressure is 30 inches WC. At a differential pressure of 40 inches WC what would the flow rate be most nearly equal to? (*answer a – 92.4 gpm*)

$$\frac{F_2}{F_1} = \sqrt{\frac{\Delta P_2}{\Delta P_1}} \quad \therefore \quad F_2 = \sqrt{\frac{\Delta P_2}{\Delta P_1}}(F_1)$$

16. Given the following data for process liquid flow, what would the orifice bore diameter be most nearly equal to?

Flow rate:  0 to 100 gpm
Pipe size:  3-inch schedule 40 carbon steel
Process fluid at 20°F and 50 psia
Differential Pressure: 100 inches WC
β ≐ 0.65

*(answer d – 2 inches, all of the given information is extraneous except the pipe size and the beta ratio where β = d/D, where D for 3-inch schedule 40 pipe is 3.068 inches)*

17. A magnetic flowmeter would NOT be a good choice to measure the flow rate for which of the following process mediums? (*answer d – glacial acetic acid, glacial acetic acid is 99.9% pure acid and such does not have sufficient conductivity to measure with a magnetic flowmeter*)

18. Water is flowing at a rate of 150 gpm with a temperature of 125°F through a 4-inch schedule 40 carbon steel pipe. What is the maximum fluid velocity most nearly equal to? (*answer a = 3.8 ft/sec*)

$$Q = AV \quad \therefore \quad V = \frac{Q}{A} = \frac{\dfrac{150 gal}{min} \times \dfrac{1 ft^3}{7.48 gal} \times \dfrac{1 min}{60 sec}}{\pi \left(\dfrac{4.026 in}{2}\right)^2 \dfrac{1 ft^2}{144 in^2}} = \frac{0.334 ft^3 / sec}{0.088 ft^2} = 3.8 ft / sec$$

19. What should the span of the transmitter be set to based upon the information given in the diagram above? [*answer b – 259 inches (18'0" x 12" x 1.2)*]

20. What should the transmitter zero be set to based upon the information given in the diagram above? (*answer b: –444*)

*Zero setting = –(h2 × 1.85) – (h1 – h2)(1.85) = –(72" × 1.85) – (240" – 72")(1.85) = –444"*

21. The tank shown below is level measured utilizing a differential pressure transmitter and bubbler tube assembly. If the vessel shown below is filled to 50% (15'-0") with a process fluid of SG = 0.9, what would the pressure gauge reading (in psig) on the bubbler assembly be most nearly equal to? *(answer d – 5.8 psig)*

$$H = L_{DT} \times SG_{fluid}$$
$$H = 15ft \times 0.9 \times \frac{12"}{1ft} \times \frac{1psig}{27.7"} = 5.85\,psig$$

22. According to API 521 what is the maximum relieving pressure, as a % of MAWP, for fire sizing a single relief valve? *(answer c – 120%)*

23. A Class IV valve shutoff classification corresponds to which of the following? *(answer b – 0.01% of rated valve capacity)*

24. Which valve body style would be best suited to control the flow of nickel slurry catalyst into a reactor? *(answer c – angle valve, because of the highly erosive nature of flowing nickel slurry)*

25. What would the maximum process liquid flow rate (in gpm) be most nearly equal to for a control valve with a CV of 15, an inlet pressure of 50 psig and an outlet pressure of 40 psig, and the process fluid SG = 1.15? *(answer b – 44 gpm)*

$$Q = C_V \sqrt{\frac{(P_1 - P_2)}{G_f}} = (15)\sqrt{\frac{(50 - 40)}{1.15}} = 44.2$$

26. Which type of large motor design should never be allowed to operate with no load connected to the motor? *(answer d – series motor)*

27. What is the synchronous speed of a 4-pole AC 60 Hz induction motor? *(answer c – 1800rpm)*

$$S_{Syn} = \frac{120f}{\#\,poles} = \frac{(120)(60)}{4} = 1800$$

28. What NEMA motor design exhibits the highest slip %? *(answer d – Design D)*

29. What Variable Frequency Drive design type would be best utilized in a centrifugal pump motor application? *(answer b – variable torque)*

30. What is the DC voltage drop most nearly equal to for a circuit which utilizes #16 AWG wiring (uncoated-stranded copper from NEC Chapter 9, Table 8) and where the field device, which draws 200 mA, is located 750 feet from the 24 VDC power supply? *(answer b – 1.50 V)*

$$V_d = \left( \frac{2 \times L}{1000} \right) \times I \times R = \left( \frac{2 \times 750}{1000} \right) \times 0.2 \times 4.99$$

31. A Safety Instrumented System having which SIL level is least likely to experience a failure of a safety function? *(answer d – SIL4)*

32. What NEMA designated enclosure type is designed to prevent the ignition of combustible dust? *(answer c – NEMA 9)*

33. According to the NEC, what is the maximum % of cross-sectional area fill allowed for a wireway? *(answer a – 20%)*

34. A Cat3 cable can support up to what maximum data rate? *(answer a – 10 Mbps)*

35. Which OSI model layer provides switching and routing technologies? *(answer c – Network Layer)*

36. Which disk array setup provides for 100% data redundancy? *(answer a – RAID 1)*

37. What is the maximum allowable FOUNDATION Fieldbus segment length utilizing a #18 AWG shielded twisted pair cable? *(answer d – 1900 meters)*

38. What is the best wavelength analytical solution to find the concentration of a gas that absorbs light at a wavelength of 5000 nanometers? *(answer b – MIR)*

    Ultraviolet (UV) Wavelength Range: 100 to 400 nanometers
    Near Infrared (NIR): 780 to 3000 nanometers
    Mid Infrared (MIR): 4000 to 50000 nanometers
    Far Infrared (FIR): 50000 nanometers to 1 mm

39. Of the following level measurement techniques, which would be the best solution for measuring phosgene in a vessel with under vacuum conditions? *(answer d – nuclear)* *Since phosgene is a lethal mixture ($COCl_2$) this eliminates D/P and conductivity, with the vessel being under vacuum conditions this eliminates ultrasonic.*

40. Of the following types of control, which will have the effect of eliminating the steady-state error? *(answer b – integral control)*

# 12. MISCELLANEOUS TABLES/INFORMATION

## 12.1 VISCOSITY EQUIVALENCY NOMOGRAPH

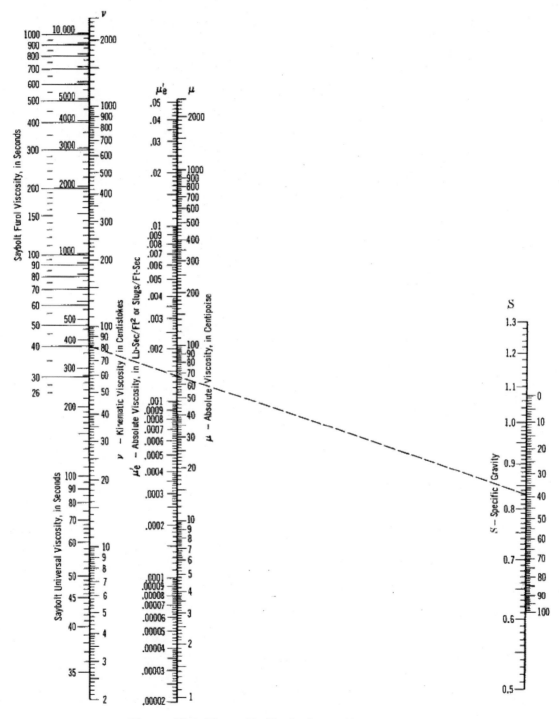

**Figure 12-1. Viscosity Equivalency Nomograph**

## 12.2 COPPER RESISTANCE TABLE (TABLE 12-1)

The resistance, R, of a length of wire is described by the expression:

$$R = \frac{\rho L}{A}$$

$\rho$ = resistivity of the material composing the wire (reference Table 12-1, note that values given in the table are ohms per 1000 feet)

$L$ = length of the wire, and

$A$ = area of the conducting cross section of the wire.

### Table 12-1. Copper Resistance Table

| Size AWG / kcmil | Area Circular Mils | Conductors | | | | DC Resistance at 75°C | | |
| | | Stranding | | Overall | | Copper | | Alum. |
| | | Qty. of Strands | Diameter Inches | Diameter Inches | Area In$^2$ | Uncoated $\Omega$/kft. | Coated $\Omega$/kft. | $\Omega$/kft. |
|---|---|---|---|---|---|---|---|---|
| 18 | 1620 | 1 | - | 0.04 | 0.001 | 7.77 | 8.08 | 12.8 |
| 18 | 1620 | 7 | 0.015 | 0.046 | 0.002 | 7.95 | 8.45 | 13.1 |
| 16 | 2580 | 1 | - | 0.051 | 0.002 | 4.89 | 5.08 | 8.05 |
| 16 | 2580 | 7 | 0.019 | 0.058 | 0.003 | 4.99 | 5.29 | 8.21 |
| 14 | 4110 | 1 | - | 0.064 | 0.003 | 3.07 | 3.19 | 5.06 |
| 14 | 4110 | 7 | 0.024 | 0.073 | 0.004 | 3.14 | 3.26 | 5.17 |
| 12 | 6530 | 1 | - | 0.081 | 0.005 | 1.93 | 2.01 | 3.18 |
| 12 | 6530 | 7 | 0.03 | 0.092 | 0.006 | 1.98 | 2.05 | 3.25 |
| 10 | 10380 | 1 | - | 0.102 | 0.008 | 1.21 | 1.26 | 2 |
| 10 | 10380 | 7 | 0.038 | 0.116 | 0.011 | 1.24 | 1.29 | 2.04 |
| 8 | 16510 | 1 | - | 0.128 | 0.013 | 0.764 | 0.786 | 1.26 |
| 8 | 16510 | 7 | 0.049 | 0.146 | 0.017 | 0.778 | 0.809 | 1.28 |
| 6 | 26240 | 7 | 0.061 | 0.184 | 0.027 | 0.491 | 0.51 | 0.808 |
| 4 | 41740 | 7 | 0.077 | 0.232 | 0.042 | 0.308 | 0.321 | 0.508 |
| 3 | 52620 | 7 | 0.087 | 0.26 | 0.053 | 0.245 | 0.254 | 0.403 |
| 2 | 66360 | 7 | 0.097 | 0.292 | 0.067 | 0.194 | 0.201 | 0.319 |
| 1 | 83690 | 19 | 0.066 | 0.332 | 0.087 | 0.154 | 0.16 | 0.253 |
| 1/0 | 105600 | 19 | 0.074 | 0.372 | 0.109 | 0.122 | 0.127 | 0.201 |
| 2/0 | 133100 | 19 | 0.084 | 0.418 | 0.137 | 0.0967 | 0.101 | 0.159 |
| 3/0 | 167800 | 19 | 0.094 | 0.47 | 0.173 | 0.0766 | 0.0797 | 0.126 |
| 4/0 | 211600 | 19 | 0.106 | 0.528 | 0.219 | 0.0608 | 0.0626 | 0.1 |
| 250 | 250000 | 37 | 0.082 | 0.575 | 0.26 | 0.0515 | 0.0535 | 0.0847 |
| 300 | 300000 | 37 | 0.09 | 0.63 | 0.312 | 0.0429 | 0.0446 | 0.0707 |
| 350 | 350000 | 37 | 0.097 | 0.681 | 0.364 | 0.0367 | 0.0382 | 0.0605 |
| 400 | 400000 | 37 | 0.104 | 0.728 | 0.416 | 0.0321 | 0.0331 | 0.0529 |
| 500 | 500000 | 37 | 0.116 | 0.813 | 0.519 | 0.0258 | 0.0265 | 0.0424 |
| 600 | 600000 | 61 | 0.099 | 0.893 | 0.626 | 0.0214 | 0.0223 | 0.0353 |
| 700 | 700000 | 61 | 0.107 | 0.964 | 0.73 | 0.0184 | 0.0189 | 0.0303 |
| 750 | 750000 | 61 | 0.111 | 0.998 | 0.782 | 0.0171 | 0.0176 | 0.0282 |

# 12.3 RTD Resistance Tables[150]

## Table 12-2. 100Ω Platinum in °F (α = 0.00385)

| °F | 0 | 1 | 2 | 3 | 4 | 5 | 6 | 7 | 8 | 9 | 10 | °F |
|---|---|---|---|---|---|---|---|---|---|---|---|---|
| | | | | | Resistance in Ohms | | | | | | | |
| -320 | 20.44 | 20.20 | 19.96 | 19.72 | 19.48 | 19.24 | 19.00 | 18.76 | 18.52 | | | -320 |
| -310 | 22.83 | 22.59 | 22.35 | 22.11 | 21.87 | 21.63 | 21.39 | 21.16 | 20.92 | 20.68 | 20.44 | -310 |
| -300 | 25.20 | 24.97 | 24.73 | 24.49 | 24.25 | 24.02 | 23.78 | 23.54 | 23.30 | 23.06 | 22.83 | -300 |
| -290 | 27.57 | 27.33 | 27.10 | 26.86 | 26.62 | 26.39 | 26.15 | 25.91 | 25.68 | 25.44 | 25.20 | -290 |
| -280 | 29.93 | 29.69 | 29.46 | 29.22 | 28.98 | 28.75 | 28.51 | 28.28 | 28.04 | 27.81 | 27.57 | -280 |
| -270 | 32.27 | 32.04 | 31.80 | 31.57 | 31.34 | 31.10 | 30.87 | 30.63 | 30.40 | 30.16 | 29.93 | -270 |
| -260 | 34.61 | 34.38 | 34.14 | 33.91 | 33.68 | 33.44 | 33.21 | 32.98 | 32.74 | 32.51 | 32.27 | -260 |
| -250 | 36.94 | 36.71 | 36.47 | 36.24 | 36.01 | 35.78 | 35.54 | 35.31 | 35.08 | 34.84 | 34.61 | -250 |
| -240 | 39.26 | 39.03 | 38.80 | 38.56 | 38.33 | 38.10 | 37.87 | 37.64 | 37.40 | 37.17 | 36.94 | -240 |
| -230 | 41.57 | 41.34 | 41.11 | 40.88 | 40.65 | 40.42 | 40.19 | 39.95 | 39.72 | 39.49 | 39.26 | -230 |
| -220 | 43.88 | 43.65 | 43.42 | 43.19 | 42.96 | 42.73 | 42.49 | 42.26 | 42.03 | 41.80 | 41.57 | -220 |
| -210 | 46.17 | 45.94 | 45.71 | 45.48 | 45.26 | 45.03 | 44.80 | 44.57 | 44.34 | 44.11 | 43.88 | -210 |
| -200 | 48.46 | 48.23 | 48.00 | 47.78 | 47.55 | 47.32 | 47.09 | 46.86 | 46.63 | 46.40 | 46.17 | -200 |
| -190 | 50.74 | 50.52 | 50.29 | 50.06 | 49.83 | 49.60 | 49.38 | 49.15 | 48.92 | 48.69 | 48.46 | -190 |
| -180 | 53.02 | 52.79 | 52.56 | 52.34 | 52.11 | 51.88 | 51.65 | 51.43 | 51.20 | 50.97 | 50.74 | -180 |
| -170 | 55.29 | 55.06 | 54.83 | 54.61 | 54.38 | 54.15 | 53.93 | 53.70 | 53.47 | 53.25 | 53.02 | -170 |
| -160 | 57.55 | 57.32 | 57.10 | 56.87 | 56.65 | 56.42 | 56.19 | 55.97 | 55.74 | 55.51 | 55.29 | -160 |
| -150 | 59.81 | 59.58 | 59.35 | 59.13 | 58.90 | 58.68 | 58.45 | 58.23 | 58.00 | 57.78 | 57.55 | -150 |
| -140 | 62.06 | 61.83 | 61.61 | 61.38 | 61.16 | 60.93 | 60.71 | 60.48 | 60.26 | 60.03 | 59.81 | -140 |
| -130 | 64.30 | 64.08 | 63.85 | 63.63 | 63.40 | 63.18 | 62.95 | 62.73 | 62.50 | 62.28 | 62.06 | -130 |
| -120 | 66.54 | 66.31 | 66.09 | 65.87 | 65.64 | 65.42 | 65.20 | 64.97 | 64.75 | 64.52 | 64.30 | -120 |
| -110 | 68.77 | 68.55 | 68.33 | 68.10 | 67.88 | 67.66 | 67.43 | 67.21 | 66.99 | 66.76 | 66.54 | -110 |
| -100 | 71.00 | 70.78 | 70.55 | 70.33 | 70.11 | 69.89 | 69.66 | 69.44 | 69.22 | 68.99 | 68.77 | -100 |
| -90 | 73.22 | 73.00 | 72.78 | 72.56 | 72.33 | 72.11 | 71.89 | 71.67 | 71.45 | 71.22 | 71.00 | -90 |
| -80 | 75.44 | 75.22 | 75.00 | 74.78 | 74.55 | 74.33 | 74.11 | 73.89 | 73.67 | 73.45 | 73.22 | -80 |
| -70 | 77.66 | 77.43 | 77.21 | 76.99 | 76.77 | 76.55 | 76.33 | 76.11 | 75.88 | 75.66 | 75.44 | -70 |
| -60 | 79.86 | 79.64 | 79.42 | 79.20 | 78.98 | 78.76 | 78.54 | 78.32 | 78.10 | 77.88 | 77.66 | -60 |
| -50 | 82.07 | 81.85 | 81.63 | 81.41 | 81.19 | 80.97 | 80.75 | 80.53 | 80.31 | 80.09 | 79.86 | -50 |
| -40 | 84.27 | 84.05 | 83.83 | 83.61 | 83.39 | 83.17 | 82.95 | 82.73 | 82.51 | 82.29 | 82.07 | -40 |
| -30 | 86.47 | 86.25 | 86.03 | 85.81 | 85.59 | 85.37 | 85.15 | 84.93 | 84.71 | 84.49 | 84.27 | -30 |
| -20 | 88.66 | 88.44 | 88.22 | 88.00 | 87.78 | 87.56 | 87.34 | 87.13 | 86.91 | 86.69 | 86.47 | -20 |
| -10 | 90.85 | 90.63 | 90.41 | 90.19 | 89.97 | 89.75 | 89.54 | 89.32 | 89.10 | 88.88 | 88.66 | -10 |
| 0 | 93.03 | 92.82 | 92.60 | 92.38 | 92.16 | 91.94 | 91.72 | 91.50 | 91.29 | 91.07 | 90.85 | 0 |
| 0 | 93.03 | 93.25 | 93.47 | 93.69 | 93.91 | 94.12 | 94.34 | 94.56 | 94.78 | 95.00 | 95.21 | 0 |
| 10 | 95.21 | 95.43 | 95.65 | 95.87 | 96.09 | 96.30 | 96.52 | 96.74 | 96.96 | 97.17 | 97.39 | 10 |
| 20 | 97.39 | 97.61 | 97.83 | 98.04 | 98.26 | 98.48 | 98.70 | 98.91 | 99.13 | 99.35 | 99.57 | 20 |
| 30 | 99.57 | 99.78 | 100.00 | 100.22 | 100.43 | 100.65 | 100.87 | 101.09 | 101.30 | 101.52 | 101.74 | 30 |
| 40 | 101.74 | 101.95 | 102.17 | 102.39 | 102.60 | 102.82 | 103.04 | 103.25 | 103.47 | 103.69 | 103.90 | 40 |
| 50 | 103.90 | 104.12 | 104.34 | 104.55 | 104.77 | 104.98 | 105.20 | 105.42 | 105.63 | 105.85 | 106.07 | 50 |
| 60 | 106.07 | 106.28 | 106.50 | 106.71 | 106.93 | 107.15 | 107.36 | 107.58 | 107.79 | 108.01 | 108.23 | 60 |
| 70 | 108.23 | 108.44 | 108.66 | 108.87 | 109.09 | 109.30 | 109.52 | 109.73 | 109.95 | 110.17 | 110.38 | 70 |
| 80 | 110.38 | 110.60 | 110.81 | 111.03 | 111.24 | 111.46 | 111.67 | 111.89 | 112.10 | 112.32 | 112.53 | 80 |
| 90 | 112.53 | 112.75 | 112.96 | 113.18 | 113.39 | 113.61 | 113.82 | 114.04 | 114.25 | 114.47 | 114.68 | 90 |
| 100 | 114.68 | 114.90 | 115.11 | 115.33 | 115.54 | 115.76 | 115.97 | 116.18 | 116.40 | 116.61 | 116.83 | 100 |
| 110 | 116.83 | 117.04 | 117.26 | 117.47 | 117.68 | 117.90 | 118.11 | 118.33 | 118.54 | 118.76 | 118.97 | 110 |
| 120 | 118.97 | 119.18 | 119.40 | 119.61 | 119.82 | 120.04 | 120.25 | 120.47 | 120.68 | 120.89 | 121.11 | 120 |
| 130 | 121.11 | 121.32 | 121.53 | 121.75 | 121.96 | 122.18 | 122.39 | 122.60 | 122.82 | 123.03 | 123.24 | 130 |
| 140 | 123.24 | 123.46 | 123.67 | 123.88 | 124.09 | 124.31 | 124.52 | 124.73 | 124.95 | 125.16 | 125.37 | 140 |

---

150. Reference www.bluelinecontrols.com.

# Table 12-2 Continued

| °F | 0 | 1 | 2 | 3 | 4 | 5 | 6 | 7 | 8 | 9 | 10 | °F |
|---|---|---|---|---|---|---|---|---|---|---|---|---|

Resistance in Ohms

| °F | 0 | 1 | 2 | 3 | 4 | 5 | 6 | 7 | 8 | 9 | 10 | °F |
|---|---|---|---|---|---|---|---|---|---|---|---|---|
| 150 | 125.37 | 125.59 | 125.80 | 126.01 | 126.22 | 126.44 | 126.65 | 126.86 | 127.08 | 127.29 | 127.50 | 150 |
| 160 | 127.50 | 127.71 | 127.93 | 128.14 | 128.35 | 128.56 | 128.78 | 128.99 | 129.20 | 129.41 | 129.62 | 160 |
| 170 | 129.62 | 129.84 | 130.05 | 130.26 | 130.47 | 130.68 | 130.90 | 131.11 | 131.32 | 131.53 | 131.74 | 170 |
| 180 | 131.74 | 131.96 | 132.17 | 132.38 | 132.59 | 132.80 | 133.01 | 133.23 | 133.44 | 133.65 | 133.86 | 180 |
| 190 | 133.86 | 134.07 | 134.28 | 134.50 | 134.71 | 134.92 | 135.13 | 135.34 | 135.55 | 135.76 | 135.97 | 190 |
| 200 | 135.97 | 136.19 | 136.40 | 136.61 | 136.82 | 137.03 | 137.24 | 137.45 | 137.66 | 137.87 | 138.08 | 200 |
| 210 | 138.08 | 138.29 | 138.51 | 138.72 | 138.93 | 139.14 | 139.35 | 139.56 | 139.77 | 139.98 | 140.19 | 210 |
| 220 | 140.19 | 140.40 | 140.61 | 140.82 | 141.03 | 141.24 | 141.45 | 141.66 | 141.87 | 142.08 | 142.29 | 220 |
| 230 | 142.29 | 142.50 | 142.71 | 142.92 | 143.13 | 143.34 | 143.55 | 143.76 | 143.97 | 144.18 | 144.39 | 230 |
| 240 | 144.39 | 144.60 | 144.81 | 145.02 | 145.23 | 145.44 | 145.65 | 145.86 | 146.07 | 146.28 | 146.49 | 240 |
| 250 | 146.49 | 146.70 | 146.91 | 147.11 | 147.32 | 147.53 | 147.74 | 147.95 | 148.16 | 148.37 | 148.58 | 250 |
| 260 | 148.58 | 148.79 | 149.00 | 149.21 | 149.41 | 149.62 | 149.83 | 150.04 | 150.25 | 150.46 | 150.67 | 260 |
| 270 | 150.67 | 150.88 | 151.08 | 151.29 | 151.50 | 151.71 | 151.92 | 152.13 | 152.33 | 152.54 | 152.75 | 270 |
| 280 | 152.75 | 152.96 | 153.17 | 153.38 | 153.58 | 153.79 | 154.00 | 154.21 | 154.42 | 154.62 | 154.83 | 280 |
| 290 | 154.83 | 155.04 | 155.25 | 155.46 | 155.66 | 155.87 | 156.08 | 156.29 | 156.49 | 156.70 | 156.91 | 290 |
| 300 | 156.91 | 157.12 | 157.33 | 157.53 | 157.74 | 157.95 | 158.15 | 158.36 | 158.57 | 158.78 | 158.98 | 300 |
| 310 | 158.98 | 159.19 | 159.40 | 159.61 | 159.81 | 160.02 | 160.23 | 160.43 | 160.64 | 160.85 | 161.05 | 310 |
| 320 | 161.05 | 161.26 | 161.47 | 161.67 | 161.88 | 162.09 | 162.29 | 162.50 | 162.71 | 162.91 | 163.12 | 320 |
| 330 | 163.12 | 163.33 | 163.53 | 163.74 | 163.95 | 164.15 | 164.36 | 164.57 | 164.77 | 164.98 | 165.18 | 330 |
| 340 | 165.18 | 165.39 | 165.60 | 165.80 | 166.01 | 166.21 | 166.42 | 166.63 | 166.83 | 167.04 | 167.24 | 340 |
| 350 | 167.24 | 167.45 | 167.66 | 167.86 | 168.07 | 168.27 | 168.48 | 168.68 | 168.89 | 169.09 | 169.30 | 350 |
| 360 | 169.30 | 169.51 | 169.71 | 169.92 | 170.12 | 170.33 | 170.53 | 170.74 | 170.94 | 171.15 | 171.35 | 360 |
| 370 | 171.35 | 171.56 | 171.76 | 171.97 | 172.17 | 172.38 | 172.58 | 172.79 | 172.99 | 173.20 | 173.40 | 370 |
| 380 | 173.40 | 173.61 | 173.81 | 174.02 | 174.22 | 174.43 | 174.63 | 174.83 | 175.04 | 175.24 | 175.45 | 380 |
| 390 | 175.45 | 175.65 | 175.86 | 176.06 | 176.26 | 176.47 | 176.67 | 176.88 | 177.08 | 177.29 | 177.49 | 390 |
| 400 | 177.49 | 177.69 | 177.90 | 178.10 | 178.30 | 178.51 | 178.71 | 178.92 | 179.12 | 179.32 | 179.53 | 400 |
| 410 | 179.53 | 179.73 | 179.93 | 180.14 | 180.34 | 180.55 | 180.75 | 180.95 | 181.16 | 181.36 | 181.56 | 410 |
| 420 | 181.56 | 181.77 | 181.97 | 182.17 | 182.38 | 182.58 | 182.78 | 182.98 | 183.19 | 183.39 | 183.59 | 420 |
| 430 | 183.59 | 183.80 | 184.00 | 184.20 | 184.40 | 184.61 | 184.81 | 185.01 | 185.22 | 185.42 | 185.62 | 430 |
| 440 | 185.62 | 185.82 | 186.03 | 186.23 | 186.43 | 186.63 | 186.84 | 187.04 | 187.24 | 187.44 | 187.65 | 440 |
| 450 | 187.65 | 187.85 | 188.05 | 188.25 | 188.45 | 188.66 | 188.86 | 189.06 | 189.26 | 189.46 | 189.67 | 450 |
| 460 | 189.67 | 189.87 | 190.07 | 190.27 | 190.47 | 190.67 | 190.88 | 191.08 | 191.28 | 191.48 | 191.68 | 460 |
| 470 | 191.68 | 191.88 | 192.09 | 192.29 | 192.49 | 192.69 | 192.89 | 193.09 | 193.29 | 193.49 | 193.70 | 470 |
| 480 | 193.70 | 193.90 | 194.10 | 194.30 | 194.50 | 194.70 | 194.90 | 195.10 | 195.30 | 195.50 | 195.71 | 480 |
| 490 | 195.71 | 195.91 | 196.11 | 196.31 | 196.51 | 196.71 | 196.91 | 197.11 | 197.31 | 197.51 | 197.71 | 490 |
| 500 | 197.71 | 197.91 | 198.11 | 198.31 | 198.51 | 198.71 | 198.91 | 199.11 | 199.31 | 199.51 | 199.71 | 500 |
| 510 | 199.71 | 199.91 | 200.11 | 200.31 | 200.51 | 200.71 | 200.91 | 201.11 | 201.31 | 201.51 | 201.71 | 510 |
| 520 | 201.71 | 201.91 | 202.11 | 202.31 | 202.51 | 202.71 | 202.91 | 203.11 | 203.31 | 203.51 | 203.71 | 520 |
| 530 | 203.71 | 203.91 | 204.11 | 204.31 | 204.51 | 204.71 | 204.90 | 205.10 | 205.30 | 205.50 | 205.70 | 530 |
| 540 | 205.70 | 205.90 | 206.10 | 206.30 | 206.50 | 206.70 | 206.89 | 207.09 | 207.29 | 207.49 | 207.69 | 540 |
| 550 | 207.69 | 207.89 | 208.09 | 208.29 | 208.48 | 208.68 | 208.88 | 209.08 | 209.28 | 209.48 | 209.67 | 550 |
| 560 | 209.67 | 209.87 | 210.07 | 210.27 | 210.47 | 210.67 | 210.86 | 211.06 | 211.26 | 211.46 | 211.66 | 560 |
| 570 | 211.66 | 211.85 | 212.05 | 212.25 | 212.45 | 212.64 | 212.84 | 213.04 | 213.24 | 213.44 | 213.63 | 570 |
| 580 | 213.63 | 213.83 | 214.03 | 214.23 | 214.42 | 214.62 | 214.82 | 215.02 | 215.21 | 215.41 | 215.61 | 580 |
| 590 | 215.61 | 215.80 | 216.00 | 216.20 | 216.40 | 216.59 | 216.79 | 216.99 | 217.18 | 217.38 | 217.58 | 590 |
| 600 | 217.58 | 217.77 | 217.97 | 218.17 | 218.37 | 218.56 | 218.76 | 218.96 | 219.15 | 219.35 | 219.55 | 600 |
| 610 | 219.55 | 219.74 | 219.94 | 220.13 | 220.33 | 220.53 | 220.72 | 220.92 | 221.12 | 221.31 | 221.51 | 610 |
| 620 | 221.51 | 221.70 | 221.90 | 222.10 | 222.29 | 222.49 | 222.68 | 222.88 | 223.08 | 223.27 | 223.47 | 620 |
| 630 | 223.47 | 223.66 | 223.86 | 224.06 | 224.25 | 224.45 | 224.64 | 224.84 | 225.03 | 225.23 | 225.42 | 630 |
| 640 | 225.42 | 225.62 | 225.82 | 226.01 | 226.21 | 226.40 | 226.60 | 226.79 | 226.99 | 227.18 | 227.38 | 640 |

## Table 12-2 Continued

| °F | 0 | 1 | 2 | 3 | 4 | 5 | 6 | 7 | 8 | 9 | 10 | °F |
|---|---|---|---|---|---|---|---|---|---|---|---|---|

Resistance in Ohms

| °F | 0 | 1 | 2 | 3 | 4 | 5 | 6 | 7 | 8 | 9 | 10 | °F |
|---|---|---|---|---|---|---|---|---|---|---|---|---|
| 650 | 227.38 | 227.57 | 227.77 | 227.96 | 228.16 | 228.35 | 228.55 | 228.74 | 228.94 | 229.13 | 229.33 | 650 |
| 660 | 229.33 | 229.52 | 229.72 | 229.91 | 230.11 | 230.30 | 230.49 | 230.69 | 230.88 | 231.08 | 231.27 | 660 |
| 670 | 231.27 | 231.47 | 231.66 | 231.86 | 232.05 | 232.24 | 232.44 | 232.63 | 232.83 | 233.02 | 233.21 | 670 |
| 680 | 233.21 | 233.41 | 233.60 | 233.80 | 233.99 | 234.18 | 234.38 | 234.57 | 234.77 | 234.96 | 235.15 | 680 |
| 690 | 235.15 | 235.35 | 235.54 | 235.73 | 235.93 | 236.12 | 236.31 | 236.51 | 236.70 | 236.89 | 237.09 | 690 |
| 700 | 237.09 | 237.28 | 237.47 | 237.67 | 237.86 | 238.05 | 238.25 | 238.44 | 238.63 | 238.83 | 239.02 | 700 |
| 710 | 239.02 | 239.21 | 239.41 | 239.60 | 239.79 | 239.98 | 240.18 | 240.37 | 240.56 | 240.75 | 240.95 | 710 |
| 720 | 240.95 | 241.14 | 241.33 | 241.52 | 241.72 | 241.91 | 242.10 | 242.29 | 242.49 | 242.68 | 242.87 | 720 |
| 730 | 242.87 | 243.06 | 243.26 | 243.45 | 243.64 | 243.83 | 244.02 | 244.22 | 244.41 | 244.60 | 244.79 | 730 |
| 740 | 244.79 | 244.98 | 245.18 | 245.37 | 245.56 | 245.75 | 245.94 | 246.13 | 246.33 | 246.52 | 246.71 | 740 |
| 750 | 246.71 | 246.90 | 247.09 | 247.28 | 247.47 | 247.67 | 247.86 | 248.05 | 248.24 | 248.43 | 248.62 | 750 |
| 760 | 248.62 | 248.81 | 249.00 | 249.20 | 249.39 | 249.58 | 249.77 | 249.96 | 250.15 | 250.34 | 250.53 | 760 |
| 770 | 250.53 | 250.72 | 250.91 | 251.10 | 251.30 | 251.49 | 251.68 | 251.87 | 252.06 | 252.25 | 252.44 | 770 |
| 780 | 252.44 | 252.63 | 252.82 | 253.01 | 253.20 | 253.39 | 253.58 | 253.77 | 253.96 | 254.15 | 254.34 | 780 |
| 790 | 254.34 | 254.53 | 254.72 | 254.91 | 255.10 | 255.29 | 255.48 | 255.67 | 255.86 | 256.05 | 256.24 | 790 |
| 800 | 256.24 | 256.43 | 256.62 | 256.81 | 257.00 | 257.19 | 257.38 | 257.57 | 257.76 | 257.95 | 258.14 | 800 |
| 810 | 258.14 | 258.33 | 258.52 | 258.70 | 258.89 | 259.08 | 259.27 | 259.46 | 259.65 | 259.84 | 260.03 | 810 |
| 820 | 260.03 | 260.22 | 260.41 | 260.60 | 260.78 | 260.97 | 261.16 | 261.35 | 261.54 | 261.73 | 261.92 | 820 |
| 830 | 261.92 | 262.11 | 262.29 | 262.48 | 262.67 | 262.86 | 263.05 | 263.24 | 263.43 | 263.61 | 263.80 | 830 |
| 840 | 263.80 | 263.99 | 264.18 | 264.37 | 264.56 | 264.74 | 264.93 | 265.12 | 265.31 | 265.50 | 265.68 | 840 |
| 850 | 265.68 | 265.87 | 266.06 | 266.25 | 266.44 | 266.62 | 266.81 | 267.00 | 267.19 | 267.37 | 267.56 | 850 |
| 860 | 267.56 | 267.75 | 267.94 | 268.12 | 268.31 | 268.50 | 268.69 | 268.87 | 269.06 | 269.25 | 269.44 | 860 |
| 870 | 269.44 | 269.62 | 269.81 | 270.00 | 270.18 | 270.37 | 270.56 | 270.75 | 270.93 | 271.12 | 271.31 | 870 |
| 880 | 271.31 | 271.49 | 271.68 | 271.87 | 272.05 | 272.24 | 272.43 | 272.61 | 272.80 | 272.99 | 273.17 | 880 |
| 890 | 273.17 | 273.36 | 273.55 | 273.73 | 273.92 | 274.11 | 274.29 | 274.48 | 274.67 | 274.85 | 275.04 | 890 |
| 900 | 275.04 | 275.22 | 275.41 | 275.60 | 275.78 | 275.97 | 276.15 | 276.34 | 276.53 | 276.71 | 276.90 | 900 |
| 910 | 276.90 | 277.08 | 277.27 | 277.46 | 277.64 | 277.83 | 278.01 | 278.20 | 278.38 | 278.57 | 278.75 | 910 |
| 920 | 278.75 | 278.94 | 279.13 | 279.31 | 279.50 | 279.68 | 279.87 | 280.05 | 280.24 | 280.42 | 280.61 | 920 |
| 930 | 280.61 | 280.79 | 280.98 | 281.16 | 281.35 | 281.53 | 281.72 | 281.90 | 282.09 | 282.27 | 282.46 | 930 |
| 940 | 282.46 | 282.64 | 282.83 | 283.01 | 283.20 | 283.38 | 283.56 | 283.75 | 283.93 | 284.12 | 284.30 | 940 |
| 950 | 284.30 | 284.49 | 284.67 | 284.86 | 285.04 | 285.22 | 285.41 | 285.59 | 285.78 | 285.96 | 286.14 | 950 |
| 960 | 286.14 | 286.33 | 286.51 | 286.70 | 286.88 | 287.06 | 287.25 | 287.43 | 287.62 | 287.80 | 287.98 | 960 |
| 970 | 287.98 | 288.17 | 288.35 | 288.53 | 288.72 | 288.90 | 289.08 | 289.27 | 289.45 | 289.64 | 289.82 | 970 |
| 980 | 289.82 | 290.00 | 290.19 | 290.37 | 290.55 | 290.73 | 290.92 | 291.10 | 291.28 | 291.47 | 291.65 | 980 |
| 990 | 291.65 | 291.83 | 292.02 | 292.20 | 292.38 | 292.56 | 292.75 | 292.93 | 293.11 | 293.30 | 293.48 | 990 |
| 1000 | 293.48 | 293.66 | 293.84 | 294.03 | 294.21 | 294.39 | 294.57 | 294.76 | 294.94 | 295.12 | 295.30 | 1000 |
| 1010 | 295.30 | 295.48 | 295.67 | 295.85 | 296.03 | 296.21 | 296.40 | 296.58 | 296.76 | 296.94 | 297.12 | 1010 |
| 1020 | 297.12 | 297.31 | 297.49 | 297.67 | 297.85 | 298.03 | 298.21 | 298.40 | 298.58 | 298.76 | 298.94 | 1020 |
| 1030 | 298.94 | 299.12 | 299.30 | 299.49 | 299.67 | 299.85 | 300.03 | 300.21 | 300.39 | 300.57 | 300.75 | 1030 |
| 1040 | 300.75 | 300.94 | 301.12 | 301.30 | 301.48 | 301.66 | 301.84 | 302.02 | 302.20 | 302.38 | 302.56 | 1040 |
| 1050 | 302.56 | 302.75 | 302.93 | 303.11 | 303.29 | 303.47 | 303.65 | 303.83 | 304.01 | 304.19 | 304.37 | 1050 |
| 1060 | 304.37 | 304.55 | 304.73 | 304.91 | 305.09 | 305.27 | 305.45 | 305.63 | 305.81 | 305.99 | 306.17 | 1060 |
| 1070 | 306.17 | 306.35 | 306.53 | 306.71 | 306.89 | 307.07 | 307.25 | 307.43 | 307.61 | 307.79 | 307.97 | 1070 |
| 1080 | 307.97 | 308.15 | 308.33 | 308.51 | 308.69 | 308.87 | 309.05 | 309.23 | 309.41 | 309.59 | 309.77 | 1080 |
| 1090 | 309.77 | 309.95 | 310.13 | 310.31 | 310.49 | 310.67 | 310.85 | 311.02 | 311.20 | 311.38 | 311.56 | 1090 |
| 1100 | 311.56 | 311.74 | 311.92 | 312.10 | 312.28 | 312.46 | 312.64 | 312.81 | 312.99 | 313.17 | 313.35 | 1100 |
| 1110 | 313.35 | 313.53 | 313.71 | 313.89 | 314.07 | 314.24 | 314.42 | 314.60 | 314.78 | 314.96 | 315.14 | 1110 |
| 1120 | 315.14 | 315.31 | 315.49 | 315.67 | 315.85 | 316.03 | 316.21 | 316.38 | 316.56 | 316.74 | 316.92 | 1120 |
| 1130 | 316.92 | 317.10 | 317.27 | 317.45 | 317.63 | 317.81 | 317.98 | 318.16 | 318.34 | 318.52 | 318.70 | 1130 |
| 1140 | 318.70 | 318.87 | 319.05 | 319.23 | 319.41 | 319.58 | 319.76 | 319.94 | 320.12 | 320.29 | 320.47 | 1140 |

# Table 12-2 Continued

| °F | 0 | 1 | 2 | 3 | 4 | 5 | 6 | 7 | 8 | 9 | 10 | °F |
|---|---|---|---|---|---|---|---|---|---|---|---|---|
| | | | | | Resistance in Ohms | | | | | | | |
| 1150 | 320.47 | 320.65 | 320.82 | 321.00 | 321.18 | 321.36 | 321.53 | 321.71 | 321.89 | 322.06 | 322.24 | 1150 |
| 1160 | 322.24 | 322.42 | 322.59 | 322.77 | 322.95 | 323.13 | 323.30 | 323.48 | 323.66 | 323.83 | 324.01 | 1160 |
| 1170 | 324.01 | 324.18 | 324.36 | 324.54 | 324.71 | 324.89 | 325.07 | 325.24 | 325.42 | 325.60 | 325.77 | 1170 |
| 1180 | 325.77 | 325.95 | 326.12 | 326.30 | 326.48 | 326.65 | 326.83 | 327.00 | 327.18 | 327.36 | 327.53 | 1180 |
| 1190 | 327.53 | 327.71 | 327.88 | 328.06 | 328.24 | 328.41 | 328.59 | 328.76 | 328.94 | 329.11 | 329.29 | 1190 |
| 1200 | 329.29 | 329.46 | 329.64 | 329.82 | 329.99 | 330.17 | 330.34 | 330.52 | 330.69 | 330.87 | 331.04 | 1200 |
| 1210 | 331.04 | 331.22 | 331.39 | 331.57 | 331.74 | 331.92 | 332.09 | 332.27 | 332.44 | 332.62 | 332.79 | 1210 |
| 1220 | 332.79 | | | | | | | | | | | 1220 |

# Table 12-3. 10Ω Copper RTD in °F (α = 0.00427)

| °F | 0 | 1 | 2 | 3 | 4 | 5 | 6 | 7 | 8 | 9 | 10 | °F |
|---|---|---|---|---|---|---|---|---|---|---|---|---|
| | | | | | Resistance in Ohms | | | | | | | |
| -320 | 1.242 | 1.219 | 1.196 | 1.173 | 1.150 | 1.127 | 1.104 | 1.081 | 1.058 | | | -320 |
| -310 | 1.472 | 1.449 | 1.426 | 1.403 | 1.380 | 1.357 | 1.334 | 1.311 | 1.288 | 1.265 | 1.242 | -310 |
| -300 | 1.701 | 1.678 | 1.655 | 1.632 | 1.609 | 1.587 | 1.564 | 1.541 | 1.518 | 1.495 | 1.472 | -300 |
| -290 | 1.930 | 1.907 | 1.884 | 1.861 | 1.839 | 1.816 | 1.793 | 1.770 | 1.747 | 1.724 | 1.701 | -290 |
| -280 | 2.158 | 2.136 | 2.113 | 2.090 | 2.067 | 2.044 | 2.021 | 1.999 | 1.976 | 1.953 | 1.930 | -280 |
| -270 | 2.386 | 2.364 | 2.341 | 2.318 | 2.295 | 2.272 | 2.250 | 2.227 | 2.204 | 2.181 | 2.158 | -270 |
| -260 | 2.614 | 2.591 | 2.568 | 2.546 | 2.523 | 2.500 | 2.477 | 2.455 | 2.432 | 2.409 | 2.386 | -260 |
| -250 | 2.841 | 2.818 | 2.795 | 2.773 | 2.750 | 2.727 | 2.705 | 2.682 | 2.659 | 2.637 | 2.614 | -250 |
| -240 | 3.067 | 3.045 | 3.022 | 2.999 | 2.977 | 2.954 | 2.931 | 2.909 | 2.886 | 2.863 | 2.841 | -240 |
| -230 | 3.293 | 3.271 | 3.248 | 3.226 | 3.203 | 3.180 | 3.158 | 3.135 | 3.113 | 3.090 | 3.067 | -230 |
| -220 | 3.519 | 3.496 | 3.474 | 3.451 | 3.429 | 3.406 | 3.384 | 3.361 | 3.338 | 3.316 | 3.293 | -220 |
| -210 | 3.744 | 3.721 | 3.699 | 3.676 | 3.654 | 3.631 | 3.609 | 3.586 | 3.564 | 3.541 | 3.519 | -210 |
| -200 | 3.968 | 3.946 | 3.923 | 3.901 | 3.879 | 3.856 | 3.834 | 3.811 | 3.789 | 3.766 | 3.744 | -200 |
| -190 | 4.192 | 4.170 | 4.148 | 4.125 | 4.103 | 4.080 | 4.058 | 4.036 | 4.013 | 3.991 | 3.968 | -190 |
| -180 | 4.416 | 4.394 | 4.371 | 4.349 | 4.327 | 4.304 | 4.282 | 4.259 | 4.237 | 4.215 | 4.192 | -180 |
| -170 | 4.639 | 4.617 | 4.594 | 4.572 | 4.550 | 4.527 | 4.505 | 4.483 | 4.461 | 4.438 | 4.416 | -170 |
| -160 | 4.862 | 4.839 | 4.817 | 4.795 | 4.773 | 4.750 | 4.728 | 4.706 | 4.684 | 4.661 | 4.639 | -160 |
| -150 | 5.084 | 5.061 | 5.039 | 5.017 | 4.995 | 4.973 | 4.950 | 4.928 | 4.906 | 4.884 | 4.862 | -150 |
| -140 | 5.305 | 5.283 | 5.261 | 5.239 | 5.217 | 5.195 | 5.172 | 5.150 | 5.128 | 5.106 | 5.084 | -140 |
| -130 | 5.526 | 5.504 | 5.482 | 5.460 | 5.438 | 5.416 | 5.394 | 5.372 | 5.350 | 5.327 | 5.305 | -130 |
| -120 | 5.747 | 5.725 | 5.703 | 5.681 | 5.659 | 5.637 | 5.615 | 5.593 | 5.571 | 5.548 | 5.526 | -120 |
| -110 | 5.967 | 5.945 | 5.923 | 5.901 | 5.879 | 5.857 | 5.835 | 5.813 | 5.791 | 5.769 | 5.747 | -110 |
| -100 | 6.187 | 6.165 | 6.143 | 6.121 | 6.099 | 6.077 | 6.055 | 6.033 | 6.011 | 5.989 | 5.967 | -100 |
| -90 | 6.406 | 6.384 | 6.362 | 6.340 | 6.318 | 6.296 | 6.275 | 6.253 | 6.231 | 6.209 | 6.187 | -90 |
| -80 | 6.625 | 6.603 | 6.581 | 6.559 | 6.537 | 6.515 | 6.494 | 6.472 | 6.450 | 6.428 | 6.406 | -80 |
| -70 | 6.843 | 6.821 | 6.799 | 6.777 | 6.756 | 6.734 | 6.712 | 6.690 | 6.668 | 6.647 | 6.625 | -70 |
| -60 | 7.061 | 7.039 | 7.017 | 6.995 | 6.974 | 6.952 | 6.930 | 6.908 | 6.886 | 6.865 | 6.843 | -60 |
| -50 | 7.276 | 7.254 | 7.233 | 7.211 | 7.190 | 7.168 | 7.147 | 7.126 | 7.104 | 7.082 | 7.061 | -50 |
| -40 | 7.490 | 7.469 | 7.447 | 7.426 | 7.404 | 7.383 | 7.362 | 7.340 | 7.319 | 7.297 | 7.276 | -40 |
| -30 | 7.705 | 7.683 | 7.662 | 7.640 | 7.619 | 7.598 | 7.576 | 7.555 | 7.533 | 7.512 | 7.490 | -30 |
| -20 | 7.919 | 7.898 | 7.876 | 7.855 | 7.834 | 7.812 | 7.791 | 7.769 | 7.748 | 7.726 | 7.705 | -20 |
| -10 | 8.134 | 8.112 | 8.091 | 8.070 | 8.048 | 8.027 | 8.005 | 7.984 | 7.962 | 7.941 | 7.919 | -10 |
| 0 | 8.348 | 8.327 | 8.306 | 8.284 | 8.263 | 8.241 | 8.220 | 8.198 | 8.177 | 8.155 | 8.134 | 0 |
| 0 | 8.348 | 8.370 | 8.391 | 8.413 | 8.434 | 8.456 | 8.477 | 8.499 | 8.520 | 8.542 | 8.563 | 0 |
| 10 | 8.563 | 8.584 | 8.606 | 8.627 | 8.649 | 8.670 | 8.692 | 8.713 | 8.735 | 8.756 | 8.778 | 10 |
| 20 | 8.778 | 8.799 | 8.820 | 8.842 | 8.863 | 8.885 | 8.906 | 8.928 | 8.949 | 8.971 | 8.992 | 20 |
| 30 | 8.992 | 9.014 | 9.035 | 9.056 | 9.078 | 9.099 | 9.121 | 9.142 | 9.164 | 9.185 | 9.207 | 30 |
| 40 | 9.207 | 9.228 | 9.250 | 9.271 | 9.292 | 9.314 | 9.335 | 9.357 | 9.378 | 9.400 | 9.421 | 40 |
| 50 | 9.421 | 9.443 | 9.464 | 9.486 | 9.507 | 9.528 | 9.550 | 9.571 | 9.593 | 9.614 | 9.636 | 50 |
| 60 | 9.636 | 9.657 | 9.679 | 9.700 | 9.722 | 9.743 | 9.764 | 9.786 | 9.807 | 9.829 | 9.850 | 60 |
| 70 | 9.850 | 9.872 | 9.893 | 9.915 | 9.936 | 9.958 | 9.979 | 10.000 | 10.022 | 10.043 | 10.065 | 70 |
| 80 | 10.065 | 10.086 | 10.108 | 10.129 | 10.151 | 10.172 | 10.194 | 10.215 | 10.236 | 10.258 | 10.279 | 80 |
| 90 | 10.279 | 10.301 | 10.322 | 10.344 | 10.365 | 10.387 | 10.408 | 10.430 | 10.451 | 10.472 | 10.494 | 90 |
| 100 | 10.494 | 10.515 | 10.537 | 10.558 | 10.580 | 10.601 | 10.623 | 10.644 | 10.666 | 10.687 | 10.708 | 100 |
| 110 | 10.708 | 10.730 | 10.751 | 10.773 | 10.794 | 10.816 | 10.837 | 10.859 | 10.880 | 10.902 | 10.923 | 110 |
| 120 | 10.923 | 10.944 | 10.966 | 10.987 | 11.009 | 11.030 | 11.052 | 11.073 | 11.095 | 11.116 | 11.138 | 120 |
| 130 | 11.138 | 11.159 | 11.180 | 11.202 | 11.223 | 11.245 | 11.266 | 11.288 | 11.309 | 11.331 | 11.352 | 130 |
| 140 | 11.352 | 11.374 | 11.395 | 11.416 | 11.438 | 11.459 | 11.481 | 11.502 | 11.524 | 11.545 | 11.567 | 140 |

# Table 12-3 Continued

| | | | | | | | | | | | | |
|---|---|---|---|---|---|---|---|---|---|---|---|---|
| 150 | 11.567 | 11.588 | 11.610 | 11.631 | 11.652 | 11.674 | 11.695 | 11.717 | 11.738 | 11.760 | 11.781 | 150 |
| 160 | 11.781 | 11.803 | 11.824 | 11.846 | 11.867 | 11.888 | 11.910 | 11.931 | 11.953 | 11.974 | 11.996 | 160 |
| 170 | 11.996 | 12.017 | 12.039 | 12.060 | 12.082 | 12.103 | 12.124 | 12.146 | 12.167 | 12.189 | 12.210 | 170 |
| 180 | 12.210 | 12.232 | 12.253 | 12.275 | 12.296 | 12.318 | 12.339 | 12.360 | 12.382 | 12.403 | 12.425 | 180 |
| 190 | 12.425 | 12.446 | 12.468 | 12.489 | 12.511 | 12.532 | 12.554 | 12.575 | 12.596 | 12.618 | 12.639 | 190 |
| 200 | 12.639 | 12.661 | 12.682 | 12.704 | 12.725 | 12.747 | 12.768 | 12.790 | 12.811 | 12.832 | 12.854 | 200 |
| 210 | 12.854 | 12.875 | 12.897 | 12.918 | 12.940 | 12.961 | 12.983 | 13.004 | 13.026 | 13.047 | 13.068 | 210 |
| 220 | 13.068 | 13.090 | 13.111 | 13.133 | 13.154 | 13.176 | 13.197 | 13.219 | 13.240 | 13.262 | 13.283 | 220 |
| 230 | 13.283 | 13.304 | 13.326 | 13.347 | 13.369 | 13.390 | 13.412 | 13.433 | 13.455 | 13.476 | 13.498 | 230 |
| 240 | 13.498 | 13.519 | 13.540 | 13.562 | 13.583 | 13.605 | 13.626 | 13.648 | 13.669 | 13.691 | 13.712 | 240 |
| 250 | 13.712 | 13.734 | 13.755 | 13.776 | 13.798 | 13.819 | 13.841 | 13.862 | 13.884 | 13.905 | 13.927 | 250 |
| 260 | 13.927 | 13.948 | 13.970 | 13.991 | 14.012 | 14.034 | 14.055 | 14.077 | 14.098 | 14.120 | 14.141 | 260 |
| 270 | 14.141 | 14.163 | 14.184 | 14.206 | 14.227 | 14.248 | 14.270 | 14.291 | 14.313 | 14.334 | 14.356 | 270 |
| 280 | 14.356 | 14.377 | 14.399 | 14.420 | 14.442 | 14.463 | 14.484 | 14.506 | 14.527 | 14.549 | 14.570 | 280 |
| 290 | 14.570 | 14.592 | 14.613 | 14.635 | 14.656 | 14.678 | 14.699 | 14.720 | 14.742 | 14.763 | 14.785 | 290 |
| 300 | 14.785 | 14.806 | 14.828 | 14.849 | 14.871 | 14.893 | 14.914 | 14.936 | 14.958 | 14.979 | 15.001 | 300 |
| 310 | 15.001 | 15.022 | 15.044 | 15.066 | 15.087 | 15.109 | 15.131 | 15.152 | 15.174 | 15.196 | 15.217 | 310 |
| 320 | 15.217 | 15.239 | 15.260 | 15.282 | 15.304 | 15.325 | 15.347 | 15.369 | 15.390 | 15.412 | 15.434 | 320 |
| 330 | 15.434 | 15.455 | 15.477 | 15.499 | 15.520 | 15.542 | 15.563 | 15.585 | 15.607 | 15.628 | 15.650 | 330 |
| 340 | 15.650 | 15.672 | 15.693 | 15.715 | 15.737 | 15.758 | 15.780 | 15.802 | 15.823 | 15.845 | 15.866 | 340 |
| 350 | 15.866 | 15.888 | 15.910 | 15.931 | 15.953 | 15.975 | 15.996 | 16.018 | 16.040 | 16.061 | 16.083 | 350 |
| 360 | 16.083 | 16.105 | 16.126 | 16.148 | 16.170 | 16.191 | 16.213 | 16.234 | 16.256 | 16.278 | 16.299 | 360 |
| 370 | 16.299 | 16.321 | 16.343 | 16.364 | 16.386 | 16.408 | 16.429 | 16.451 | 16.473 | 16.494 | 16.516 | 370 |
| 380 | 16.516 | 16.538 | 16.559 | 16.581 | 16.603 | 16.624 | 16.646 | 16.667 | 16.689 | 16.711 | 16.732 | 380 |
| 390 | 16.732 | 16.754 | 16.776 | 16.797 | 16.819 | 16.841 | 16.862 | 16.884 | 16.906 | 16.927 | 16.949 | 390 |
| 400 | 16.949 | 16.971 | 16.992 | 17.014 | 17.036 | 17.057 | 17.079 | 17.101 | 17.122 | 17.144 | 17.166 | 400 |
| 410 | 17.166 | 17.187 | 17.209 | 17.231 | 17.252 | 17.274 | 17.296 | 17.317 | 17.339 | 17.360 | 17.382 | 410 |
| 420 | 17.382 | 17.404 | 17.425 | 17.447 | 17.469 | 17.490 | 17.512 | 17.534 | 17.555 | 17.577 | 17.599 | 420 |
| 430 | 17.599 | 17.620 | 17.642 | 17.664 | 17.685 | 17.707 | 17.729 | 17.750 | 17.772 | 17.794 | 17.815 | 430 |
| 440 | 17.815 | 17.837 | 17.859 | 17.880 | 17.902 | 17.924 | 17.945 | 17.967 | 17.989 | 18.010 | 18.032 | 440 |
| 450 | 18.032 | 18.054 | 18.075 | 18.097 | 18.119 | 18.140 | 18.162 | 18.184 | 18.205 | 18.227 | 18.249 | 450 |
| 460 | 18.249 | 18.270 | 18.292 | 18.314 | 18.335 | 18.357 | 18.379 | 18.400 | 18.422 | 18.444 | 18.465 | 460 |
| 470 | 18.465 | 18.487 | 18.509 | 18.530 | 18.552 | 18.574 | 18.595 | 18.617 | 18.639 | 18.661 | 18.682 | 470 |
| 480 | 18.682 | 18.704 | 18.726 | 18.747 | 18.769 | 18.791 | 18.812 | 18.834 | 18.856 | 18.877 | 18.899 | 480 |
| 490 | 18.899 | 18.921 | 18.942 | 18.964 | 18.986 | 19.007 | 19.029 | 19.051 | 19.072 | 19.094 | 19.116 | 490 |
| 500 | 19.116 | | | | | | | | | | | 500 |

# Table 12-4. 120Ω Nickel RTD in °F (α = 0.00672)

| °F | 0 | 1 | 2 | 3 | 4 | 5 | 6 | 7 | 8 | 9 | 10 | °F |
|---|---|---|---|---|---|---|---|---|---|---|---|---|
| | | | | | Resistance in Ohms | | | | | | | |
| -110 | 67.32 | 66.96 | 66.60 | | | | | | | | | -110 |
| -100 | 70.93 | 70.57 | 70.21 | 69.85 | 69.49 | 69.13 | 68.77 | 68.41 | 68.04 | 67.68 | 67.32 | -100 |
| -90 | 74.55 | 74.19 | 73.83 | 73.46 | 73.10 | 72.74 | 72.38 | 72.02 | 71.66 | 71.30 | 70.93 | -90 |
| -80 | 78.17 | 77.81 | 77.45 | 77.08 | 76.72 | 76.36 | 76.00 | 75.64 | 75.27 | 74.91 | 74.55 | -80 |
| -70 | 81.80 | 81.43 | 81.07 | 80.71 | 80.35 | 79.98 | 79.62 | 79.26 | 78.89 | 78.53 | 78.17 | -70 |
| -60 | 85.44 | 85.07 | 84.71 | 84.34 | 83.98 | 83.62 | 83.25 | 82.89 | 82.52 | 82.16 | 81.80 | -60 |
| -50 | 89.09 | 88.72 | 88.36 | 87.99 | 87.62 | 87.26 | 86.89 | 86.53 | 86.16 | 85.80 | 85.44 | -50 |
| -40 | 92.76 | 92.39 | 92.02 | 91.65 | 91.29 | 90.92 | 90.55 | 90.19 | 89.82 | 89.45 | 89.09 | -40 |
| -30 | 96.44 | 96.07 | 95.70 | 95.33 | 94.97 | 94.60 | 94.23 | 93.86 | 93.49 | 93.12 | 92.76 | -30 |
| -20 | 100.15 | 99.78 | 99.41 | 99.04 | 98.67 | 98.30 | 97.92 | 97.55 | 97.18 | 96.81 | 96.44 | -20 |
| -10 | 103.89 | 103.52 | 103.14 | 102.77 | 102.39 | 102.02 | 101.65 | 101.27 | 100.90 | 100.53 | 100.15 | -10 |
| 0 | 107.66 | 107.28 | 106.91 | 106.53 | 106.15 | 105.77 | 105.40 | 105.02 | 104.64 | 104.27 | 103.89 | 0 |
| 0 | 107.66 | 108.04 | 108.42 | 108.80 | 109.18 | 109.56 | 109.94 | 110.32 | 110.70 | 111.09 | 111.47 | 0 |
| 10 | 111.47 | 111.85 | 112.23 | 112.62 | 113.00 | 113.39 | 113.77 | 114.16 | 114.54 | 114.93 | 115.32 | 10 |
| 20 | 115.32 | 115.71 | 116.09 | 116.48 | 116.87 | 117.26 | 117.65 | 118.04 | 118.43 | 118.82 | 119.21 | 20 |
| 30 | 119.21 | 119.61 | 120.00 | 120.39 | 120.79 | 121.18 | 121.58 | 121.97 | 122.37 | 122.77 | 123.16 | 30 |
| 40 | 123.16 | 123.56 | 123.96 | 124.36 | 124.76 | 125.16 | 125.56 | 125.96 | 126.36 | 126.76 | 127.17 | 40 |
| 50 | 127.17 | 127.57 | 127.98 | 128.38 | 128.79 | 129.19 | 129.60 | 130.00 | 130.41 | 130.82 | 131.23 | 50 |
| 60 | 131.23 | 131.64 | 132.05 | 132.46 | 132.87 | 133.28 | 133.69 | 134.11 | 134.52 | 134.93 | 135.35 | 60 |
| 70 | 135.35 | 135.76 | 136.18 | 136.59 | 137.01 | 137.43 | 137.85 | 138.26 | 138.68 | 139.10 | 139.52 | 70 |
| 80 | 139.52 | 139.95 | 140.37 | 140.79 | 141.21 | 141.64 | 142.06 | 142.48 | 142.91 | 143.34 | 143.76 | 80 |
| 90 | 143.76 | 144.19 | 144.62 | 145.05 | 145.48 | 145.90 | 146.34 | 146.77 | 147.20 | 147.63 | 148.06 | 90 |
| 100 | 148.06 | 148.50 | 148.93 | 149.37 | 149.80 | 150.24 | 150.67 | 151.11 | 151.55 | 151.99 | 152.43 | 100 |
| 110 | 152.43 | 152.87 | 153.31 | 153.75 | 154.19 | 154.63 | 155.07 | 155.52 | 155.96 | 156.41 | 156.85 | 110 |
| 120 | 156.85 | 157.30 | 157.75 | 158.19 | 158.64 | 159.09 | 159.54 | 159.99 | 160.44 | 160.89 | 161.34 | 120 |
| 130 | 161.34 | 161.80 | 162.25 | 162.70 | 163.16 | 163.61 | 164.07 | 164.53 | 164.98 | 165.44 | 165.90 | 130 |
| 140 | 165.90 | 166.36 | 166.82 | 167.28 | 167.74 | 168.20 | 168.67 | 169.13 | 169.59 | 170.06 | 170.52 | 140 |
| 150 | 170.52 | 170.99 | 171.46 | 171.92 | 172.39 | 172.86 | 173.33 | 173.80 | 174.27 | 174.74 | 175.21 | 150 |
| 160 | 175.21 | 175.68 | 176.16 | 176.63 | 177.10 | 177.58 | 178.06 | 178.53 | 179.01 | 179.49 | 179.97 | 160 |
| 170 | 179.97 | 180.44 | 180.92 | 181.40 | 181.89 | 182.37 | 182.85 | 183.33 | 183.82 | 184.30 | 184.78 | 170 |
| 180 | 184.78 | 185.27 | 185.76 | 186.24 | 186.73 | 187.22 | 187.71 | 188.20 | 188.69 | 189.18 | 189.67 | 180 |
| 190 | 189.67 | 190.16 | 190.65 | 191.15 | 191.64 | 192.13 | 192.63 | 193.12 | 193.62 | 194.12 | 194.62 | 190 |
| 200 | 194.62 | 195.11 | 195.61 | 196.11 | 196.61 | 197.11 | 197.62 | 198.12 | 198.62 | 199.12 | 199.63 | 200 |
| 210 | 199.63 | 200.13 | 200.64 | 201.15 | 201.65 | 202.16 | 202.67 | 203.18 | 203.69 | 204.20 | 204.71 | 210 |
| 220 | 204.71 | 205.22 | 205.73 | 206.24 | 206.76 | 207.27 | 207.79 | 208.30 | 208.82 | 209.34 | 209.85 | 220 |
| 230 | 209.85 | 210.37 | 210.89 | 211.41 | 211.93 | 212.45 | 212.97 | 213.50 | 214.02 | 214.54 | 215.07 | 230 |
| 240 | 215.07 | 215.59 | 216.12 | 216.65 | 217.17 | 217.70 | 218.23 | 218.76 | 219.29 | 219.82 | 220.35 | 240 |
| 250 | 220.35 | 220.88 | 221.42 | 221.95 | 222.48 | 223.02 | 223.56 | 224.09 | 224.63 | 225.17 | 225.70 | 250 |
| 260 | 225.70 | 226.24 | 226.78 | 227.32 | 227.87 | 228.41 | 228.95 | 229.49 | 230.04 | 230.58 | 231.13 | 260 |
| 270 | 231.13 | 231.67 | 232.22 | 232.77 | 233.32 | 233.86 | 234.41 | 234.96 | 235.52 | 236.07 | 236.62 | 270 |
| 280 | 236.62 | 237.17 | 237.73 | 238.28 | 238.84 | 239.39 | 239.95 | 240.51 | 241.06 | 241.62 | 242.18 | 280 |
| 290 | 242.18 | 242.74 | 243.30 | 243.86 | 244.43 | 244.99 | 245.55 | 246.12 | 246.68 | 247.25 | 247.82 | 290 |
| 300 | 247.82 | 248.38 | 248.95 | 249.52 | 250.09 | 250.66 | 251.23 | 251.80 | 252.37 | 252.94 | 253.52 | 300 |
| 310 | 253.52 | 254.09 | 254.67 | 255.24 | 255.82 | 256.40 | 256.98 | 257.55 | 258.13 | 258.71 | 259.30 | 310 |
| 320 | 259.30 | 259.88 | 260.46 | 261.04 | 261.63 | 262.21 | 262.80 | 263.39 | 263.97 | 264.56 | 265.15 | 320 |
| 330 | 265.15 | 265.74 | 266.33 | 266.92 | 267.52 | 268.11 | 268.70 | 269.30 | 269.89 | 270.49 | 271.09 | 330 |
| 340 | 271.09 | 271.69 | 272.29 | 272.89 | 273.49 | 274.09 | 274.69 | 275.30 | 275.90 | 276.51 | 277.11 | 340 |

# Table 12-4 Continued

| °F | 0 | 1 | 2 | 3 | 4 | 5 | 6 | 7 | 8 | 9 | 10 | °F |
|----|----|----|----|----|----|----|----|----|----|----|----|----|
| | | | | | Resistance in Ohms | | | | | | | |
| 350 | 277.11 | 277.72 | 278.33 | 278.94 | 279.55 | 280.16 | 280.77 | 281.38 | 282.00 | 282.61 | 283.23 | 350 |
| 360 | 283.23 | 283.84 | 284.46 | 285.08 | 285.70 | 286.32 | 286.94 | 287.56 | 288.19 | 288.81 | 289.44 | 360 |
| 370 | 289.44 | 290.06 | 290.69 | 291.32 | 291.95 | 292.58 | 293.21 | 293.84 | 294.48 | 295.11 | 295.75 | 370 |
| 380 | 295.75 | 296.38 | 297.02 | 297.66 | 298.30 | 298.94 | 299.58 | 300.22 | 300.87 | 301.51 | 302.16 | 380 |
| 390 | 302.16 | 302.81 | 303.45 | 304.10 | 304.75 | 305.41 | 306.06 | 306.71 | 307.37 | 308.02 | 308.68 | 390 |
| 400 | 308.68 | 309.34 | 310.00 | 310.66 | 311.32 | 311.98 | 312.64 | 313.31 | 313.97 | 314.64 | 315.31 | 400 |
| 410 | 315.31 | 315.98 | 316.65 | 317.32 | 317.99 | 318.67 | 319.34 | 320.02 | 320.70 | 321.38 | 322.06 | 410 |
| 420 | 322.06 | 322.74 | 323.42 | 324.10 | 324.79 | 325.47 | 326.16 | 326.85 | 327.54 | 328.23 | 328.92 | 420 |
| 430 | 328.92 | 329.61 | 330.30 | 331.00 | 331.70 | 332.39 | 333.09 | 333.79 | 334.49 | 335.19 | 335.90 | 430 |
| 440 | 335.90 | 336.60 | 337.31 | 338.02 | 338.72 | 339.43 | 340.14 | 340.86 | 341.57 | 342.28 | 343.00 | 440 |
| 450 | 343.00 | 343.71 | 344.43 | 345.15 | 345.87 | 346.59 | 347.32 | 348.04 | 348.76 | 349.49 | 350.22 | 450 |
| 460 | 350.22 | 350.95 | 351.68 | 352.41 | 353.14 | 353.87 | 354.61 | 355.34 | 356.08 | 356.82 | 357.56 | 460 |
| 470 | 357.56 | 358.30 | 359.04 | 359.79 | 360.53 | 361.28 | 362.02 | 362.77 | 363.52 | 364.27 | 365.02 | 470 |
| 480 | 365.02 | 365.78 | 366.53 | 367.29 | 368.04 | 368.80 | 369.56 | 370.32 | 371.08 | 371.84 | 372.61 | 480 |
| 490 | 372.61 | 373.37 | 374.14 | 374.91 | 375.67 | 376.44 | 377.21 | 377.99 | 378.76 | 379.53 | 380.31 | 490 |
| 500 | 380.31 | | | | | | | | | | | 500 |

## Table 12-5. 604Ω Nickel-Iron (Balco) RTD in °F (α = 0.00518)

| °F | 0 | 1 | 2 | 3 | 4 | 5 | 6 | 7 | 8 | 9 | 10 | °F |
|---|---|---|---|---|---|---|---|---|---|---|---|---|
| | | | | | | Resistance in Ohms | | | | | | |
| -320 | 248.80 | 248.36 | 247.92 | 247.48 | 247.05 | 246.62 | 246.19 | 245.77 | 245.34 | | | -320 |
| -310 | 253.42 | 252.94 | 252.47 | 252.00 | 251.53 | 251.07 | 250.61 | 250.16 | 249.70 | 249.25 | 248.80 | -310 |
| -300 | 258.35 | 257.85 | 257.34 | 256.84 | 256.34 | 255.85 | 255.35 | 254.87 | 254.38 | 253.90 | 253.42 | -300 |
| -290 | 263.61 | 263.07 | 262.53 | 262.00 | 261.47 | 260.94 | 260.42 | 259.90 | 259.38 | 258.86 | 258.35 | -290 |
| -280 | 269.18 | 268.61 | 268.04 | 267.48 | 266.91 | 266.36 | 265.80 | 265.25 | 264.70 | 264.15 | 263.61 | -280 |
| -270 | 275.08 | 274.47 | 273.87 | 273.28 | 272.68 | 272.09 | 271.50 | 270.92 | 270.34 | 269.76 | 269.18 | -270 |
| -260 | 281.29 | 280.66 | 280.03 | 279.40 | 278.77 | 278.15 | 277.53 | 276.91 | 276.30 | 275.69 | 275.08 | -260 |
| -250 | 287.83 | 287.16 | 286.50 | 285.84 | 285.18 | 284.52 | 283.87 | 283.22 | 282.58 | 281.93 | 281.29 | -250 |
| -240 | 294.69 | 293.99 | 293.29 | 292.60 | 291.91 | 291.22 | 290.53 | 289.85 | 289.18 | 288.50 | 287.83 | -240 |
| -230 | 301.86 | 301.13 | 300.40 | 299.68 | 298.95 | 298.23 | 297.52 | 296.81 | 296.10 | 295.39 | 294.69 | -230 |
| -220 | 309.36 | 308.59 | 307.83 | 307.08 | 306.32 | 305.57 | 304.82 | 304.08 | 303.34 | 302.60 | 301.86 | -220 |
| -210 | 317.18 | 316.38 | 315.59 | 314.80 | 314.01 | 313.23 | 312.45 | 311.67 | 310.90 | 310.13 | 309.36 | -210 |
| -200 | 325.31 | 324.48 | 323.66 | 322.84 | 322.02 | 321.20 | 320.39 | 319.58 | 318.78 | 317.98 | 317.18 | -200 |
| -190 | 333.77 | 332.91 | 332.05 | 331.20 | 330.35 | 329.50 | 328.66 | 327.82 | 326.98 | 326.14 | 325.31 | -190 |
| -180 | 342.55 | 341.66 | 340.77 | 339.88 | 339.00 | 338.12 | 337.24 | 336.37 | 335.50 | 334.63 | 333.77 | -180 |
| -170 | 351.65 | 350.72 | 349.80 | 348.88 | 347.97 | 347.06 | 346.15 | 345.24 | 344.34 | 343.44 | 342.55 | -170 |
| -160 | 361.06 | 360.11 | 359.16 | 358.21 | 357.26 | 356.32 | 355.37 | 354.44 | 353.50 | 352.57 | 351.65 | -160 |
| -150 | 370.80 | 369.81 | 368.83 | 367.85 | 366.87 | 365.89 | 364.92 | 363.95 | 362.99 | 362.02 | 361.06 | -150 |
| -140 | 380.86 | 379.84 | 378.82 | 377.81 | 376.80 | 375.79 | 374.79 | 373.79 | 372.79 | 371.79 | 370.80 | -140 |
| -130 | 391.24 | 390.19 | 389.14 | 388.09 | 387.05 | 386.01 | 384.97 | 383.94 | 382.91 | 381.89 | 380.86 | -130 |
| -120 | 401.94 | 400.86 | 399.77 | 398.70 | 397.62 | 396.55 | 395.48 | 394.42 | 393.35 | 392.30 | 391.24 | -120 |
| -110 | 412.96 | 411.84 | 410.73 | 409.62 | 408.51 | 407.41 | 406.31 | 405.21 | 404.12 | 403.03 | 401.94 | -110 |
| -100 | 424.30 | 423.15 | 422.01 | 420.86 | 419.72 | 418.59 | 417.46 | 416.33 | 415.20 | 414.08 | 412.96 | -100 |
| -90 | 435.96 | 434.78 | 433.60 | 432.43 | 431.26 | 430.09 | 428.92 | 427.76 | 426.61 | 425.45 | 424.30 | -90 |
| -80 | 447.94 | 446.73 | 445.52 | 444.31 | 443.11 | 441.91 | 440.71 | 439.52 | 438.33 | 437.14 | 435.96 | -80 |
| -70 | 460.24 | 459.00 | 457.75 | 456.52 | 455.28 | 454.05 | 452.82 | 451.60 | 450.37 | 449.16 | 447.94 | -70 |
| -60 | 472.86 | 471.58 | 470.31 | 469.04 | 467.77 | 466.51 | 465.25 | 463.99 | 462.74 | 461.49 | 460.24 | -60 |
| -50 | 485.80 | 484.49 | 483.19 | 481.89 | 480.59 | 479.29 | 478.00 | 476.71 | 475.42 | 474.14 | 472.86 | -50 |
| -40 | 499.06 | 497.72 | 496.39 | 495.05 | 493.72 | 492.39 | 491.07 | 489.75 | 488.43 | 487.11 | 485.80 | -40 |
| -30 | 512.65 | 511.27 | 509.90 | 508.54 | 507.17 | 505.81 | 504.46 | 503.10 | 501.75 | 500.41 | 499.06 | -30 |
| -20 | 526.55 | 525.14 | 523.74 | 522.34 | 520.95 | 519.56 | 518.17 | 516.78 | 515.40 | 514.02 | 512.65 | -20 |
| -10 | 540.77 | 539.33 | 537.90 | 536.47 | 535.04 | 533.62 | 532.20 | 530.78 | 529.37 | 527.96 | 526.55 | -10 |
| 0 | 555.31 | 553.84 | 552.38 | 550.92 | 549.46 | 548.00 | 546.55 | 545.10 | 543.65 | 542.21 | 540.77 | 0 |
| 0 | 555.31 | 556.78 | 558.26 | 559.74 | 561.22 | 562.70 | 564.19 | 565.68 | 567.18 | 568.67 | 570.17 | 0 |
| 10 | 570.17 | 571.68 | 573.19 | 574.70 | 576.21 | 577.73 | 579.25 | 580.77 | 582.30 | 583.82 | 585.36 | 10 |
| 20 | 585.36 | 586.89 | 588.43 | 589.97 | 591.52 | 593.07 | 594.62 | 596.18 | 597.73 | 599.30 | 600.86 | 20 |
| 30 | 600.86 | 602.43 | 604.00 | 605.54 | 607.09 | 608.64 | 610.19 | 611.74 | 613.30 | 614.85 | 616.41 | 30 |
| 40 | 616.41 | 617.98 | 619.54 | 621.11 | 622.67 | 624.24 | 625.82 | 627.39 | 628.97 | 630.55 | 632.13 | 40 |
| 50 | 632.13 | 633.71 | 635.30 | 636.89 | 638.48 | 640.07 | 641.66 | 643.26 | 644.86 | 646.46 | 648.06 | 50 |
| 60 | 648.06 | 649.67 | 651.28 | 652.89 | 654.50 | 656.11 | 657.73 | 659.35 | 660.97 | 662.59 | 664.22 | 60 |
| 70 | 664.22 | 665.85 | 667.48 | 669.11 | 670.74 | 672.38 | 674.02 | 675.66 | 677.30 | 678.95 | 680.59 | 70 |
| 80 | 680.59 | 682.24 | 683.89 | 685.55 | 687.20 | 688.86 | 690.52 | 692.19 | 693.85 | 695.52 | 697.19 | 80 |
| 90 | 697.19 | 698.86 | 700.53 | 702.21 | 703.89 | 705.57 | 707.25 | 708.93 | 710.62 | 712.31 | 714.00 | 90 |
| 100 | 714.00 | 715.69 | 717.39 | 719.09 | 720.79 | 722.49 | 724.19 | 725.90 | 727.61 | 729.32 | 731.03 | 100 |
| 110 | 731.03 | 732.75 | 734.47 | 736.19 | 737.91 | 739.63 | 741.36 | 743.09 | 744.82 | 746.55 | 748.29 | 110 |
| 120 | 748.29 | 750.03 | 751.77 | 753.51 | 755.25 | 757.00 | 758.75 | 760.50 | 762.25 | 764.00 | 765.76 | 120 |
| 130 | 765.76 | 767.52 | 769.28 | 771.05 | 772.81 | 774.58 | 776.35 | 778.12 | 779.90 | 781.67 | 783.45 | 130 |
| 140 | 783.45 | 785.24 | 787.02 | 788.80 | 790.59 | 792.38 | 794.18 | 795.97 | 797.77 | 799.57 | 801.37 | 140 |

# Table 12-5 Continued

| °F | 0 | 1 | 2 | 3 | 4 | 5 | 6 | 7 | 8 | 9 | 10 | °F |
|---|---|---|---|---|---|---|---|---|---|---|---|---|
| | | | | | Resistance in Ohms | | | | | | | |
| 150 | 801.37 | 803.17 | 804.98 | 806.78 | 808.59 | 810.41 | 812.22 | 814.04 | 815.86 | 817.68 | 819.50 | 150 |
| 160 | 819.50 | 821.32 | 823.15 | 824.98 | 826.81 | 828.65 | 830.48 | 832.32 | 834.16 | 836.01 | 837.85 | 160 |
| 170 | 837.85 | 839.70 | 841.55 | 843.40 | 845.25 | 847.11 | 848.97 | 850.83 | 852.69 | 854.56 | 856.42 | 170 |
| 180 | 856.42 | 858.29 | 860.16 | 862.04 | 863.91 | 865.79 | 867.67 | 869.55 | 871.44 | 873.33 | 875.21 | 180 |
| 190 | 875.21 | 877.11 | 879.00 | 880.90 | 882.79 | 884.69 | 886.60 | 888.50 | 890.41 | 892.32 | 894.23 | 190 |
| 200 | 894.23 | 896.14 | 898.05 | 899.97 | 901.89 | 903.81 | 905.74 | 907.67 | 909.59 | 911.52 | 913.46 | 200 |
| 210 | 913.46 | 915.39 | 917.33 | 919.27 | 921.21 | 923.16 | 925.10 | 927.05 | 929.00 | 930.95 | 932.91 | 210 |
| 220 | 932.91 | 934.87 | 936.82 | 938.79 | 940.75 | 942.72 | 944.68 | 946.65 | 948.63 | 950.60 | 952.58 | 220 |
| 230 | 952.58 | 954.56 | 956.54 | 958.52 | 960.51 | 962.50 | 964.49 | 966.48 | 968.47 | 970.47 | 972.47 | 230 |
| 240 | 972.47 | 974.47 | 976.47 | 978.48 | 980.49 | 982.50 | 984.51 | 986.52 | 988.54 | 990.56 | 992.58 | 240 |
| 250 | 992.58 | 994.60 | 996.63 | 998.66 | 1000.7 | 1002.7 | 1004.8 | 1006.8 | 1008.8 | 1010.9 | 1012.9 | 250 |
| 260 | 1012.9 | 1015.0 | 1017.0 | 1019.1 | 1021.1 | 1023.2 | 1025.2 | 1027.3 | 1029.3 | 1031.4 | 1033.5 | 260 |
| 270 | 1033.5 | 1035.5 | 1037.6 | 1039.7 | 1041.7 | 1043.8 | 1045.9 | 1048.0 | 1050.1 | 1052.1 | 1054.2 | 270 |
| 280 | 1054.2 | 1056.3 | 1058.4 | 1060.5 | 1062.6 | 1064.7 | 1066.8 | 1068.9 | 1071.0 | 1073.1 | 1075.2 | 280 |
| 290 | 1075.2 | 1077.3 | 1079.4 | 1081.6 | 1083.7 | 1085.8 | 1087.9 | 1090.0 | 1092.2 | 1094.3 | 1096.4 | 290 |
| 300 | 1096.4 | 1098.6 | 1100.7 | 1102.8 | 1105.0 | 1107.1 | 1109.3 | 1111.4 | 1113.6 | 1115.7 | 1117.9 | 300 |
| 310 | 1117.9 | 1120.0 | 1122.2 | 1124.3 | 1126.5 | 1128.7 | 1130.8 | 1133.0 | 1135.2 | 1137.3 | 1139.5 | 310 |
| 320 | 1139.5 | 1141.7 | 1143.9 | 1146.0 | 1148.2 | 1150.4 | 1152.6 | 1154.8 | 1157.0 | 1159.2 | 1161.4 | 320 |
| 330 | 1161.4 | 1163.6 | 1165.8 | 1168.0 | 1170.2 | 1172.4 | 1174.6 | 1176.8 | 1179.0 | 1181.2 | 1183.5 | 330 |
| 340 | 1183.5 | 1185.7 | 1187.9 | 1190.1 | 1192.4 | 1194.6 | 1196.8 | 1199.1 | 1201.3 | 1203.5 | 1205.8 | 340 |
| 350 | 1205.8 | 1208.0 | 1210.3 | 1212.5 | 1214.8 | 1217.0 | 1219.3 | 1221.5 | 1223.8 | 1226.0 | 1228.3 | 350 |
| 360 | 1228.3 | 1230.6 | 1232.8 | 1235.1 | 1237.4 | 1239.6 | 1241.9 | 1244.2 | 1246.5 | 1248.8 | 1251.0 | 360 |
| 370 | 1251.0 | 1253.3 | 1255.6 | 1257.9 | 1260.2 | 1262.5 | 1264.8 | 1267.1 | 1269.4 | 1271.7 | 1274.0 | 370 |
| 380 | 1274.0 | 1276.3 | 1278.6 | 1280.9 | 1283.3 | 1285.6 | 1287.9 | 1290.2 | 1292.5 | 1294.9 | 1297.2 | 380 |
| 390 | 1297.2 | 1299.5 | 1301.9 | 1304.2 | 1306.5 | 1308.9 | 1311.2 | 1313.6 | 1315.9 | 1318.3 | 1320.6 | 390 |
| 400 | 1320.6 | | | | | | | | | | | 400 |

# 12.4 THERMOCOUPLE MILLIVOLT TABLES[151]

## Table 12-6. Type J in °C

ITS-90 Table for Type J Thermocouple (Ref Junction 0°C)

| °C | 0 | -1 | -2 | -3 | -4 | -5 | -6 | -7 | -8 | -9 | -10 |
|---|---|---|---|---|---|---|---|---|---|---|---|
| | | | | | | | | | | | |

**Thermoelectric Voltage in mV**

| °C | 0 | -1 | -2 | -3 | -4 | -5 | -6 | -7 | -8 | -9 | -10 |
|---|---|---|---|---|---|---|---|---|---|---|---|
| 350 | 19.090 | 19.146 | 19.201 | 19.256 | 19.311 | 19.366 | 19.422 | 19.477 | 19.532 | 19.587 | 19.642 |
| 360 | 19.642 | 19.697 | 19.753 | 19.808 | 19.863 | 19.918 | 19.973 | 20.028 | 20.083 | 20.139 | 20.194 |
| 370 | 20.194 | 20.249 | 20.304 | 20.359 | 20.414 | 20.469 | 20.525 | 20.580 | 20.635 | 20.690 | 20.745 |
| 380 | 20.745 | 20.800 | 20.855 | 20.911 | 20.966 | 21.021 | 21.076 | 21.131 | 21.186 | 21.241 | 21.297 |
| 390 | 21.297 | 21.352 | 21.407 | 21.462 | 21.517 | 21.572 | 21.627 | 21.683 | 21.738 | 21.793 | 21.848 |
| 400 | 21.848 | 21.903 | 21.958 | 22.014 | 22.069 | 22.124 | 22.179 | 22.234 | 22.289 | 22.345 | 22.400 |
| 410 | 22.400 | 22.455 | 22.510 | 22.565 | 22.620 | 22.676 | 22.731 | 22.786 | 22.841 | 22.896 | 22.952 |
| 420 | 22.952 | 23.007 | 23.062 | 23.117 | 23.172 | 23.228 | 23.283 | 23.338 | 23.393 | 23.449 | 23.504 |
| 430 | 23.504 | 23.559 | 23.614 | 23.670 | 23.725 | 23.780 | 23.835 | 23.891 | 23.946 | 24.001 | 24.057 |
| 440 | 24.057 | 24.112 | 24.167 | 24.223 | 24.278 | 24.333 | 24.389 | 24.444 | 24.499 | 24.555 | 24.610 |
| 450 | 24.610 | 24.665 | 24.721 | 24.776 | 24.832 | 24.887 | 24.943 | 24.998 | 25.053 | 25.109 | 25.164 |
| 460 | 25.164 | 25.220 | 25.275 | 25.331 | 25.386 | 25.442 | 25.497 | 25.553 | 25.608 | 25.664 | 25.720 |
| 470 | 25.720 | 25.775 | 25.831 | 25.886 | 25.942 | 25.998 | 26.053 | 26.109 | 26.165 | 26.220 | 26.276 |
| 480 | 26.276 | 26.332 | 26.387 | 26.443 | 26.499 | 26.555 | 26.610 | 26.666 | 26.722 | 26.778 | 26.834 |
| 490 | 26.834 | 26.889 | 26.945 | 27.001 | 27.057 | 27.113 | 27.169 | 27.225 | 27.281 | 27.337 | 27.393 |
| 500 | 27.393 | 27.449 | 27.505 | 27.561 | 27.617 | 27.673 | 27.729 | 27.785 | 27.841 | 27.897 | 27.953 |
| 510 | 27.953 | 28.010 | 28.066 | 28.122 | 28.178 | 28.234 | 28.291 | 28.347 | 28.403 | 28.460 | 28.516 |
| 520 | 28.516 | 28.572 | 28.629 | 28.685 | 28.741 | 28.798 | 28.854 | 28.911 | 28.967 | 29.024 | 29.080 |
| 530 | 29.080 | 29.137 | 29.194 | 29.250 | 29.307 | 29.363 | 29.420 | 29.477 | 29.534 | 29.590 | 29.647 |
| 540 | 29.647 | 29.704 | 29.761 | 29.818 | 29.874 | 29.931 | 29.988 | 30.045 | 30.102 | 30.159 | 30.216 |
| 550 | 30.216 | 30.273 | 30.330 | 30.387 | 30.444 | 30.502 | 30.559 | 30.616 | 30.673 | 30.730 | 30.788 |
| 560 | 30.788 | 30.845 | 30.902 | 30.960 | 31.017 | 31.074 | 31.132 | 31.189 | 31.247 | 31.304 | 31.362 |
| 570 | 31.362 | 31.419 | 31.477 | 31.535 | 31.592 | 31.650 | 31.708 | 31.766 | 31.823 | 31.881 | 31.939 |
| 580 | 31.939 | 31.997 | 32.055 | 32.113 | 32.171 | 32.229 | 32.287 | 32.345 | 32.403 | 32.461 | 32.519 |
| 590 | 32.519 | 32.577 | 32.636 | 32.694 | 32.752 | 32.810 | 32.869 | 32.927 | 32.985 | 33.044 | 33.102 |
| 600 | 33.102 | 33.161 | 33.219 | 33.278 | 33.337 | 33.395 | 33.454 | 33.513 | 33.571 | 33.630 | 33.689 |
| 610 | 33.689 | 33.748 | 33.807 | 33.866 | 33.925 | 33.984 | 34.043 | 34.102 | 34.161 | 34.220 | 34.279 |
| 620 | 34.279 | 34.338 | 34.397 | 34.457 | 34.516 | 34.575 | 34.635 | 34.694 | 34.754 | 34.813 | 34.873 |
| 630 | 34.873 | 34.932 | 34.992 | 35.051 | 35.111 | 35.171 | 35.230 | 35.290 | 35.350 | 35.410 | 35.470 |
| 640 | 35.470 | 35.530 | 35.590 | 35.650 | 35.710 | 35.770 | 35.830 | 35.890 | 35.950 | 36.010 | 36.071 |
| 650 | 36.071 | 36.131 | 36.191 | 36.252 | 36.312 | 36.373 | 36.433 | 36.494 | 36.554 | 36.615 | 36.675 |
| 660 | 36.675 | 36.736 | 36.797 | 36.858 | 36.918 | 36.979 | 37.040 | 37.101 | 37.162 | 37.223 | 37.284 |
| 670 | 37.284 | 37.345 | 37.406 | 37.467 | 37.528 | 37.590 | 37.651 | 37.712 | 37.773 | 37.835 | 37.896 |
| 680 | 37.896 | 37.958 | 38.019 | 38.081 | 38.142 | 38.204 | 38.265 | 38.327 | 38.389 | 38.450 | 38.512 |
| 690 | 38.512 | 38.574 | 38.636 | 38.698 | 38.760 | 38.822 | 38.884 | 38.946 | 39.008 | 39.070 | 39.132 |
| 700 | 39.132 | 39.194 | 39.256 | 39.318 | 39.381 | 39.443 | 39.505 | 39.568 | 39.630 | 39.693 | 39.755 |
| 710 | 39.755 | 39.818 | 39.880 | 39.943 | 40.005 | 40.068 | 40.131 | 40.193 | 40.256 | 40.319 | 40.382 |
| 720 | 40.382 | 40.445 | 40.508 | 40.570 | 40.633 | 40.696 | 40.759 | 40.822 | 40.886 | 40.949 | 41.012 |
| 730 | 41.012 | 41.075 | 41.138 | 41.201 | 41.265 | 41.328 | 41.391 | 41.455 | 41.518 | 41.581 | 41.645 |
| 740 | 41.645 | 41.708 | 41.772 | 41.835 | 41.899 | 41.962 | 42.026 | 42.090 | 42.153 | 42.217 | 42.281 |

---

151. Reference www.instrumentation-central.com.

# Table 12-6 Continued

ITS-90 Table for Type J Thermocouple (Ref Junction 0°C)

| °C | 0 | -1 | -2 | -3 | -4 | -5 | -6 | -7 | -8 | -9 | -10 |
|-----|---|----|----|----|----|----|----|----|----|----|-----|

Thermoelectric Voltage in mV

| °C | 0 | -1 | -2 | -3 | -4 | -5 | -6 | -7 | -8 | -9 | -10 |
|-----|--------|--------|--------|--------|--------|--------|--------|--------|--------|--------|--------|
| 350 | 19.090 | 19.146 | 19.201 | 19.256 | 19.311 | 19.366 | 19.422 | 19.477 | 19.532 | 19.587 | 19.642 |
| 360 | 19.642 | 19.697 | 19.753 | 19.808 | 19.863 | 19.918 | 19.973 | 20.028 | 20.083 | 20.139 | 20.194 |
| 370 | 20.194 | 20.249 | 20.304 | 20.359 | 20.414 | 20.469 | 20.525 | 20.580 | 20.635 | 20.690 | 20.745 |
| 380 | 20.745 | 20.800 | 20.855 | 20.911 | 20.966 | 21.021 | 21.076 | 21.131 | 21.186 | 21.241 | 21.297 |
| 390 | 21.297 | 21.352 | 21.407 | 21.462 | 21.517 | 21.572 | 21.627 | 21.683 | 21.738 | 21.793 | 21.848 |
| 400 | 21.848 | 21.903 | 21.958 | 22.014 | 22.069 | 22.124 | 22.179 | 22.234 | 22.289 | 22.345 | 22.400 |
| 410 | 22.400 | 22.455 | 22.510 | 22.565 | 22.620 | 22.676 | 22.731 | 22.786 | 22.841 | 22.896 | 22.952 |
| 420 | 22.952 | 23.007 | 23.062 | 23.117 | 23.172 | 23.228 | 23.283 | 23.338 | 23.393 | 23.449 | 23.504 |
| 430 | 23.504 | 23.559 | 23.614 | 23.670 | 23.725 | 23.780 | 23.835 | 23.891 | 23.946 | 24.001 | 24.057 |
| 440 | 24.057 | 24.112 | 24.167 | 24.223 | 24.278 | 24.333 | 24.389 | 24.444 | 24.499 | 24.555 | 24.610 |
| 450 | 24.610 | 24.665 | 24.721 | 24.776 | 24.832 | 24.887 | 24.943 | 24.998 | 25.053 | 25.109 | 25.164 |
| 460 | 25.164 | 25.220 | 25.275 | 25.331 | 25.386 | 25.442 | 25.497 | 25.553 | 25.608 | 25.664 | 25.720 |
| 470 | 25.720 | 25.775 | 25.831 | 25.886 | 25.942 | 25.998 | 26.053 | 26.109 | 26.165 | 26.220 | 26.276 |
| 480 | 26.276 | 26.332 | 26.387 | 26.443 | 26.499 | 26.555 | 26.610 | 26.666 | 26.722 | 26.778 | 26.834 |
| 490 | 26.834 | 26.889 | 26.945 | 27.001 | 27.057 | 27.113 | 27.169 | 27.225 | 27.281 | 27.337 | 27.393 |
| 500 | 27.393 | 27.449 | 27.505 | 27.561 | 27.617 | 27.673 | 27.729 | 27.785 | 27.841 | 27.897 | 27.953 |
| 510 | 27.953 | 28.010 | 28.066 | 28.122 | 28.178 | 28.234 | 28.291 | 28.347 | 28.403 | 28.460 | 28.516 |
| 520 | 28.516 | 28.572 | 28.629 | 28.685 | 28.741 | 28.798 | 28.854 | 28.911 | 28.967 | 29.024 | 29.080 |
| 530 | 29.080 | 29.137 | 29.194 | 29.250 | 29.307 | 29.363 | 29.420 | 29.477 | 29.534 | 29.590 | 29.647 |
| 540 | 29.647 | 29.704 | 29.761 | 29.818 | 29.874 | 29.931 | 29.988 | 30.045 | 30.102 | 30.159 | 30.216 |
| 550 | 30.216 | 30.273 | 30.330 | 30.387 | 30.444 | 30.502 | 30.559 | 30.616 | 30.673 | 30.730 | 30.788 |
| 560 | 30.788 | 30.845 | 30.902 | 30.960 | 31.017 | 31.074 | 31.132 | 31.189 | 31.247 | 31.304 | 31.362 |
| 570 | 31.362 | 31.419 | 31.477 | 31.535 | 31.592 | 31.650 | 31.708 | 31.766 | 31.823 | 31.881 | 31.939 |
| 580 | 31.939 | 31.997 | 32.055 | 32.113 | 32.171 | 32.229 | 32.287 | 32.345 | 32.403 | 32.461 | 32.519 |
| 590 | 32.519 | 32.577 | 32.636 | 32.694 | 32.752 | 32.810 | 32.869 | 32.927 | 32.985 | 33.044 | 33.102 |
| 600 | 33.102 | 33.161 | 33.219 | 33.278 | 33.337 | 33.395 | 33.454 | 33.513 | 33.571 | 33.630 | 33.689 |
| 610 | 33.689 | 33.748 | 33.807 | 33.866 | 33.925 | 33.984 | 34.043 | 34.102 | 34.161 | 34.220 | 34.279 |
| 620 | 34.279 | 34.338 | 34.397 | 34.457 | 34.516 | 34.575 | 34.635 | 34.694 | 34.754 | 34.813 | 34.873 |
| 630 | 34.873 | 34.932 | 34.992 | 35.051 | 35.111 | 35.171 | 35.230 | 35.290 | 35.350 | 35.410 | 35.470 |
| 640 | 35.470 | 35.530 | 35.590 | 35.650 | 35.710 | 35.770 | 35.830 | 35.890 | 35.950 | 36.010 | 36.071 |
| 650 | 36.071 | 36.131 | 36.191 | 36.252 | 36.312 | 36.373 | 36.433 | 36.494 | 36.554 | 36.615 | 36.675 |
| 660 | 36.675 | 36.736 | 36.797 | 36.858 | 36.918 | 36.979 | 37.040 | 37.101 | 37.162 | 37.223 | 37.284 |
| 670 | 37.284 | 37.345 | 37.406 | 37.467 | 37.528 | 37.590 | 37.651 | 37.712 | 37.773 | 37.835 | 37.896 |
| 680 | 37.896 | 37.958 | 38.019 | 38.081 | 38.142 | 38.204 | 38.265 | 38.327 | 38.389 | 38.450 | 38.512 |
| 690 | 38.512 | 38.574 | 38.636 | 38.698 | 38.760 | 38.822 | 38.884 | 38.946 | 39.008 | 39.070 | 39.132 |
| 700 | 39.132 | 39.194 | 39.256 | 39.318 | 39.381 | 39.443 | 39.505 | 39.568 | 39.630 | 39.693 | 39.755 |
| 710 | 39.755 | 39.818 | 39.880 | 39.943 | 40.005 | 40.068 | 40.131 | 40.193 | 40.256 | 40.319 | 40.382 |
| 720 | 40.382 | 40.445 | 40.508 | 40.570 | 40.633 | 40.696 | 40.759 | 40.822 | 40.886 | 40.949 | 41.012 |
| 730 | 41.012 | 41.075 | 41.138 | 41.201 | 41.265 | 41.328 | 41.391 | 41.455 | 41.518 | 41.581 | 41.645 |
| 740 | 41.645 | 41.708 | 41.772 | 41.835 | 41.899 | 41.962 | 42.026 | 42.090 | 42.153 | 42.217 | 42.281 |

# Table 12-6 Continued

ITS-90 Table for Type J Thermocouple (Ref Junction 0°C)

| °C | 0 | -1 | -2 | -3 | -4 | -5 | -6 | -7 | -8 | -9 | -10 |
|---|---|---|---|---|---|---|---|---|---|---|---|
| | | | | | Thermoelectric Voltage in mV | | | | | | |
| 750 | 42.281 | 42.344 | 42.408 | 42.472 | 42.536 | 42.599 | 42.663 | 42.727 | 42.791 | 42.855 | 42.919 |
| 760 | 42.919 | 42.983 | 43.047 | 43.111 | 43.175 | 43.239 | 43.303 | 43.367 | 43.431 | 43.495 | 43.559 |
| 770 | 43.559 | 43.624 | 43.688 | 43.752 | 43.817 | 43.881 | 43.945 | 44.010 | 44.074 | 44.139 | 44.203 |
| 780 | 44.203 | 44.267 | 44.332 | 44.396 | 44.461 | 44.525 | 44.590 | 44.655 | 44.719 | 44.784 | 44.848 |
| 790 | 44.848 | 44.913 | 44.977 | 45.042 | 45.107 | 45.171 | 45.236 | 45.301 | 45.365 | 45.430 | 45.494 |
| 800 | 45.494 | 45.559 | 45.624 | 45.688 | 45.753 | 45.818 | 45.882 | 45.947 | 46.011 | 46.076 | 46.141 |
| 810 | 46.141 | 46.205 | 46.270 | 46.334 | 46.399 | 46.464 | 46.528 | 46.593 | 46.657 | 46.722 | 46.786 |
| 820 | 46.786 | 46.851 | 46.915 | 46.980 | 47.044 | 47.109 | 47.173 | 47.238 | 47.302 | 47.367 | 47.431 |
| 830 | 47.431 | 47.495 | 47.560 | 47.624 | 47.688 | 47.753 | 47.817 | 47.881 | 47.946 | 48.010 | 48.074 |
| 840 | 48.074 | 48.138 | 48.202 | 48.267 | 48.331 | 48.395 | 48.459 | 48.523 | 48.587 | 48.651 | 48.715 |
| 850 | 48.715 | 48.779 | 48.843 | 48.907 | 48.971 | 49.034 | 49.098 | 49.162 | 49.226 | 49.290 | 49.353 |
| 860 | 49.353 | 49.417 | 49.481 | 49.544 | 49.608 | 49.672 | 49.735 | 49.799 | 49.862 | 49.926 | 49.989 |
| 870 | 49.989 | 50.052 | 50.116 | 50.179 | 50.243 | 50.306 | 50.369 | 50.432 | 50.495 | 50.559 | 50.622 |
| 880 | 50.622 | 50.685 | 50.748 | 50.811 | 50.874 | 50.937 | 51.000 | 51.063 | 51.126 | 51.188 | 51.251 |
| 890 | 51.251 | 51.314 | 51.377 | 51.439 | 51.502 | 51.565 | 51.627 | 51.690 | 51.752 | 51.815 | 51.877 |
| 900 | 51.877 | 51.940 | 52.002 | 52.064 | 52.127 | 52.189 | 52.251 | 52.314 | 52.376 | 52.438 | 52.500 |
| 910 | 52.500 | 52.562 | 52.624 | 52.686 | 52.748 | 52.810 | 52.872 | 52.934 | 52.996 | 53.057 | 53.119 |
| 920 | 53.119 | 53.181 | 53.243 | 53.304 | 53.366 | 53.427 | 53.489 | 53.550 | 53.612 | 53.673 | 53.735 |
| 930 | 53.735 | 53.796 | 53.857 | 53.919 | 53.980 | 54.041 | 54.102 | 54.164 | 54.225 | 54.286 | 54.347 |
| 940 | 54.347 | 54.408 | 54.469 | 54.530 | 54.591 | 54.652 | 54.713 | 54.773 | 54.834 | 54.895 | 54.956 |
| 950 | 54.956 | 55.016 | 55.077 | 55.138 | 55.198 | 55.259 | 55.319 | 55.380 | 55.440 | 55.501 | 55.561 |
| 960 | 55.561 | 55.622 | 55.682 | 55.742 | 55.803 | 55.863 | 55.923 | 55.983 | 56.043 | 56.104 | 56.164 |
| 970 | 56.164 | 56.224 | 56.284 | 56.344 | 56.404 | 56.464 | 56.524 | 56.584 | 56.643 | 56.703 | 56.763 |
| 980 | 56.763 | 56.823 | 56.883 | 56.942 | 57.002 | 57.062 | 57.121 | 57.181 | 57.240 | 57.300 | 57.360 |
| 990 | 57.360 | 57.419 | 57.479 | 57.538 | 57.597 | 57.657 | 57.716 | 57.776 | 57.835 | 57.894 | 57.953 |
| 1000 | 57.953 | 58.013 | 58.072 | 58.131 | 58.190 | 58.249 | 58.309 | 58.368 | 58.427 | 58.486 | 58.545 |
| 1010 | 58.545 | 58.604 | 58.663 | 58.722 | 58.781 | 58.840 | 58.899 | 58.957 | 59.016 | 59.075 | 59.134 |
| 1020 | 59.134 | 59.193 | 59.252 | 59.310 | 59.369 | 59.428 | 59.487 | 59.545 | 59.604 | 59.663 | 59.721 |
| 1030 | 59.721 | 59.780 | 59.838 | 59.897 | 59.956 | 60.014 | 60.073 | 60.131 | 60.190 | 60.248 | 60.307 |
| 1040 | 60.307 | 60.365 | 60.423 | 60.482 | 60.540 | 60.599 | 60.657 | 60.715 | 60.774 | 60.832 | 60.890 |
| 1050 | 60.890 | 60.949 | 61.007 | 61.065 | 61.123 | 61.182 | 61.240 | 61.298 | 61.356 | 61.415 | 61.473 |
| 1060 | 61.473 | 61.531 | 61.589 | 61.647 | 61.705 | 61.763 | 61.822 | 61.880 | 61.938 | 61.996 | 62.054 |
| 1070 | 62.054 | 62.112 | 62.170 | 62.228 | 62.286 | 62.344 | 62.402 | 62.460 | 62.518 | 62.576 | 62.634 |
| 1080 | 62.634 | 62.692 | 62.750 | 62.808 | 62.866 | 62.924 | 62.982 | 63.040 | 63.098 | 63.156 | 63.214 |
| 1090 | 63.214 | 63.271 | 63.329 | 63.387 | 63.445 | 63.503 | 63.561 | 63.619 | 63.677 | 63.734 | 63.792 |
| 1100 | 63.792 | 63.850 | 63.908 | 63.966 | 64.024 | 64.081 | 64.139 | 64.197 | 64.255 | 64.313 | 64.370 |
| 1110 | 64.370 | 64.428 | 64.486 | 64.544 | 64.602 | 64.659 | 64.717 | 64.775 | 64.833 | 64.890 | 64.948 |
| 1120 | 64.948 | 65.006 | 65.064 | 65.121 | 65.179 | 65.237 | 65.295 | 65.352 | 65.410 | 65.468 | 65.525 |
| 1130 | 65.525 | 65.583 | 65.641 | 65.699 | 65.756 | 65.814 | 65.872 | 65.929 | 65.987 | 66.045 | 66.102 |
| 1140 | 66.102 | 66.160 | 66.218 | 66.275 | 66.333 | 66.391 | 66.448 | 66.506 | 66.564 | 66.621 | 66.679 |
| 1150 | 66.679 | 66.737 | 66.794 | 66.852 | 66.910 | 66.967 | 67.025 | 67.082 | 67.140 | 67.198 | 67.255 |
| 1160 | 67.255 | 67.313 | 67.370 | 67.428 | 67.486 | 67.543 | 67.601 | 67.658 | 67.716 | 67.773 | 67.831 |
| 1170 | 67.831 | 67.888 | 67.946 | 68.003 | 68.061 | 68.119 | 68.176 | 68.234 | 68.291 | 68.348 | 68.406 |
| 1180 | 68.406 | 68.463 | 68.521 | 68.578 | 68.636 | 68.693 | 68.751 | 68.808 | 68.865 | 68.923 | 68.980 |
| 1190 | 68.980 | 69.037 | 69.095 | 69.152 | 69.209 | 69.267 | 69.324 | 69.381 | 69.439 | 69.496 | 69.553 |

# Table 12-7. Type K in °C

ITS-90 Table for Type K Thermocouple (Ref Junction 0°C)

| °C | 0 | 1 | 2 | 3 | 4 | 5 | 6 | 7 | 8 | 9 | 10 |
|---|---|---|---|---|---|---|---|---|---|---|---|
| | | | | | | Thermoelectric Voltage in mV | | | | | |
| 0 | 0.000 | 0.039 | 0.079 | 0.119 | 0.158 | 0.198 | 0.238 | 0.277 | 0.317 | 0.357 | 0.397 |
| 10 | 0.397 | 0.437 | 0.477 | 0.517 | 0.557 | 0.597 | 0.637 | 0.677 | 0.718 | 0.758 | 0.798 |
| 20 | 0.798 | 0.838 | 0.879 | 0.919 | 0.960 | 1.000 | 1.041 | 1.081 | 1.122 | 1.163 | 1.203 |
| 30 | 1.203 | 1.244 | 1.285 | 1.326 | 1.366 | 1.407 | 1.448 | 1.489 | 1.530 | 1.571 | 1.612 |
| 40 | 1.612 | 1.653 | 1.694 | 1.735 | 1.776 | 1.817 | 1.858 | 1.899 | 1.941 | 1.982 | 2.023 |
| 50 | 2.023 | 2.064 | 2.106 | 2.147 | 2.188 | 2.230 | 2.271 | 2.312 | 2.354 | 2.395 | 2.436 |
| 60 | 2.436 | 2.478 | 2.519 | 2.561 | 2.602 | 2.644 | 2.685 | 2.727 | 2.768 | 2.810 | 2.851 |
| 70 | 2.851 | 2.893 | 2.934 | 2.976 | 3.017 | 3.059 | 3.100 | 3.142 | 3.184 | 3.225 | 3.267 |
| 80 | 3.267 | 3.308 | 3.350 | 3.391 | 3.433 | 3.474 | 3.516 | 3.557 | 3.599 | 3.640 | 3.682 |
| 90 | 3.682 | 3.723 | 3.765 | 3.806 | 3.848 | 3.889 | 3.931 | 3.972 | 4.013 | 4.055 | 4.096 |
| 100 | 4.096 | 4.138 | 4.179 | 4.220 | 4.262 | 4.303 | 4.344 | 4.385 | 4.427 | 4.468 | 4.509 |
| 110 | 4.509 | 4.550 | 4.591 | 4.633 | 4.674 | 4.715 | 4.756 | 4.797 | 4.838 | 4.879 | 4.920 |
| 120 | 4.920 | 4.961 | 5.002 | 5.043 | 5.084 | 5.124 | 5.165 | 5.206 | 5.247 | 5.288 | 5.328 |
| 130 | 5.328 | 5.369 | 5.410 | 5.450 | 5.491 | 5.532 | 5.572 | 5.613 | 5.653 | 5.694 | 5.735 |
| 140 | 5.735 | 5.775 | 5.815 | 5.856 | 5.896 | 5.937 | 5.977 | 6.017 | 6.058 | 6.098 | 6.138 |
| 150 | 6.138 | 6.179 | 6.219 | 6.259 | 6.299 | 6.339 | 6.380 | 6.420 | 6.460 | 6.500 | 6.540 |
| 160 | 6.540 | 6.580 | 6.620 | 6.660 | 6.701 | 6.741 | 6.781 | 6.821 | 6.861 | 6.901 | 6.941 |
| 170 | 6.941 | 6.981 | 7.021 | 7.060 | 7.100 | 7.140 | 7.180 | 7.220 | 7.260 | 7.300 | 7.340 |
| 180 | 7.340 | 7.380 | 7.420 | 7.460 | 7.500 | 7.540 | 7.579 | 7.619 | 7.659 | 7.699 | 7.739 |
| 190 | 7.739 | 7.779 | 7.819 | 7.859 | 7.899 | 7.939 | 7.979 | 8.019 | 8.059 | 8.099 | 8.138 |
| 200 | 8.138 | 8.178 | 8.218 | 8.258 | 8.298 | 8.338 | 8.378 | 8.418 | 8.458 | 8.499 | 8.539 |
| 210 | 8.539 | 8.579 | 8.619 | 8.659 | 8.699 | 8.739 | 8.779 | 8.819 | 8.860 | 8.900 | 8.940 |
| 220 | 8.940 | 8.980 | 9.020 | 9.061 | 9.101 | 9.141 | 9.181 | 9.222 | 9.262 | 9.302 | 9.343 |
| 230 | 9.343 | 9.383 | 9.423 | 9.464 | 9.504 | 9.545 | 9.585 | 9.626 | 9.666 | 9.707 | 9.747 |
| 240 | 9.747 | 9.788 | 9.828 | 9.869 | 9.909 | 9.950 | 9.991 | 10.031 | 10.072 | 10.113 | 10.153 |
| 250 | 10.153 | 10.194 | 10.235 | 10.276 | 10.316 | 10.357 | 10.398 | 10.439 | 10.480 | 10.520 | 10.561 |
| 260 | 10.561 | 10.602 | 10.643 | 10.684 | 10.725 | 10.766 | 10.807 | 10.848 | 10.889 | 10.930 | 10.971 |
| 270 | 10.971 | 11.012 | 11.053 | 11.094 | 11.135 | 11.176 | 11.217 | 11.259 | 11.300 | 11.341 | 11.382 |
| 280 | 11.382 | 11.423 | 11.465 | 11.506 | 11.547 | 11.588 | 11.630 | 11.671 | 11.712 | 11.753 | 11.795 |
| 290 | 11.795 | 11.836 | 11.877 | 11.919 | 11.960 | 12.001 | 12.043 | 12.084 | 12.126 | 12.167 | 12.209 |
| 300 | 12.209 | 12.250 | 12.291 | 12.333 | 12.374 | 12.416 | 12.457 | 12.499 | 12.540 | 12.582 | 12.624 |
| 310 | 12.624 | 12.665 | 12.707 | 12.748 | 12.790 | 12.831 | 12.873 | 12.915 | 12.956 | 12.998 | 13.040 |
| 320 | 13.040 | 13.081 | 13.123 | 13.165 | 13.206 | 13.248 | 13.290 | 13.331 | 13.373 | 13.415 | 13.457 |
| 330 | 13.457 | 13.498 | 13.540 | 13.582 | 13.624 | 13.665 | 13.707 | 13.749 | 13.791 | 13.833 | 13.874 |
| 340 | 13.874 | 13.916 | 13.958 | 14.000 | 14.042 | 14.084 | 14.126 | 14.167 | 14.209 | 14.251 | 14.293 |
| 350 | 14.293 | 14.335 | 14.377 | 14.419 | 14.461 | 14.503 | 14.545 | 14.587 | 14.629 | 14.671 | 14.713 |
| 360 | 14.713 | 14.755 | 14.797 | 14.839 | 14.881 | 14.923 | 14.965 | 15.007 | 15.049 | 15.091 | 15.133 |
| 370 | 15.133 | 15.175 | 15.217 | 15.259 | 15.301 | 15.343 | 15.385 | 15.427 | 15.469 | 15.511 | 15.554 |
| 380 | 15.554 | 15.596 | 15.638 | 15.680 | 15.722 | 15.764 | 15.806 | 15.849 | 15.891 | 15.933 | 15.975 |
| 390 | 15.975 | 16.017 | 16.059 | 16.102 | 16.144 | 16.186 | 16.228 | 16.270 | 16.313 | 16.355 | 16.397 |
| 400 | 16.397 | 16.439 | 16.482 | 16.524 | 16.566 | 16.608 | 16.651 | 16.693 | 16.735 | 16.778 | 16.820 |
| 410 | 16.820 | 16.862 | 16.904 | 16.947 | 16.989 | 17.031 | 17.074 | 17.116 | 17.158 | 17.201 | 17.243 |
| 420 | 17.243 | 17.285 | 17.328 | 17.370 | 17.413 | 17.455 | 17.497 | 17.540 | 17.582 | 17.624 | 17.667 |
| 430 | 17.667 | 17.709 | 17.752 | 17.794 | 17.837 | 17.879 | 17.921 | 17.964 | 18.006 | 18.049 | 18.091 |
| 440 | 18.091 | 18.134 | 18.176 | 18.218 | 18.261 | 18.303 | 18.346 | 18.388 | 18.431 | 18.473 | 18.516 |

# Table 12-7 Continued

ITS-90 Table for Type K Thermocouple (Ref Junction 0°C)

| °C | 0 | 1 | 2 | 3 | 4 | 5 | 6 | 7 | 8 | 9 | 10 |
|---|---|---|---|---|---|---|---|---|---|---|---|
| | | | | | Thermoelectric Voltage in mV | | | | | | |
| 450 | 18.516 | 18.558 | 18.601 | 18.643 | 18.686 | 18.728 | 18.771 | 18.813 | 18.856 | 18.898 | 18.941 |
| 460 | 18.941 | 18.983 | 19.026 | 19.068 | 19.111 | 19.154 | 19.196 | 19.239 | 19.281 | 19.324 | 19.366 |
| 470 | 19.366 | 19.409 | 19.451 | 19.494 | 19.537 | 19.579 | 19.622 | 19.664 | 19.707 | 19.750 | 19.792 |
| 480 | 19.792 | 19.835 | 19.877 | 19.920 | 19.962 | 20.005 | 20.048 | 20.090 | 20.133 | 20.175 | 20.218 |
| 490 | 20.218 | 20.261 | 20.303 | 20.346 | 20.389 | 20.431 | 20.474 | 20.516 | 20.559 | 20.602 | 20.644 |
| 500 | 20.644 | 20.687 | 20.730 | 20.772 | 20.815 | 20.857 | 20.900 | 20.943 | 20.985 | 21.028 | 21.071 |
| 510 | 21.071 | 21.113 | 21.156 | 21.199 | 21.241 | 21.284 | 21.326 | 21.369 | 21.412 | 21.454 | 21.497 |
| 520 | 21.497 | 21.540 | 21.582 | 21.625 | 21.668 | 21.710 | 21.753 | 21.796 | 21.838 | 21.881 | 21.924 |
| 530 | 21.924 | 21.966 | 22.009 | 22.052 | 22.094 | 22.137 | 22.179 | 22.222 | 22.265 | 22.307 | 22.350 |
| 540 | 22.350 | 22.393 | 22.435 | 22.478 | 22.521 | 22.563 | 22.606 | 22.649 | 22.691 | 22.734 | 22.776 |
| 550 | 22.776 | 22.819 | 22.862 | 22.904 | 22.947 | 22.990 | 23.032 | 23.075 | 23.117 | 23.160 | 23.203 |
| 560 | 23.203 | 23.245 | 23.288 | 23.331 | 23.373 | 23.416 | 23.458 | 23.501 | 23.544 | 23.586 | 23.629 |
| 570 | 23.629 | 23.671 | 23.714 | 23.757 | 23.799 | 23.842 | 23.884 | 23.927 | 23.970 | 24.012 | 24.055 |
| 580 | 24.055 | 24.097 | 24.140 | 24.182 | 24.225 | 24.267 | 24.310 | 24.353 | 24.395 | 24.438 | 24.480 |
| 590 | 24.480 | 24.523 | 24.565 | 24.608 | 24.650 | 24.693 | 24.735 | 24.778 | 24.820 | 24.863 | 24.905 |
| 600 | 24.905 | 24.948 | 24.990 | 25.033 | 25.075 | 25.118 | 25.160 | 25.203 | 25.245 | 25.288 | 25.330 |
| 610 | 25.330 | 25.373 | 25.415 | 25.458 | 25.500 | 25.543 | 25.585 | 25.627 | 25.670 | 25.712 | 25.755 |
| 620 | 25.755 | 25.797 | 25.840 | 25.882 | 25.924 | 25.967 | 26.009 | 26.052 | 26.094 | 26.136 | 26.179 |
| 630 | 26.179 | 26.221 | 26.263 | 26.306 | 26.348 | 26.390 | 26.433 | 26.475 | 26.517 | 26.560 | 26.602 |
| 640 | 26.602 | 26.644 | 26.687 | 26.729 | 26.771 | 26.814 | 26.856 | 26.898 | 26.940 | 26.983 | 27.025 |
| 650 | 27.025 | 27.067 | 27.109 | 27.152 | 27.194 | 27.236 | 27.278 | 27.320 | 27.363 | 27.405 | 27.447 |
| 660 | 27.447 | 27.489 | 27.531 | 27.574 | 27.616 | 27.658 | 27.700 | 27.742 | 27.784 | 27.826 | 27.869 |
| 670 | 27.869 | 27.911 | 27.953 | 27.995 | 28.037 | 28.079 | 28.121 | 28.163 | 28.205 | 28.247 | 28.289 |
| 680 | 28.289 | 28.332 | 28.374 | 28.416 | 28.458 | 28.500 | 28.542 | 28.584 | 28.626 | 28.668 | 28.710 |
| 690 | 28.710 | 28.752 | 28.794 | 28.835 | 28.877 | 28.919 | 28.961 | 29.003 | 29.045 | 29.087 | 29.129 |
| 700 | 29.129 | 29.171 | 29.213 | 29.255 | 29.297 | 29.338 | 29.380 | 29.422 | 29.464 | 29.506 | 29.548 |
| 710 | 29.548 | 29.589 | 29.631 | 29.673 | 29.715 | 29.757 | 29.798 | 29.840 | 29.882 | 29.924 | 29.965 |
| 720 | 29.965 | 30.007 | 30.049 | 30.090 | 30.132 | 30.174 | 30.216 | 30.257 | 30.299 | 30.341 | 30.382 |
| 730 | 30.382 | 30.424 | 30.466 | 30.507 | 30.549 | 30.590 | 30.632 | 30.674 | 30.715 | 30.757 | 30.798 |
| 740 | 30.798 | 30.840 | 30.881 | 30.923 | 30.964 | 31.006 | 31.047 | 31.089 | 31.130 | 31.172 | 31.213 |
| 750 | 31.213 | 31.255 | 31.296 | 31.338 | 31.379 | 31.421 | 31.462 | 31.504 | 31.545 | 31.586 | 31.628 |
| 760 | 31.628 | 31.669 | 31.710 | 31.752 | 31.793 | 31.834 | 31.876 | 31.917 | 31.958 | 32.000 | 32.041 |
| 770 | 32.041 | 32.082 | 32.124 | 32.165 | 32.206 | 32.247 | 32.289 | 32.330 | 32.371 | 32.412 | 32.453 |
| 780 | 32.453 | 32.495 | 32.536 | 32.577 | 32.618 | 32.659 | 32.700 | 32.742 | 32.783 | 32.824 | 32.865 |
| 790 | 32.865 | 32.906 | 32.947 | 32.988 | 33.029 | 33.070 | 33.111 | 33.152 | 33.193 | 33.234 | 33.275 |
| 800 | 33.275 | 33.316 | 33.357 | 33.398 | 33.439 | 33.480 | 33.521 | 33.562 | 33.603 | 33.644 | 33.685 |
| 810 | 33.685 | 33.726 | 33.767 | 33.808 | 33.848 | 33.889 | 33.930 | 33.971 | 34.012 | 34.053 | 34.093 |
| 820 | 34.093 | 34.134 | 34.175 | 34.216 | 34.257 | 34.297 | 34.338 | 34.379 | 34.420 | 34.460 | 34.501 |
| 830 | 34.501 | 34.542 | 34.582 | 34.623 | 34.664 | 34.704 | 34.745 | 34.786 | 34.826 | 34.867 | 34.908 |
| 840 | 34.908 | 34.948 | 34.989 | 35.029 | 35.070 | 35.110 | 35.151 | 35.192 | 35.232 | 35.273 | 35.313 |
| 850 | 35.313 | 35.354 | 35.394 | 35.435 | 35.475 | 35.516 | 35.556 | 35.596 | 35.637 | 35.677 | 35.718 |
| 860 | 35.718 | 35.758 | 35.798 | 35.839 | 35.879 | 35.920 | 35.960 | 36.000 | 36.041 | 36.081 | 36.121 |
| 870 | 36.121 | 36.162 | 36.202 | 36.242 | 36.282 | 36.323 | 36.363 | 36.403 | 36.443 | 36.484 | 36.524 |
| 880 | 36.524 | 36.564 | 36.604 | 36.644 | 36.685 | 36.725 | 36.765 | 36.805 | 36.845 | 36.885 | 36.925 |
| 890 | 36.925 | 36.965 | 37.006 | 37.046 | 37.086 | 37.126 | 37.166 | 37.206 | 37.246 | 37.286 | 37.326 |

# Table 12-7 Continued

ITS-90 Table for Type K Thermocouple (Ref Junction 0°C)

| °C | 0 | 1 | 2 | 3 | 4 | 5 | 6 | 7 | 8 | 9 | 10 |
|---|---|---|---|---|---|---|---|---|---|---|---|
| | | | | Thermoelectric Voltage in mV | | | | | | | |
| 900 | 37.326 | 37.366 | 37.406 | 37.446 | 37.486 | 37.526 | 37.566 | 37.606 | 37.646 | 37.686 | 37.725 |
| 910 | 37.725 | 37.765 | 37.805 | 37.845 | 37.885 | 37.925 | 37.965 | 38.005 | 38.044 | 38.084 | 38.124 |
| 920 | 38.124 | 38.164 | 38.204 | 38.243 | 38.283 | 38.323 | 38.363 | 38.402 | 38.442 | 38.482 | 38.522 |
| 930 | 38.522 | 38.561 | 38.601 | 38.641 | 38.680 | 38.720 | 38.760 | 38.799 | 38.839 | 38.878 | 38.918 |
| 940 | 38.918 | 38.958 | 38.997 | 39.037 | 39.076 | 39.116 | 39.155 | 39.195 | 39.235 | 39.274 | 39.314 |
| 950 | 39.314 | 39.353 | 39.393 | 39.432 | 39.471 | 39.511 | 39.550 | 39.590 | 39.629 | 39.669 | 39.708 |
| 960 | 39.708 | 39.747 | 39.787 | 39.826 | 39.866 | 39.905 | 39.944 | 39.984 | 40.023 | 40.062 | 40.101 |
| 970 | 40.101 | 40.141 | 40.180 | 40.219 | 40.259 | 40.298 | 40.337 | 40.376 | 40.415 | 40.455 | 40.494 |
| 980 | 40.494 | 40.533 | 40.572 | 40.611 | 40.651 | 40.690 | 40.729 | 40.768 | 40.807 | 40.846 | 40.885 |
| 990 | 40.885 | 40.924 | 40.963 | 41.002 | 41.042 | 41.081 | 41.120 | 41.159 | 41.198 | 41.237 | 41.276 |
| 1000 | 41.276 | 41.315 | 41.354 | 41.393 | 41.431 | 41.470 | 41.509 | 41.548 | 41.587 | 41.626 | 41.665 |
| 1010 | 41.665 | 41.704 | 41.743 | 41.781 | 41.820 | 41.859 | 41.898 | 41.937 | 41.976 | 42.014 | 42.053 |
| 1020 | 42.053 | 42.092 | 42.131 | 42.169 | 42.208 | 42.247 | 42.286 | 42.324 | 42.363 | 42.402 | 42.440 |
| 1030 | 42.440 | 42.479 | 42.518 | 42.556 | 42.595 | 42.633 | 42.672 | 42.711 | 42.749 | 42.788 | 42.826 |
| 1040 | 42.826 | 42.865 | 42.903 | 42.942 | 42.980 | 43.019 | 43.057 | 43.096 | 43.134 | 43.173 | 43.211 |
| 1050 | 43.211 | 43.250 | 43.288 | 43.327 | 43.365 | 43.403 | 43.442 | 43.480 | 43.518 | 43.557 | 43.595 |
| 1060 | 43.595 | 43.633 | 43.672 | 43.710 | 43.748 | 43.787 | 43.825 | 43.863 | 43.901 | 43.940 | 43.978 |
| 1070 | 43.978 | 44.016 | 44.054 | 44.092 | 44.130 | 44.169 | 44.207 | 44.245 | 44.283 | 44.321 | 44.359 |
| 1080 | 44.359 | 44.397 | 44.435 | 44.473 | 44.512 | 44.550 | 44.588 | 44.626 | 44.664 | 44.702 | 44.740 |
| 1090 | 44.740 | 44.778 | 44.816 | 44.853 | 44.891 | 44.929 | 44.967 | 45.005 | 45.043 | 45.081 | 45.119 |
| 1100 | 45.119 | 45.157 | 45.194 | 45.232 | 45.270 | 45.308 | 45.346 | 45.383 | 45.421 | 45.459 | 45.497 |
| 1110 | 45.497 | 45.534 | 45.572 | 45.610 | 45.647 | 45.685 | 45.723 | 45.760 | 45.798 | 45.836 | 45.873 |
| 1120 | 45.873 | 45.911 | 45.948 | 45.986 | 46.024 | 46.061 | 46.099 | 46.136 | 46.174 | 46.211 | 46.249 |
| 1130 | 46.249 | 46.286 | 46.324 | 46.361 | 46.398 | 46.436 | 46.473 | 46.511 | 46.548 | 46.585 | 46.623 |
| 1140 | 46.623 | 46.660 | 46.697 | 46.735 | 46.772 | 46.809 | 46.847 | 46.884 | 46.921 | 46.958 | 46.995 |
| 1150 | 46.995 | 47.033 | 47.070 | 47.107 | 47.144 | 47.181 | 47.218 | 47.256 | 47.293 | 47.330 | 47.367 |
| 1160 | 47.367 | 47.404 | 47.441 | 47.478 | 47.515 | 47.552 | 47.589 | 47.626 | 47.663 | 47.700 | 47.737 |
| 1170 | 47.737 | 47.774 | 47.811 | 47.848 | 47.884 | 47.921 | 47.958 | 47.995 | 48.032 | 48.069 | 48.105 |
| 1180 | 48.105 | 48.142 | 48.179 | 48.216 | 48.252 | 48.289 | 48.326 | 48.363 | 48.399 | 48.436 | 48.473 |
| 1190 | 48.473 | 48.509 | 48.546 | 48.582 | 48.619 | 48.656 | 48.692 | 48.729 | 48.765 | 48.802 | 48.838 |
| 1200 | 48.838 | 48.875 | 48.911 | 48.948 | 48.984 | 49.021 | 49.057 | 49.093 | 49.130 | 49.166 | 49.202 |
| 1210 | 49.202 | 49.239 | 49.275 | 49.311 | 49.348 | 49.384 | 49.420 | 49.456 | 49.493 | 49.529 | 49.565 |
| 1220 | 49.565 | 49.601 | 49.637 | 49.674 | 49.710 | 49.746 | 49.782 | 49.818 | 49.854 | 49.890 | 49.926 |
| 1230 | 49.926 | 49.962 | 49.998 | 50.034 | 50.070 | 50.106 | 50.142 | 50.178 | 50.214 | 50.250 | 50.286 |
| 1240 | 50.286 | 50.322 | 50.358 | 50.393 | 50.429 | 50.465 | 50.501 | 50.537 | 50.572 | 50.608 | 50.644 |
| 1250 | 50.644 | 50.680 | 50.715 | 50.751 | 50.787 | 50.822 | 50.858 | 50.894 | 50.929 | 50.965 | 51.000 |
| 1260 | 51.000 | 51.036 | 51.071 | 51.107 | 51.142 | 51.178 | 51.213 | 51.249 | 51.284 | 51.320 | 51.355 |
| 1270 | 51.355 | 51.391 | 51.426 | 51.461 | 51.497 | 51.532 | 51.567 | 51.603 | 51.638 | 51.673 | 51.708 |
| 1280 | 51.708 | 51.744 | 51.779 | 51.814 | 51.849 | 51.885 | 51.920 | 51.955 | 51.990 | 52.025 | 52.060 |
| 1290 | 52.060 | 52.095 | 52.130 | 52.165 | 52.200 | 52.235 | 52.270 | 52.305 | 52.340 | 52.375 | 52.410 |
| 1300 | 52.410 | 52.445 | 52.480 | 52.515 | 52.550 | 52.585 | 52.620 | 52.654 | 52.689 | 52.724 | 52.759 |
| 1310 | 52.759 | 52.794 | 52.828 | 52.863 | 52.898 | 52.932 | 52.967 | 53.002 | 53.037 | 53.071 | 53.106 |
| 1320 | 53.106 | 53.140 | 53.175 | 53.210 | 53.244 | 53.279 | 53.313 | 53.348 | 53.382 | 53.417 | 53.451 |
| 1330 | 53.451 | 53.486 | 53.520 | 53.555 | 53.589 | 53.623 | 53.658 | 53.692 | 53.727 | 53.761 | 53.795 |
| 1340 | 53.795 | 53.830 | 53.864 | 53.898 | 53.932 | 53.967 | 54.001 | 54.035 | 54.069 | 54.104 | 54.138 |

MISCELLANEOUS TABLES AND INFORMATION

## 12.5 INSTRUMENT AIR QUALITY

Reference ISA-7.0.01-1996 Standard

### Dewpoint

The pressure dewpoint (the dewpoint of the air under pressure, not at atmsopheric) as measured at the dryer outlet shall be at least 10°C (18°F) below the minimum temperature to which any part of the instrument air system will be exposed. The pressure dewpoint shall not exceed 4°C (39°F) at line pressure.

### Particle Size

A maximum 40 µmeter particle size in the instrument air system is acceptable for the majority of pneumatic devices.

### Lubricant Content

The lubricant content should be as close to zero as possible, and under no circumstances shall it exceed one (1) ppm w/w or v/v. Any lubricant in the compressed air system shall be evaluated for compatibility with the use of pneumatic devices connected to the instrument air system.

### Contaminants

Instrument air shall be free of corrosive contaminants and hazardous gases which could be drawn into the instrument air supply. The air intake should be monitored for contaminants (e.g., paint vapor, chemical vapors, engine exhaust).

Traditionally, regenerative desiccant air dryers for instrument air systems are designed to provide –40°F dewpoint air at pressure. However, in extremely cold climates, instrument air applications may require dewpoints as low as –100°F at operating pressure to prevent the formation of ice within the instrument air system.

### Composition of Air

| | |
|---|---|
| Nitrogen | 78.084% |
| Oxygen | 20.948% |
| Argon | 0.934% |
| Carbon Dioxide | 0.0314% |
| Neon | 0.00182% |
| Helium | 0.000524% |
| Methane | 0.00015% |
| Hydrogen | 0.00005% |
| Trace Gases | 0.000056% |

# 12.6 THEVENIN & NORTON EQUIVALENCIES

## 12.6.1 Thevenin

Thevenin: Defined as any linear circuit containing several voltages and resistances that can be replaced by just a single voltage in series with a single resistor.

*DC – Thevenin*

Basic steps for solving a DC Thevenin analysis:

1.  Remove the portion of the network across which the Thevenin equivalent circuit is to be found.

2.  Mark the terminals of the remaining two terminal network

3.  Calculate $R_{TH}$ by first setting all sources to zero (voltage sources are replaced by short circuits and current sources are replaced by open circuits). Then find the resultant resistance between the two marked terminals.

4.  Calculate $E_{TH}$ by first replacing the voltage and current sources and then find the short circuit current between the two marked terminals.

5.  Draw the Thevenin equivalent circuit with the portion of the circuit previously removed replaced between the terminals of the equivalent circuit.

**Examples:**

*   Find the Thevenin equivalent circuit for the network in the shaded area of the network below. Then find the current through RL for values of 2, 10 and 100 Ω.

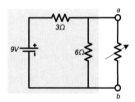

Solution:

Steps 1 & 2 produce the following:

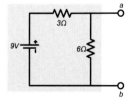

Step 3: Short the voltage source:

$$R_{TH} = \frac{(3)(6)}{3+6} = 2\Omega$$

Step 4: Replace the voltage source

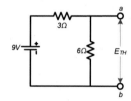

$$E_{TH} = \frac{(6)(9)}{6+3} = 6V$$

Step 5: Draw the Equivalent Circuit

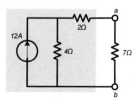

$$R_L = 2\Omega \quad I = \frac{6V}{2\Omega + 2\Omega} = 1.5A$$
$$R_L = 10\Omega \quad I = \frac{6V}{2\Omega + 10\Omega} = 0.5A$$
$$R_L = 100\Omega \quad I = \frac{6V}{2\Omega + 100\Omega} = 0.059A$$

- Find the Thevenin equivalent circuit for the network in the shaded area of the network below.

Solution:

Steps 1 & 2 produce the following:

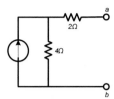

Step 3: Open the current source:

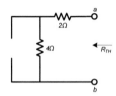

$$R_{TH} = 4\Omega + 2\Omega = 6\Omega$$

Step 4: Replace the current source In this case. Since there exists an open circuit between the two marked terminals, the current is zero between these terminals and the 2 $\Omega$ resistor, thus no voltage drop across the 2 $\Omega$ resistor. So $E_{TH}$ is simply the voltage drop across the 4 $\Omega$ resistor.

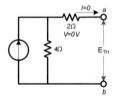

$$E_{TH} = (4\Omega)(12A) = 48V$$

Step 5: Draw the Equivalent Circuit

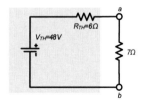

## AC – Thevenin

Basic steps for solving an AC Thevenin analysis:

1.  Remove the portion of the network across which the Thevenin equivalent circuit is to be found.

2.  Mark the terminals of the remaining two terminal network

3.  Calculate $Z_{TH}$ by first setting all sources to zero (voltage sources are replaced by short circuits and current sources are replaced by open circuits). Then find the resultant impedance between the two marked terminals.

4.  Calculate $E_{TH}$ by first replacing the voltage and current sources and then find the open circuit voltage between the two marked terminals.

5.  Draw the Thevenin equivalent circuit with the portion of the circuit previously removed replaced between the terminals of the equivalent circuit.

**Examples:**

*   Find the Thevenin equivalent circuit for the network external to resistor $R$ as shown below:

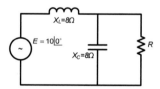

Solution:

Steps 1 & 2 produce the following:

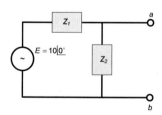

Z1 = j8 and Z2 = – j2

Step 3: Short the voltage source:

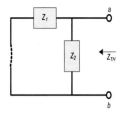

$$Z_{TH} = \frac{Z_1 Z_2}{Z_1 + Z_2} = \frac{(j8)(-j2)}{(j8 - j20)} = \frac{-j^2 16}{j6} = \frac{16}{6\underline{|90°}} = 2.67\underline{|-90°}$$

Step 4: Replace the voltage source

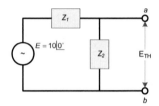

$$E_{TH} = \frac{Z_2 E}{Z_1 + Z_2} = \frac{(-j2)(10V)}{(j8 - j2)} = \frac{-j20}{j6} = 3.33\underline{|-180°}$$

Step 5: Draw the Equivalent Circuit

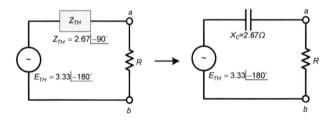

## 12.6.2 Norton's Theorem

Named after Bell Labs engineer Edward L. Norton: Any two terminal bilateral network can be replaced by an equivalent circuit consisting of a current source and a parallel impedance.

### DC Norton

Basic steps for solving a DC Norton analysis:

1. Remove the portion of the network across which the Norton equivalent circuit is to be found.

2. Mark the terminals of the remaining two terminal network

3. Calculate $R_N$ by first setting all sources to zero (voltage sources are replaced by short circuits and current sources are replaced by open circuits). Then find the resultant resistance between the two marked terminals.

4. Calculate $I_N$ by first replacing the voltage and current sources and then find the short circuit voltage between the two marked terminals.

5. Draw the Norton equivalent circuit with the portion of the circuit previously removed replaced between the terminals of the equivalent circuit.

**Examples:**

- Find the Norton equivalent circuit for the network in the shaded area of the network below.

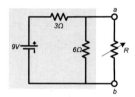

Solution:

Steps 1 & 2 produce the following:

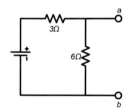

Step 3: Short the voltage source:

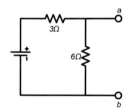

$$R_N = \frac{(3)(6)}{3+6} = 2\Omega$$

Step 4: Replace the voltage source & short-circuit the marked terminals

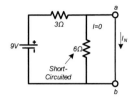

$$I_N = \frac{9V}{3\Omega} = 3A$$

Step 5: Draw the Equivalent Circuit

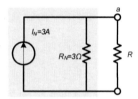

- Find the Norton equivalent circuit for the 9 Ω network in the shaded area below.

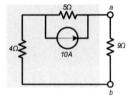

Solution:

Steps 1 & 2 produce the following:

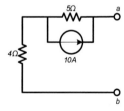

Step 3: Open the current source:

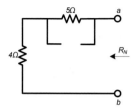

$$R_N = 4\Omega + 5\Omega = 9\Omega$$

Step 4: Replace the current source & short-circuit the marked terminals

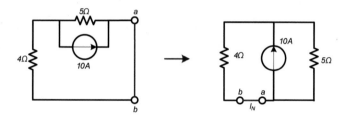

$$I_N = \frac{(5\Omega)(10A)}{4\Omega + 5\Omega} = 5.556A$$

Step 5: Draw the Equivalent Circuit

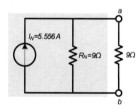

## AC Norton

Basic steps for solving an AC Norton analysis:

1.  Remove the portion of the network across which the Norton equivalent circuit is to be found.

2.  Mark the terminals of the remaining two terminal network

3.  Calculate $Z_N$ by first setting all sources to zero (voltage sources are replaced by short circuits and current sources are replaced by open circuits). Then find the resultant impedance between the two marked terminals.

4.  Calculate $I_N$ by first replacing the voltage and current sources and then find the short circuit voltage between the two marked terminals.

5. Draw the Norton equivalent circuit with the portion of the circuit previously removed replaced between the terminals of the equivalent circuit.

**Example:**

- Determine the Norton equivalent circuit for the network external to the 6 Ω resistor in the shaded area below.

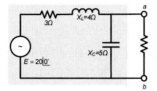

Solution:

Steps 1 & 2 produce the following:

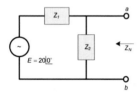

$$Z_1 = 3 + J4 = 5 \underline{|53.13°}$$
$$Z_2 = -j5 = 5 \underline{|-90}$$

Step 3: Short the voltage source:

$$Z_N = \frac{Z_1 Z_2}{Z_1 + Z_2} = \frac{(5\underline{|53.13°})(5\underline{|-90°})}{(3+j4)+(-j5)} = \frac{25\underline{|-36.87°}}{3-j1} =$$

$$\frac{25\underline{|-36.87°}}{3.16\underline{|-18.43°}} = 7.91\underline{|-18.44°} = 7.5 - j2.5$$

## Step 4: Replace the voltage source & short-circuit the marked terminals

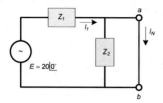

$$I_N = I_1 = \frac{E}{Z_1} = \frac{20\lfloor 0^\circ}{5\lfloor 53.13^\circ} = 4\lfloor -53.13^\circ$$

## Step 5: Draw the Equivalent Circuit

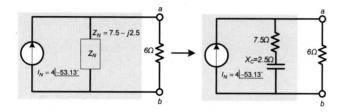

# 13. Uninterruptible Power Supply (UPS)

## 13.1 UPS Topologies

### 13.1.1 Single-Conversion

*Standby (off-line):* Allows equipment to run on utility power until the UPS detects a problem, at which point the UPS switches to battery power to protect against sags, surges or outages.

The *standby UPS (Figure 13-1)* is the simplest and least expensive UPS design. **Caution:** this configuration switches to battery mode in 5–12ms, so there may be an interruption of power during switching with the use of this design.

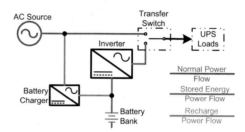

**Figure 13-1. Standby Single-Conversion UPS (Schematic)**

*Line-Interactive:* Regulates voltage by boosting input voltage or modulating (bucking) it down as necessary before allowing it to pass to the protected equipment, or resorting to battery power. This design (Figure 13-2) doesn't have to resort to batteries as often as a standby system, although it may use some battery power to support the transition between normal mode and voltage regulation mode. Battery usage is lower than a standby UPS but still higher than a double-conversion topology. This design switches to battery mode with a typical transfer time of 3–8ms; there still may be an interruption of power during switching with the use of this design.

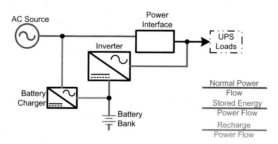

**Figure 13-2. Line-Interactive Single-Conversion UPS (Schematic)**

### 13.1.2 Double-Conversion

This design isolates equipment from raw utility power, converting power from AC to DC and back to AC again, to deliver the cleanest power and highest protection (Figure 13-3). If the AC input supply is out of predefined limits, the rectifier turns off, and the UPS draws current from the battery. Battery power passes through the output inverter and then to the load. The UPS will stay on battery power until the AC input returns to the normal tolerances (or until the battery runs out, whichever is sooner). This design has zero interruption transfer time.

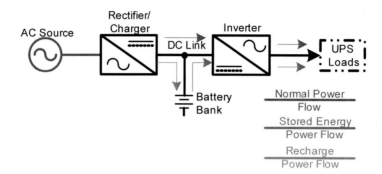

**Figure 13-3. Double-Conversion UPS (Schematic)**

## 13.2 INVERTER TECHNOLOGIES

### 13.2.1 Ferro-resonant

This type converts DC to AC by generating a square wave filtered by a ferroresonant transformer, creating a true sine wave output. The ferroresonant transformer (also called an "electric flywheel" because it rides through a one-cycle loss of input) is a "tuned" nonlinear transformer. Magnetic storage of electrical energy in the transformer supplies one cycle of ride-through energy (upon loss of input to the transformer) to allow zero-break operation of the static transfer switch.

### 13.2.2 PWM (Pulse Width Modulation)

This type converts DC to AC by using power switching at a 20 kHz to 50 kHz rate. A linear feedback loop is part of the circuitry. The output is a pulse-width modulated positive and negative square wave. A simple output low-pass filter removes the high frequency carrier for a smoothed sine wave.

### 13.2.3 Step-Wave

With this type, power semiconductors and phase shifting networks create a 6 or 12-step "staircase" waveform that is filtered into a sinusoidal shape. It requires a semiconductor device plus a phase shift for each step in the output voltage waveform.

## 13.3 MECHANICAL FLYWHEEL

A DC battery-free flywheel energy storage technology, this is an integrated motor-generator-flywheel that stores kinetic energy in a constantly spinning, quiet, low friction steel disc (Figure 13-4). This system acts as a short term energy source that allows time for generator systems to start and come on line. The main difference between a conventional battery type UPS and a mechanical flywheel is the amount of time that output power will be available. Unlike the battery type UPS, the mechanical flywheel design will only output power while the flywheel assembly is rotating and expending its kinetic energy, thus the use of the term 'short-term' energy source.

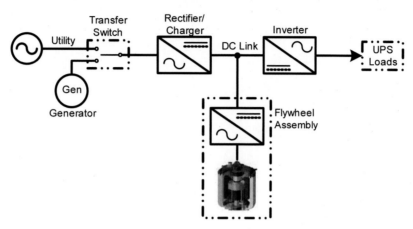

**Figure 13-4. Mechanical Flywheel UPS (Schematic)**

# 14. RECOMMENDED RESOURCES

The following is a list of useful resource material for the Control Systems Engineer:

- Instrument Engineers' Handbook, 4th Edition, Vol. 1: Process Measurement and Analysis, Béla G. Lipták, ISBN: 978-0-8493-1083-0, Published by ISA/CRC Press, 2003

- Instrument Engineers' Handbook, 4th Edition, Vol. 1: Process Software and Digital Networks, Béla G. Lipták, ISBN: 978-1-43981-776-6, Published by ISA/CRC Press, 2011

- Instrument Engineers' Handbook, 4th Edition, Vol. 2: Process Control and Optimization, Béla G. Lipták, ISBN: 0-8493-1081-4, Published by ISA/CRC Press, 2005

- Analytical Instrumentation – R.E. Sherman (ISBN: 978-1-55617-581-7) Published by ISA, 1996

- Control Valves – Guy Borden, Jr. (ISBN: 978-1-55617-565-7) Published by ISA, 1998

- Omega Engineering website (http://www.omega.com/techref/)

- Cameron Hydraulic Book (19th edition) (http://www.flowserve.com/Products/Pumps/Cameron-Hydraulic-Data-Book,en_US) Published by Flowserve with the latest reprint in 2002

- Crane Flow of Fluids Tech Paper 410 (http://www.flowoffluids.com/publications/crane-tp-410.aspx) (ISBN: 1-40052-712-0) Published by Crane Valves North America in The Woodlands, TX with the latest reprint in 2009

- Principles and Practice of Flow Meter Engineering (9th edition) – L.K. Spink no longer in print, was originally published by The Foxboro Company with the last reprint being 1978. Old copies are still available via the web on sites such as amazon, goggle, alibris, etc.

- Flow Measurement Engineering Handbook – Richard W. Miller (ISBN: 0070423660) McGraw-Hill Publishing, New York, NY © 1996

- Fisher Control Valve Handbook (available from Emerson free of charge) (http://www.documentation.emersonprocess.com/groups/public/documents/book/cvh99.pdf)

- Crosby Pressure Relief Valve Engineering Handbook TP-V300 (available from Tyco free of charge) (http://www.tycovalves-na.com/ld/CROMC-0296-US.pdf)

- Crouse-Hinds 2008 Code Digest (available from Cooper Crouse Hinds free of charge) (http://www.crouse-hinds.com/smart/PDFs/Code_Digest_2008.pdf)

- 2011 edition of NEC Code Handbook (available from NFPA, Quincy, MA) NFPA book number 70HB11 (ISBN: 978-0-877-65916-7)

- 2012 edition NFPA 70E (available from NFPA, Quincy, MA), ISBN 978-145590096-1

- Controller Tuning & Control Loop Performance (2nd edition) – David St. Clair (ISBN: 0-9669703-0-6) Published by Straight-Line Control Company in Newark, DE, Inc. with the latest printing being 2007 ([www.straightlinecontrol.com](www.straightlinecontrol.com))